T0326366

FLOW AND HEAT TRANSFER IN GEOTHERMAL SYSTEMS

FLOW AND HEAT TRANSFER IN GEOTHERMAL SYSTEMS

Basic Equations for Describing and Modeling Geothermal Phenomena and Technologies

ANIKO TOTH
ELEMER BOBOK

ELSEVIER

AMSTERDAM • BOSTON • HEIDELBERG • LONDON
NEW YORK • OXFORD • PARIS • SAN DIEGO
SAN FRANCISCO • SINGAPORE • SYDNEY • TOKYO

Elsevier
Radarweg 29, PO Box 211, 1000 AE Amsterdam, Netherlands
The Boulevard, Langford Lane, Kidlington, Oxford OX5 1GB, United Kingdom
50 Hampshire Street, 5th Floor, Cambridge, MA 02139, United States

Copyright © 2017 Elsevier Inc. All rights reserved.

No part of this publication may be reproduced or transmitted in any form or by any means,
electronic or mechanical, including photocopying, recording, or any information storage and
retrieval system, without permission in writing from the publisher. Details on how to seek
permission, further information about the Publisher's permissions policies and our
arrangements with organizations such as the Copyright Clearance Center and the Copyright
Licensing Agency, can be found at our website: www.elsevier.com/permissions.

This book and the individual contributions contained in it are protected under copyright by
the Publisher (other than as may be noted herein).

Notices
Knowledge and best practice in this field are constantly changing. As new research and
experience broaden our understanding, changes in research methods, professional practices,
or medical treatment may become necessary.

Practitioners and researchers must always rely on their own experience and knowledge in
evaluating and using any information, methods, compounds, or experiments described
herein. In using such information or methods they should be mindful of their own safety and
the safety of others, including parties for whom they have a professional responsibility.

To the fullest extent of the law, neither the Publisher nor the authors, contributors, or editors,
assume any liability for any injury and/or damage to persons or property as a matter of
products liability, negligence or otherwise, or from any use or operation of any methods,
products, instructions, or ideas contained in the material herein.

Library of Congress Cataloging-in-Publication Data
A catalog record for this book is available from the Library of Congress

British Library Cataloguing-in-Publication Data
A catalogue record for this book is available from the British Library

ISBN: 978-0-12-800277-3

For information on all Elsevier publications
visit our website at https://www.elsevier.com/

Working together
to grow libraries in
developing countries

www.elsevier.com • www.bookaid.org

Publisher: Candice G. Janco
Acquisition Editor: Amy Shapiro
Editorial Project Manager: Tasha Frank
Production Project Manager: Mohanapriyan Rajendran
Designer: Maria Ines Cruz

Typeset by TNQ Books and Journals

Contents

Preface

Flow and Heat Transfer in Geothermal Systems is intended as a systematic and analytical exploration of the most important geothermal principles. Understanding the physical principles of fluid flow and heat transfer, in both natural and artificial systems, is essential to understanding how every stage of the geothermal cycle affects geothermal production wells, injection wells, drilling operations, surface equipment, energy-conversion systems, and the geothermal reservoir itself.

Although we assume a basic knowledge of mathematics and some familiarity with the geothermal industry, our book should be accessible to beginning engineering students and even well-educated laymen who wish to understand a bit more about this promising alternative to fossil energy. We expect that *Flow and Heat Transfer in Geothermal Systems* will be especially valuable as a handbook for geologists, hydrogeologists, reservoir engineers, geophysicists, geochemists, drilling engineers, and production engineers, all of whose collaborative work is vital in creating and maintaining successful geothermal operations.

Chapter 1 is an introduction to the basic idea of a geothermal reservoir along with a brief history of early geothermal development. In Chapter 2 we explore the basic laws of fluid mechanics and thermodynamics. Chapter 3 deals with transport processes in geothermal reservoirs, based on the complex continuum model, and introduces the geothermally useful Darcy's Law.

Chapter 4 studies the different boundary and initial conditions within rock masses, and the various means of measuring how heat is conducted.

Chapter 5 looks at those important natural processes which obtain in undisturbed geothermal reservoirs: consolidation, natural convection, and the development of overpressured reservoirs.

Chapter 6 uses analytic complex functions to explain two-dimensional underground flows, including the Hele—Shaw flow. More specifically, we look at geothermal reservoirs and their producing wells, which form a serially connected synergetic flow system: a radially inward Darcy flow toward the well in the reservoir, and a turbulent upflow through the tubing.

The flow within wells is the subject of Chapter 7, which also examines homogeneous water upflow and two-phase flows induced by dissolving gas and flashing. In this chapter, our examination of the energy transfer process assumes an unsteady flow of inviscid fluid.

Chapter 8 deals with the use of submersible pumps to induce artificial lifting, describing the most important types of centrifugal pumps along with their construction, their operation, their performance curves and how they affect cavitation. This chapter describes the phenomenon by which, as production continues, heat transfer causes a gradual rise of temperature in the surrounding rock, decreasing the temperature difference and the heat flux.

Chapter 9 investigates borehole heat transfer, and how to determine the temperature distribution of the flowing fluid both in production and injection wells. In this chapter the flow patterns of two-phase water–steam mixture flows are analyzed. Temperature distribution along the pipe axis is also determined. The chapter ends with an examination of the heat transfer process around a buried horizontal hot-water transporting pipeline.

Chapter 10 looks at how gathering pipelines work in geothermal energy production systems, introducing the basic equations used to analyze one-dimensional pipe flow for both laminar and turbulent flows. We demonstrate how to assess the loss of superheated steam, assuming an isothermal case.

Chapter 11 describes the process of geothermal power generation, briefly outlining the power generation cycle, analyzing the basic thermodynamic process of wet steam generation and showing how energy is converted from thermal to mechanical energy in the steam turbines. This chapter introduces several of the most important types of geothermal power plants: single flash, dry steam, and binary plan.

In Chapters 12, 13, and 14 we investigate the following topics: propagation of the cooled region between injection and production wells in fractured reservoirs; flow and heat transfer in a borehole heat exchanger (both in shallow and in deeper regions); flow and heat transfer during drilling operations; laminar and turbulent flows of non-Newtonian fluids through annuli; and temperature distribution in the circulating drilling mud.

Chapter 15 is a case study of how much environmental damage can be caused by high-enthalpy geothermal reservoirs. This chapter relates the history of a serious industrial accident which occurred in Hungary when workers, while tapping an overpressured 200°C reservoir, provoked a steam blowout from a depth of 4000 m. As part of the resulting hydrodynamic/thermodynamic reconstruction, certain inconsistent phenomena observed during the blowout are explained with the help of thermodynamical calculations.

The book's final chapter, Chapter 16, describes two nontrivial forms of geothermal energy production: the first highlights the substantial geothermal potential of an abandoned copper ore mine, where the roadways and the shafts were flooded by mine water; the second explores another unusual application, a deicing system located at the entrance of a mine tunnel.

Acknowledgments

Over the years the following associations have provided us with an invaluable forum for the investigation of geothermal topics: the Stanford Geothermal Workshop, one of the geothermal world's longest-running technical workshops; the Geothermal Research Council (GRC); the International Geothermal Association (IGA); and the European Geothermal Energy Council (EGEC). Among the individuals we would like to thank are Prof. John Lund of the Oregon Geo Heat Center, who gave us unstinting advice and encouragement, and Prof. Roland Horne of Stanford University, who showed by his personal example the high level of academic rigor needed in this field. In the same light, special thanks are due to Prof. Burkhard Sanner of the European Geothermal Energy Council.

We would also like to thank Andras Dianovszky and Mark Zsemko for recreating several important diagrams which had gotten lost in the shuffle. Last but not least, our special thanks to David Fenerty for his editing and proofreading suggestions.

Aniko Toth
Miskolc, Hungary
September, 2016

What Is Geothermal Energy?

1.1 INTRODUCTION

Geothermal energy is energy contained within the high temperature mass of the Earth's crust, mantle, and core. Since the Earth's interior is much hotter than its surface, energy flows continuously from the deep, hot interior up to the surface. This is the so-called terrestrial heat-flow. The temperature of the Earth's crust increases with depth in accordance with Fourier's law of heat conduction. Thus the energy content of a unit of mass also increases with depth.

All of the Earth's crust contains geothermal energy, but geothermal energy can only be recovered by means of a suitable energy-bearing medium. To be practical, the energy-bearing media must be: hot enough (high-specific energy content), abundant enough, easily recoverable, inexpensive, manageable, and safe. Water satisfies these requirements perfectly. The specific heat of water is 4.187 kJ/kg°C. In the steam phase, latent heat is added to it. Hot water and steam can be recovered easily through deep, rotary-drilled wells. Through the use of a suitable designed heat exchanger, heat can be efficiently transferred from

Copyright © 2017 Elsevier Inc. All rights reserved.

the water or steam. Steam is an especially suitable working fluid for energy conversion cycles.

Nowadays, geothermal energy production is mainly achieved from hot water and steam production via deep boreholes. Another rapidly-growing production technology involves exploiting the energy content of near-surface regions by using shallow borehole heat exchangers and heat pumps.

It is likely that the natural heat of volcanoes and other geothermal sources were already being used in the remote Paleolithic era, but concrete evidence only dates from 8000 to 10,000 years ago. We are therefore forced to use indirect methods when speculating on mankind's earliest relationship with geothermal phenomena and products of the Earth's heat.

The uses of natural hot water for balneology and the exploitation of hydrothermal products for a wide range of practical applications increased remarkably during the millennium preceding the Christian era. This use eventually extended to the boundaries of ancient Rome, achieving maximum use during the 3rd century A.D., the Roman Empire's apex. After Rome's decline in the 6th century, geothermal exploitation also declined throughout Southern Europe, a period of disuse which lasted until the beginning of the second millennium. There is evidence that geothermal resources were still being exploited in the centuries that followed, in China and many other countries, but on a very limited scale and only in rudimentary forms.

Deep in the Remontalou River valley, at the south edge of Auvergne in the Central French massif, the town of Chaudes—Aigues has an 82°C hot spring, one of the hottest natural thermo-mineral springs in Europe. The region has been inhabited since prehistoric times. The main spring, called *le par*, is one of about 30 gushing springs concentrated in a small area. From mid-October to the end of April, a 5-km network of pipes brings the hot water from five of these springs to heat 150 homes. Houses built above the springs use the hot water directly below for heating, and have done so since the 14th century (Cataldi et al., 1999).

Geothermal water was first used for boric acid production in Larderello, Italy, in 1827. Boric acid production was an Italian monopoly in Europe, and became a large-scale industry in the middle of the 19th century.

Other countries also began to develop their geothermal resources on an industrial scale. V. Zsigmondy, for example, became a legend in Hungarian geothermal history after he drilled Europe's deepest well (971 m) in Budapest in 1877. Since that date, the resulting geothermal water has been used for balneology in the famous Szechenyi Spa. In 1892, the first geothermal district-heating system began operations in Boise, Idaho, USA. In 1928, Iceland, another pioneer in the utilization of geothermal energy, also began exploiting its geothermal fluids (mainly hot waters) for domestic heating purposes.

In 1904, Larderello again became famous as the first place where electricity was generated from geothermal steam. The scientific and commercial success of this experiment demonstrated the industrial value of geothermal energy, the exploitation of which would then develop more significantly. By 1942, Larderello's installed geothermoelectric capacity had reached 127,650 kW$_e$. Several countries soon followed Italy's example. In 1919, the first geothermal wells in Japan were drilled at Beppu. In 1921, geothermal wells were drilled at the Geysers, California, USA.

Between the two World Wars, oil prospectors found huge geothermal water reservoirs all over the world, usually by accident. In 1958, based on similar exploration data, and after extensively mapping variations in the Pannonian Basin's terrestrial heat-flow 15 years earlier, the Hungarian mining engineer T. Boldizsár composed the world's first regional heat-flow map of Hungary (Boldizsar, 1964). That same year, a small geothermal power plant began operating in New Zealand. Another started in 1959 in Mexico, and another in the United States in 1960. Many other countries would then follow suit in the years to come.

As of 2015, 28 nations currently use geothermal energy to generate electricity (geothermal power). There has been a significant increase since 1995. By that year, the world's installed capacity was 6833 MWe (Bertrani, 2015). By 2005, it was 8934 MWe. By 2015, it was 12,635 MWe (or 73,549 GWh/year).

As of 2015, 78 countries have direct utilization of geothermal energy, a significant increase from the 28 reported in 1995, 58 in 2000, and 72 in 2005. For 2015 the reported amount of geothermal energy used is 438,071 TJ/year (121,696 GWh/year). Approximate geothermal energy use by category is 49.0% for ground-source heat pumps, 24.9% for bathing and swimming (including balneology), 14.4% for space heating (of which 85% is for district-heating), 5.3% for greenhouses and uncovered surface heating, 2.7% for industrial process heating, 2.6% for aquaculture pond and raceway heating, 0.4% for agricultural drying, 0.5% for snow melting and cooling, and 0.2% for other uses.

1.2 THE NATURE AND ORIGIN OF GEOTHERMAL ENERGY

There was a time, not so long ago, when the high temperature of the Earth's interior was not known. Kelvin solved first the differential equation of the heat conduction in a spherical coordinate system. The spherical symmetry of Earth's shape suggested the idea of the spherically symmetrical temperature distribution around the world. The temperature distribution along the depth is monotonically increasing. In accordance the Fourier's law of heat conduction, a radial outward heat flux occurs.

This is the so-called terrestrial heat-flow. The terrestrial heat-flow is mainly conduction but can be convection also. Kelvin collected surface heat-flow data from Russia, Australia, South Africa, Deccan Plateau, and Labrador. Unfortunately, these places are geothermally similar with a relatively low heat-flow. Kelvin's measurements confirmed the idea of spherically symmetrical temperature distribution obtaining an average heat-flow value of 0.0556 W/m^2. This thermostatic model of the Earth was proven false on the basis of Boldizsár's (1943) terrestrial heat-flow measurements, especially after the discovery of the regional geothermal anomaly in the Carpathian Basin. Boldizsar's heat-flow map of Hungary was the first in the world in 1944. It was proven by the convincing evidence of the regionally varying heat-flow distribution. He got the name "Father of geothermal." Boldizsar's early results were confirmed by extended investigations of Bullard (1954), exploring the extremely high heat-flow distribution along the mid-oceanic ridges (Fig. 1.1).

As a result of international scientific cooperation, large-scale continental heat-flow maps demonstrate the varying heat-flow intensity belonging to certain tectonic structures. Along the displacing mid-ocean ridges the terrestrial heat-flow attains the value of 0.2 W/m^2. The average heat-flow in the Carpathian Basin is 0.1 W/m^2 (Toth, 2010). On the continental shields or the oceanic crust, heat-flow density hardly attains the value of 0.02 W/m^2. All these are connected as the result of the plate tectonics, the movement of the lithosphere plates.

The generally accepted model of the Earth's structure posits an outer, spherical shell, known as the crust. Its two parts can be distinguished as the continental crust, with an average thickness of 35 km, and the oceanic crust, with a thickness of about 8 km. Beneath the crust lies a boundary

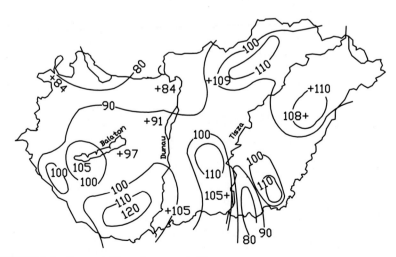

FIGURE 1.1 Terrestrial heat-flow map in Hungary.

known as the Mohorovicic discontinuity, where the speed of propagation of seismic waves suddenly increases from 7 km/s to 8.1 km/s. The Mohorovicic discontinuity can be found beneath the crust and above the mantle. The mantle extends to a depth of 2900 km, where there it changes into the much denser liquid core. The core is composed largely of molten iron. Within this liquid core is a solidified iron inner core with a radius of about 1350 km. On a large scale, these are the main components of the Earth's structure, as shown in Fig. 1.2.

From the geothermal point of view, only the crust and the upper mantle are of importance. Direct information about the mantle is available from deep boreholes only. The three deepest are in Sakhalin, Qatar, and the Kola Peninsula. They have a bottomhole depth of about 12 km. All other data derive from indirect gravimetric, seismic, dipole-resistivity, and other geophysical information.

The crust is not a homogeneous spherical shell. The continental crust beneath the continents and the enclosed seas is mainly granite composite, rich in silica with a density of 2670 kg/m^3. The oceanic crust is mainly basaltic. It is poor in silica with a density 2950 kg/m^3. The thickness of the continental crust is variable. Beneath the high ranges it can be 70–75 km thick, but beneath the sinking sedimentary basins its thickness is only 20–25 km.

Beneath the crust, the upper mantle is rigid. This is the so-called lithosphere. Its thickness is approximately 80–100 km. Under the lithosphere, the propagation speed of the seismic waves decreases in a

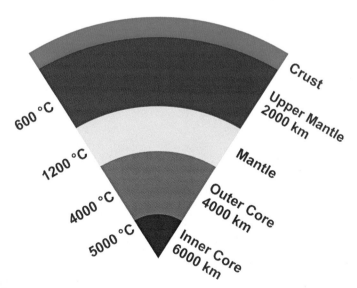

FIGURE 1.2   Structure of Earth's interior.

spherical shell, which has a thickness of 150 km. This is the so-called asthenosphere. Its temperature is possibly higher than the lithosphere or the mantle beneath it. The temperature of the asthenosphere is about 800−850°C. At this temperature, the mantle is in a plastic state; it can be flowing. Since the density of the asthenosphere is 3350 kg/m^3 on average, the lighter lithosphere is floating upon it. In accordance to Archimedes' law, beneath the high mountains the lithosphere becomes immersed deeper into the asthenosphere, while the oceanic crust is thinned. The lithosphere is not a unique rigid shell, but it consists of six large and some smaller plate pieces, which are in continuously moving relative to each other and the rotation axis of the earth as can be seen in Fig. 1.3.

Today the magmatic and tectonic activity of the Earth happens along the plate boundaries. It can be seen in Fig. 1.4, where the seismic belts coincide with the plate boundaries.

The heat can be transferred across the plastic mantle not only by conduction, but convective currents can also develop. At the lower boundary of the lithosphere, the temperature is lower than in the deepest region. At the deeper, less dense mantle, material occurs. The denser material under the influence of the gravity will sink, displacing the hotter and lighter mass, which upflows to the boundary of the lithosphere conveying its heat content. This motion caused by the temperature difference is the so-called thermal convection. Its streamline-pattern shows

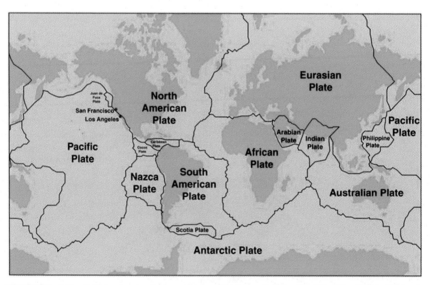

FIGURE 1.3 Lithosphere plates. *Source: Dickson, M.H., Fanelli, M., 2004. What is geothermal energy?. In: Website of the International Geothermal Association. https://www. geothermal-energy.org/what_is_geothermal_energy.html.*

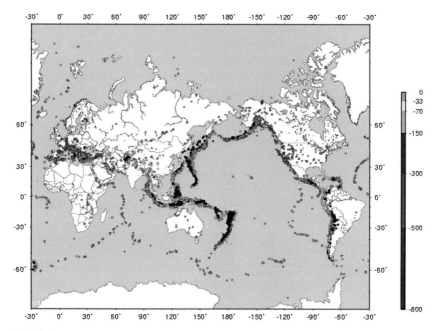

FIGURE 1.4 Large earthquakes at the plate boundaries. *Source: https://www.e-education. psu.edu/files/earth501/image/lesson2/neic_2007janjun.jpg*

characteristic convection cells. The plastic viscosity is about 10^{23} times of the water viscosity, thus the velocities of the developing thermal convection are extremely small.

Thus the lithosphere is in mechanical and thermal interaction. The convection currents cause the lithosphere plates to drift apart, while on the other hand the upflowing mantle heats the crust with great intensity. Around the stagnation point, the heat transfer is the most intense; the crust is heated here the most considerably. The tensile stresses of the flowing plastic mantle rift the weakened lithosphere plate, while the magma continuously fills the split accreting to it. This process is the so-called sea floor spreading at the mid-oceanic ridges. The opening of the lithosphere plate can also occur in the continental crust, in this way forming the East-African rift valley.

As the lithosphere plates move off each other, they can collide with other plates. In this case, the oceanic plate of greater density is creased under the lighter but thicker continental plate, submerging along the so-called Benioff plane, which has an inclination angle of about 45 degrees. This oceanic lithosphere plate submerging to the plastic mantle is warming up gradually, its strength decreases, then it attains its melting temperature, liquefies, and flows up. The submerging rigid lithosphere plate can be followed by seismic tools to the depth of 600–700 km.

The density difference between the mantle material and the melted lithosphere plate is substantially greater ($600 \, \text{kg/m}^3$) than the density difference caused by the thermal expansion which is $50 \, \text{kg/m}^3$. Thus the Archimedean lifting force can induce a more intensive uplifting flow in the region of the dissolved lithosphere plate. The re-melted intermediary and acidic magma is accreted from below to the continental crust. Thus it will be raised. At the same time, the extremely strong convective heat-flow propagates further in the solid crust as a very intensive conductive heat flux. Thus the orogenic areas are more active geothermally, and their terrestrial heat-flow is substantial. The solid crust may even be melted here, and volcanic areas can develop. This occurs typically along the plate boundaries of the Pacific coast.

There are other regions outside the plate boundaries where the terrestrial heat-flow is anomalously high. Such regions are the Carpathian Basin, the Paris Basin, or the Kuban region at the northern side of the Caucasus Mountains. The reason for the high heat-flow originates in the thinning continental crust due to the tension stresses and subcrustal erosion. The crust in the Carpathian Basin may be as thin as only 20 km. This window leads necessarily to the higher terrestrial heat-flow. It is obvious that the thin crust enables a higher heat flux since:

$$q = \frac{k}{\delta}(T_m - T_o) \tag{1.1}$$

where δ is the thickness of the crust, k is the average heat conductivity, T_m is the temperature at the top of the mantle, and T_0 is the surface temperature. The geothermal gradient is also higher:

$$\gamma = \frac{dT}{dz} = \frac{T_m - T_0}{\delta} \tag{1.2}$$

The geothermal gradient γ in the Carpathian Basin is higher than the continental mean value which is $0.045-0.065°C/m$. Thus relatively hot rock masses can be found in relatively shallow depth. This means favorable natural conditions to access geothermal reservoirs, to recover the geothermal energy.

1.3 GEOTHERMAL RESERVOIRS

The geothermal field is a geographical notion designating any region on the Earth's surface where such surface manifestations as geysers, fumaroles, or boiling mud-ponds indicate an active geothermal domain underground. These phenomena are characteristic of active volcanic regions. Where geothermal fields exist but have no spectacular surface manifestations, high terrestrial heat-flow and above-average geothermal gradient show that geothermal energy production would still be possible. Boldizsár (1943) was the first to recognize that terrestrial heat-flow is the most important indicator of geothermal activity.

The geothermal system is any arbitrarily but suitably demarcated region of the Earth's crust where geothermal phenomena are being investigated. The interaction of the system with its surroundings causes various characteristically geothermal processes.

The geothermal reservoir is part of a geothermal field from which the field's internal energy content can be recovered through the use of some reservoir fluid: steam, hot water, or a mixture of both. The geothermal reservoir is a natural or man-made sub-system demarcated to fit the demands of energy production. A *natural* hydrothermal geothermal reservoir is an extended, porous, and permeable formation saturated with hot water or steam, and possessing both a sufficiently large heat supply and a reliable recharge mechanism. The best geothermal reservoirs have a fractured rock matrix with high permeability, leaving a relatively unimpeded vertical path for thermal convection.

The geothermal system can be divided into two interacting sub-systems. One is a natural system; an existing reservoir with its own heat supply and recharge system. The other is an artificial system, characterized by such components as wells, pumps, gathering pipes, and heat exchangers. The two sub-systems operate together, exerting a mutual influence.

It is also possible to create artificial reservoirs in hot dry rocks (HDR), which have neither water nor pore/fracture type permeability. In this case, the fracture system can be formed by hydraulic fracturing, with water then being circulated through an injection and production well. The injected surface water warms as it flows through the fracture system, and is then delivered via the production well.

Whether the geothermal system is natural or man-made, heat transfer is the most important phenomenon. Heat transfer occurs partly through heat conduction and partly through conductive transport, as the heat-bearing fluid flows up to the surface. For the most part, geothermal reservoirs are classified according to how heat transfer occurs between the geothermal fluid and the reservoir rock through which it moves.

Geothermal reservoirs are continuously heated by terrestrial heat-flow. Many geothermal reservoirs are only heated by conductive heat-flow, which has a relatively moderate conductive heat flux; $0.0556 \, W/m^2$, on average (the average geothermal gradient is $30°C/km$). Although such an environment is unsuitable for recoverable hydrothermal systems, the reservoir's internal energy can still be extracted by means of geothermal heat pumps.

Regional thinning and sinking of the crust causes more intense heat-flow and a higher geothermal gradient. Recoverable hydrothermal systems become suitable as soon as the terrestrial heat flux reaches $0.1 \, W/m^2$, with a geothermal gradient of at least 40–50°C/km. It should be obvious that along with satisfactory recharge of the reservoir, a sufficiently porous (20%) and permeable (500–1000 mD) formation is also necessary. In Europe, such conditions exist mainly in sunken sedimentary basins like the Carpathian or Paris basins.

In a geothermal system, temperature increases linearly with depth. Porosity, however, decreases exponentially with depth as sediment underneath is compacted by the lithostatic pressure of sediment above it:

$$\phi = \phi_0 e^{-Az} \tag{1.3}$$

The expression Eq. (1.3) approximates this distribution quite well, in which ϕ_0 is the porosity at the surface, z is the depth, and A depends on the type of the sediment. The tendency is shown in Fig. 1.5. The porosity and the temperature distributions determine the region where recoverable hot water reservoirs can be found.

These medium-temperature hot water fields are quite common worldwide, and may be confined, artesian aquifers or open, hydrostatic systems. Their temperature is typically lower than 150°C. They may be worth direct use as district-heating, agricultural, and industrial process heat sources.

The conceptual model of such a medium-temperature hot water reservoir is shown in Fig. 1.6. Porous and permeable sedimentary layers are settled mainly horizontally on the impermeable bedrock. There are hardly permeable mainly clayey layers between the permeable

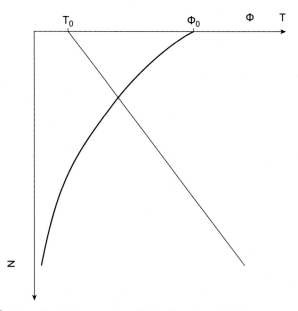

FIGURE 1.5   Porosity and temperature distribution along depth.

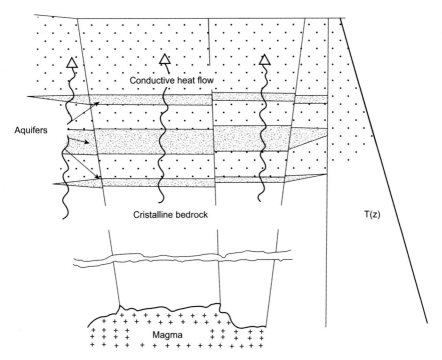

FIGURE 1.6   Conceptual model of medium temperature aquifer.

formations. These hardly permeable formations are also saturated by geothermal fluid, which is immobile-bounded pore water. Any cross-flow cannot be developed thorough these weak-permeability layers, but the continuous mass of the pore water can transmit the pressure across them. Thus the vertical pressure distribution is a continuous function along the depth, independent of the permeability of the different layers. Therefore the law of hydrostatics is valid for the whole saturated sedimentary mass.

One of the greatest conductive heated sedimentary aquifers is the Upper Pannonian formation in Hungary. This sandy and sandstone formation extends horizontally more than 40,000 km^2, while then average thickness is about 300 m, in which the sandy and clayey layers successive follow each other. The proportion of the permeable layers in the whole aquifer is about 35%.

Beneath the sedimentary layers, fractured or karstic regions of the bedrock can be found containing geothermal reservoirs connecting to the Pannonian aquifer. These carboniferous reservoirs are deeper and hotter than the Upper Pannonian formations, and their pressures are hydrostatic but somewhere weak thermal convection currents can occur. Such systems are developed in Zalakaros, Iklódbördöce, and Mélykút.

It's generally accepted that the best source of geothermal heat is a magmatic intrusion into the Earth's crust. In an active volcano, molten magma flows to the surface through a large fault system. Faults in a hard rock may provide an open channel for the upward flow of molten magma, which can overcome gravity to penetrate the rigid-rock fault system, but plastic rocks such as clay can fill the fault space and block the flow upward. In these cases the magmatic intrusion arrives at the boundary between hard and plastic rocks, but does not reach the surface. This so-called crypto-volcanism, without eruption, is common in acidic volcanoes but can also occur in basic volcanoes. In these cases the intrusion remains below the surface at a depth of 5–15 km. Above the intrusion, through the crystalline bedrock of good thermal conductivity, a very intensive terrestrial heat-flow develops.

Above the bedrock is a porous or fissured aquifer. The aquifer's rock matrix is discontinuous relative to the solid bedrock, and has weaker heat conductivity. This weak conductivity cannot transfer the intense subsurface heat, but the existence of reservoir fluid allows for thermal convection, a more efficient heat transfer mechanism than conduction. A low-permeability rock layer, the so-called cap rock, may overlay the aquifer. All steam-producing reservoirs have a cap rock. This almost impermeable cap rock is created by the deposition of dissolved minerals, or through kaolin formation brought about by the rock's hydrothermal alteration (Facca and Tonani, 1964). Cap rock may also develop through the cumulative compaction of sedimentary rocks, causing substantial overpressure. In the latter case, the compacted under-layer forms a hydraulic seal. The conceptual model of a convective heated reservoir is shown in Fig. 1.7.

In order to develop thermal convection, certain conditions must be fulfilled. The motion of a fluid, which is caused solely by the density differences brought about by temperature gradients, is called thermal or natural convection. A fluid body may be in mechanical equilibrium without being in thermal equilibrium. The hydrostatic equation may be satisfied even though the temperature distribution is not uniform. There are certain temperature gradients which allow the mechanical equilibrium to be maintained. Other temperature distributions may induce movements in the fluid. It is known that convective heat flux is a much more intensive heat transfer mechanism than conduction. The increasing temperature along the depth decreases the fluid density with the depth. The Archimedean lifting force acts to the fluid mass of higher temperature and lower density. It is superimposed to the gravity force inducing the flow of the fluid. From the point of view of heat transfer, thermal convection is equivalent with a formation of extreme high-heat conductivity. The heat-flow is high even very low temperature gradient across the zone of convection.

A high geothermal gradient is necessary to develop thermal convection. Its necessary value is typically higher than 0.2°C/m. The reason for this high

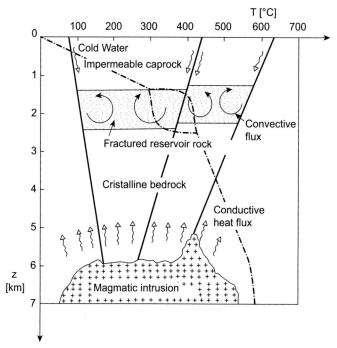

FIGURE 1.7 Conceptual model of a convective heated reservoir.

gradient is a near-surface (<5 km) fresh magmatic intrusion, with a temperature of 650–1000°C. This intensive heating produces a high (>1 W/m²) heat-flow. There are additional necessary conditions. The permeability should be greater than 1 darcy. This may possible in a fractured carboniferous reservoir or in coarse deltaic sediments. A sufficiently large vertical thickness of the aquifer is also necessary. Thus convection transfers the heat along a longer path, more efficiently. The cold-water recharge to the deeper formations can also increase the heat transfer intensity.

The temperature distribution of the conductive heated reservoir is not linear. Inside the aquifer, the temperature change is very small; the temperature gradient is large between the top of the reservoir and the surface only. Thus high temperature reservoir fluid can be found at shallow depths, which can be attained by shallow, large diameter boreholes more economically.

Both convective and conductive heated reservoirs contain hot water in liquid phase. The pressure of the water increases along the depth in accordance of the law of hydrostatics:

$$\frac{dp}{dz} = \rho g \qquad (1.4)$$

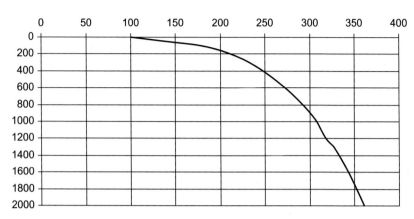

FIGURE 1.8   Boiling curve of water along depth.

where z in the depth, ρ is the density, and g is the gravity acceleration. Since ρ decreases with temperature, the expression Eq. (1.4) is nonlinear. The boiling curve is the locus of saturation temperatures that correspond to the local fluid hydrostatic pressure, which can be seen in Fig. 1.8. The actual formation temperatures are substantially lower than the saturation temperatures determined by the boiling curve. Thus the water remains in liquid state in the reservoirs and can be flashed in the near-surface section of the well as its pressure decreases more rapidly than its temperature.

The most valuable geothermal resources are dry-steam reservoirs. Large dry-steam reservoirs are found only in two places: the Geysers in California and Larderello in Italy. In dry-steam reservoirs, the reservoir's fractured rock matrix is filled with superheated steam. The steam also contains non-condensable gases, but no hot water. The steam is of meteoric origins, and its relative amounts of various water isotopes of water are $0^{16}/0^{18}$. The steam's H/D ratio is almost the same as for the isotopic composition of meteoric waters.0

The conceptual model of a dry-steam reservoir is shown in Fig. 1.9. The lateral boundaries of this type of reservoir must be impermeable, or liquid would flood sideways into the steam-filled region and collapse the superheated-steam reservoir. In a steam-filled area, the only liquid should be the condensate which forms in the reservoir's cooler lateral and upper boundaries.

As production wells tap the reservoir to provide a steam outflow, a cone-shaped depression forms around the lower-pressure production zone. This lower pressure causes the condensate to evaporate, which in turn causes further steam generation. This may lead to the reservoir eventually drying out completely. The liquid content of the deeper zone beneath the reservoir remains the only possible source for additional

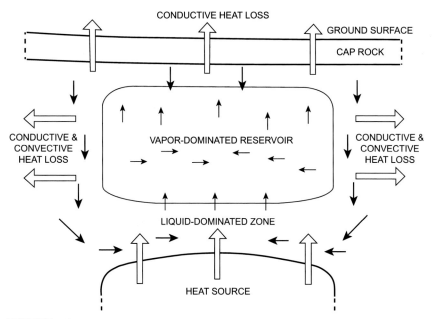

FIGURE 1.9 Conceptual model of a dry-steam reservoir (D'Amore and Truesdell, 1979).

steam. Natural recharge is possible primarily through the lateral boundaries, which are demarcated by major faults with significant offsets.
Surface water then can percolate the deepest liquid-filled region.

If the production rate exceeds the rate of recharge, the deep liquid
region will sink even deeper as decreasing pressure causes more and
more evaporation. A steam zone must have the initial pressure needed to
develop this unbalanced state. This is possible as the natural steam loss
through the upper layer before their occlusion larger than the natural
recharge. In their natural pre-production states both Larderello and the
Geysers were characterized by such surface thermal manifestations as
geysers, fumaroles, hot mud-pools, and acid-altered rocks. Over time,
permeability in the shallow formations decreases, as mineral deposits seal
the fractures and block the upflowing steam.

There is a very close correlation between the temperature of dry-steam
reservoirs and the saturated steam's temperature at maximum enthalpy.
This thermodynamic state can be seen in Fig. 1.10, in the enthalpy–
entropy diagram. At the point M, where the temperature is 235°C and the
pressure 30.6 bar, the enthalpy is 2904 kJ/kg. According to the first law of
thermodynamics, if there is no heat transfer and no work being performed
during the expansion process, the enthalpy must be constant. As the
saturated steam expands and its pressure decreases, above 235°C it turns
to wet steam, as a two-phase water-steam mixture (1-2). Below 235°C, the
expanding saturated steam becomes superheated (3-4).

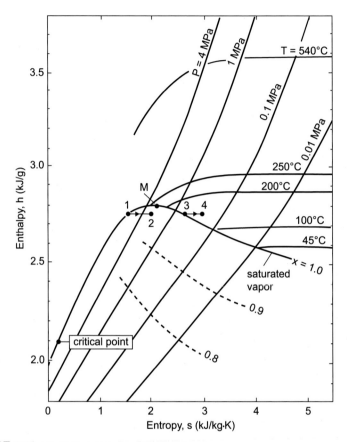

FIGURE 1.10   Enthalpy–entropy diagram of water.

The geothermal fluid is obviously at its hottest in the deepest part of the reservoir. As its lower density causes it to rise up, and as it passes through the fractures or pore channels, its pressure decreases. Meanwhile, a fraction of the steam will separate out of the water and flow upward, with the liquid flowing back down. This brings the steam to a lower pressure region, where its expansion is increased and the process is intensified. McNitt (1977) explained this phenomenon. His basic idea was to compare the amount of heat needed to evaporate a unit mass of water with the amount of heat released as the steam bubble condenses.

It is evident that the steam phase begins as soon as the reservoir temperature approximates the temperature for the enthalpy maximum on the saturation curve. The pressure should change with depth hydrostatically, where the temperature is less than 235°C level. Above that level, temperature and pressure remain almost constant throughout the dry-steam reservoir. Fig. 1.11 shows pressure distribution with increasing

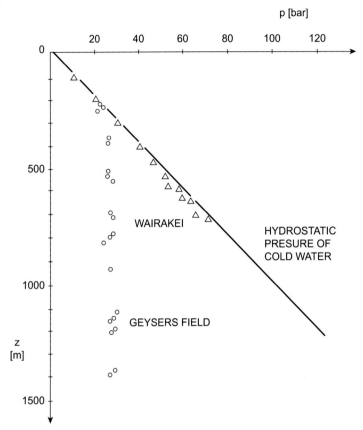

FIGURE 1.11 Pressure distribution along the depth of a dry-steam and a hot water reservoir.

depth, comparing a dry-steam reservoir in the Geysers (USA) with a hot water aquifer in Wairakei (New Zealand).

Huge regions of the Earth's crust have high temperatures, but lack permeable rock matrix and geofluid. In those cases, artificial geothermal reservoirs can be created. The necessary interconnecting fracture system can be produced by hydraulic fracturing, a routinely used technology in the petroleum industry.

The first step is to drill a sufficiently deep, mainly inclined well into the HDR. Then the so-called fracturing fluid is injected under very high pressure, at a pre-determined depth. The high pressure opens up a large crack, in a circular plane with a diameter of a few hundred meters. The aperture created is only a few millimeters wide. The plane of the crack tilts in the direction of the greatest principal stress σ_1, the overburden stress, and runs perpendicular to the direction of the least principal stress.

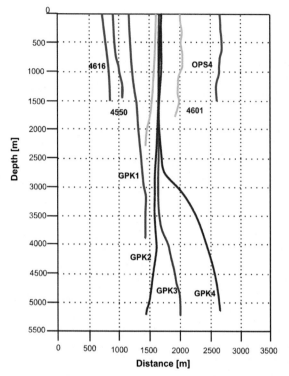

FIGURE 1.12   Position of boreholes in Soultz.

The fracturing fluid is a two-phase fluid-solid mixture, containing solid particles known as proppants. As fracturing fluid is pumped into the opening crack, proppants spread along the rock surface. The crack remains open as long as the pump produces the necessary high pressure. Once the pump stops, the fracturing fluid's pressure decreases to hydrostatic pressure level and the fracture begins to close. The fluid's proppants, typically nut-shell fragments of glass beads, then "prop" the crack open.

When the fracture is big enough, another well is drilled to intercept it. This creates a closed loop whereby cold water is pumped down through the injection well, then flows through the fracture, is warmed up, and finally returns via the production well to the surface. In this way, the energy of the deep hot formations is recovered.

The first experiment was carried out in Fenton Hill, New Mexico, where a single fracture of 40,000 m^2 was made. Although the heat-transfer area proved insufficient, a small portable unit was able to generate electric power. Successful international research efforts have since gone on to develop the HDR concept. In HDR projects, multi-fractured, inter-connecting flow systems are produced. Where very low-permeability formations are enhanced by hydraulic fracturing to create artificial reservoirs, the resulting systems are known as EGS (enhanced geothermal system). The most successful HDR/EGS project to date was implemented in Soultz sous Forets, France, where a 1.5 MW power plant currently operates (see Fig. 1.12).

References

Bertrani, R., 2015. Geothermal power generation in the world 2010-2014 update report. In: Proceedings World Geothermal Congress 2015. Melbourne, Australia.

Boldizsár, T., 1943. Aspects of Geothermal Gradient in Mining Industry (In Hungarian) BKL, 20. Budapest.

Boldizsár, T., 1964. Heat Flow Map of Hungary (In Hungarian). In: Proceedings of Hungarian Academic of Sciences, Budapest, vol. 33, pp. 307–327.

Bullard, E., 1954. The flow of heat through the floor of the Atlantic Ocean. Proc. Royal Soc. London 222 (1150), 408–429.

Cataldi, R., Hodgson, F.S., Lund, W.J., 1999. Stories from a heated Earth. In: Our Geothermal Heritage. Geothermal Resource Council, Sacramento, California, ISBN 0-934412-19-7.

D'Amore, F., Truesdel, A.H., 1979. Models for steam chemistry at Larderello and the geysers. In: Proc. 5th Workshop on Geothermal Reservoir Engineering Stanford University, Stanford, CA, pp. 262–276.

Facca, G., Tonani, F., 1964. Theory and technology of a geothermal field. Bull. Volcanol. 27 (1), 143–189.

McNitt, J.R., 1977. Origin of Steam in Geothermal Reservoirs. SPE Paper 6764.

Tóth, A., 2010. Hungarian country update 2005–2009. In: Proceedings of World Geothermal Congress, Bali, Indonesia.

Basic Equations of Fluid Mechanics and Thermodynamics

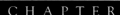
2.1 ELEMENTS OF TRANSPORT THEORY

Like any other material, a fluid is discrete at the microscopic level. The dimensions typically involved in engineering problems are, however, much greater than molecular distance. A mathematical description of fluid flow consequently requires that the actual molecular structure be replaced by a hypothetical continuous medium, called the continuum. This continuum continuously occupies a three-dimensional space, and can be infinitely subdivided while still preserving its original properties.

Copyright © 2017 Elsevier Inc. All rights reserved.

In the continuum model, the actual material is replaced by continuous functions of physical quantities. The actual physical processes are represented by the evolution of these continuous functions in space and time. This spatial- and time-evolution can be expressed by means of partial differential equations. The value of such a function at a certain point is obtained as an average value in the infinitesimal volume dV surrounding the point. The essential mathematical simplification of the continuum model is that this average value is, in the limiting case, assigned to the point itself.

This infinitesimal fluid element of volume dV must contain a sufficient number of molecules to allow a statistical interpretation of the continuum. When the molecular dimensions are very small compared with any characteristic dimension of the flow system, the average values obtained in this fluid element may be considered as variables at a given point. This infinitesimal fluid element is called a fluid particle. This notion of a fluid particle is not to be confused with any particle of the molecular theory.

If any material system is left alone for a sufficient length of time, it will achieve a state of equilibrium. In this equilibrium state, all macroscopically measurable quantities are independent of time. Variables that depend only on the state of the system are called variables of state. The pressure p and the density ρ are obviously variables of the mechanical state of the system. The thermal state of the system can also be characterized by the pressure and density, but it is an experimental observation that the density of the system is not solely a function of its pressure; a third variable of state, the temperature T, must be considered. The relation between the variables of state can be expressed by the equation of state:

$$p = p(\rho, T) \tag{2.1}$$

In the equation of state, the density is often replaced by its reciprocal called the specific volume:

$$\vartheta = \frac{1}{\rho} \tag{2.2}$$

In general, it is not possible to express the equation of state in a simple analytic form (Batchelor, 1967). It can only be tabulated or plotted against the variables of state. It is obvious that an equation such as:

$$p = p(\vartheta, T) \tag{2.3}$$

can be represented by a surface in the coordinate system ϑ, T, p. This surface of state consists of piecewise continuous surface parts as shown later in Chapter 11, Fig. 2.1. It is customary to plot this relation as projected onto any one of the three planes, $p - \vartheta$, $p - T$, or $\vartheta - T$. In such a projection, the third variable of state is treated as a parameter. The most

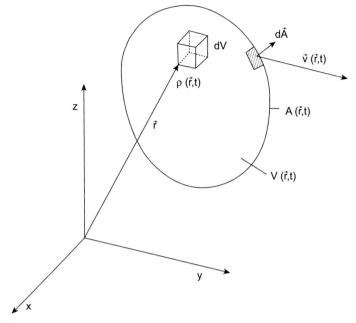

FIGURE 2.1 Material volume of the flowing fluid.

common diagram of state is that of water as shown in Chapter 11, Fig. 2.2. Thus we obtain isothermal curves (isotherms) in the p, ϑ plane, the isobars in the ϑ, T plane, and the isochores in the p, T plane. The shape of the surface of state is characteristic of a particular material.

When the energy relationships, and the exchange of work and heat between a system and its surroundings are considered, it seems to be necessary to define two further variables of state: the internal energy e, and the enthalpy i. For a simple system, e and i are functions of any two. Thus the so-called caloric equations of state express these relationships. They are obtained experimentally; their graphical representations are the so-called Mollier diagrams.

Variables of state can be classified as either intensive or extensive variables.

Intensive variables are related to the points of a system, thus they are point functions. Velocity, pressure, and temperature are typical intensive variables. Intensive variables can relate to a finite region only if their distributions are uniform. In this case, both the values of a variable relating to the whole region, and that relating to a part of it must be the same.

Extensive variables are related to a finite region of a system, thus they are set functions. The mass of the system is such an extensive variable, and so are the volume, momentum, energy, entropy, etc. Set functions are

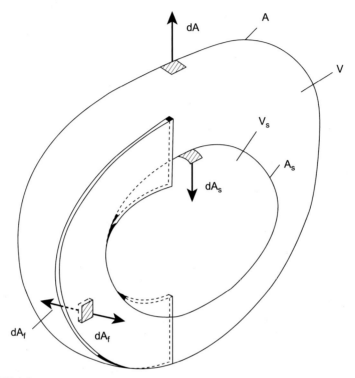

FIGURE 2.2 Multiple connected material volumes.

additive; for example the momentum of a certain body is equal to the sum of the momentums of its parts. Specific quantities such as density or specific enthalpy would seem to be intensive variables since their values vary from point to point in the manner of point functions. There is, however, an essential difference between such specific quantities and real intensive variables. The specific quantities may be integrated over a region thus producing an extensive variable; for a real intensive variable, such an integration has no physical meaning.

Intensive and extensive variables play different roles in physical processes. The state of a material system can be determined by as many extensive variables as there are types of interactions between the system and its surroundings. An interaction of certain type may induce either equilibrium or a process of change. A characteristic pair of intensive and extensive variables are associated with any interaction. The characteristic intensive variable is that one which is uniformly distributed at equilibrium. For instance, experimental observations show that thermal equilibrium can exist only if the temperature has a uniform distribution throughout the whole system. A homogeneous velocity distribution is associated with the condition of mechanical equilibrium. A system with a uniformly distributed velocity U is at rest in a coordinate system moving with the velocity U. Thus temperature and velocity have particular significance as the characteristic variables of thermal and mechanical interaction.

If the distribution of the characteristic intensive variable is not homogeneous, the equilibrium condition is terminated and certain processes are induced accompanied by changes in the extensive variables. These changes are directed so as to neutralize the inhomogeneity. Fluxes of extensive quantities may be convective, i.e., transferred by macroscopic motion; or conductive, i.e., transferred by molecular motion only. Conductive fluxes can be expressed by the product of a conductivity coefficient and the gradient of the characteristic intensive variable. These conductivity coefficients are important material properties.

2.2 BALANCE EQUATIONS

2.2.1 The Principle of Conservation of Mass

The principle of conservation of mass states that the mass of a body is constant during its motion. This can be stated in the rate form as the rate of change with time of the mass of a body being zero. It is obvious that for a material system the above statement can be expressed mathematically.

Consider the volume $V(\vec{r}, t)$ flowing with the fluid. It is constituted of the same particles of fixed identity. The volume $V(\vec{r}, t)$ and its bounding surface $A(\vec{r}, t)$ vary in time representing successive configurations of the same fluid particles. This is illustrated in Fig. 2.1.

Let an infinitesimal volume element be located at a point P, characterized by the position vector $\vec{r}$, within the flowing volume under consideration. The scalar density point function $\rho(\vec{r}, t)$ is the sum of the infinitesimal mass elements, thus it is the volume integral of the density over the volume $V(\vec{r}, t)$. The principle of conservation of mass can be expressed as the material derivative of this volume integral:

$$\frac{d}{dt} \int_{V_t} \rho dV = 0 \tag{2.4}$$

Applying Euler's transport theorem, this expression becomes:

$$\int_V \frac{\partial \rho}{\partial t} dV + \int_{(A)} \rho \vec{v} d\vec{A} = 0 \tag{2.5}$$

Therefore, the sum of rate of change of mass within the fixed volume V, which is an instantaneous configuration of V(t), and the mass flux $\rho \vec{v}$

across the bounding surface of V is zero. The first term represents the local rate of change of mass within the fixed volume V. The surface integral represents the mass which crosses the bounding surface A. This convective mass flux equals the local rate of change of mass.

A conductive mass flux may also occur, primarily as the result of a concentration gradient. This is the so-called ordinary diffusion. The conductive mass flux $\vec{j}$ is given by Fick's law:

$$\vec{j} = -D \operatorname{grad} \rho \qquad (2.6)$$

where D is the so-called diffusivity. This is one of the characteristic physical properties of the fluid, with dimensions (m^2/s).

Within the volume V under consideration, there may also be mass sources or sinks; for instance, the rate of mass produced within V by chemical reactions. (Note that conductive mass flux, sources and sinks relate only to some species of mass.)

If this is the case, the mass balance equation becomes:

$$\int_V \frac{\partial \rho_i}{\partial t} dV + \int_{(A)} \left(\rho_i \vec{v_i} - D_i \operatorname{grad} \rho_i \right) d\vec{A} = \int_V \xi_i dV; \quad i = 1, 2, \ldots n \quad (2.7)$$

where ξ is the strength of the sources or sinks per unit volume. Either of these equations represents the mass balance equation for the i-th species. When all n equations are added together, one obtains:

$$\int_V \frac{\partial \rho}{\partial t} dV + \int_{(A)} \rho \vec{v} d\vec{A} = 0 \qquad (2.8)$$

Since the total mass of the body is always constant, it is necessary for the conductive fluxes (the sum of the sources and the sinks), to vanish.

The conservation of mass equation may be rewritten in differential form. The term representing the surface integral may be transformed into a volume integral by means of Gauss's divergence theorem:

$$\int_V \left[\frac{\partial \rho}{\partial t} + \operatorname{div}(\rho \vec{v}) \right] dV = 0 \qquad (2.9)$$

Since the limit of integration is arbitrary, and ρ and $\vec{v}$ are continuous functions with continuous derivatives, the integrand must be zero. Removing the integral signs, we obtain the well-known continuity equation; the differential equation form of the law of conservation of mass:

$$\frac{\partial \rho}{\partial t} + \operatorname{div}(\rho \vec{v}) = 0 \qquad (2.10)$$

This equation can be expanded both for Cartesian and cylindrical coordinates:

$$\frac{\partial \rho}{\partial t} + \frac{\partial}{\partial x}(\rho v_x) + \frac{\partial}{\partial y}(\rho v_y) + \frac{\partial}{\partial z}(\rho v_z) = 0 \tag{2.11}$$

$$\frac{\partial \rho}{\partial t} + \frac{1}{r}\frac{\partial}{\partial r}(r\rho v_r) + \frac{1}{r}\frac{\partial}{\partial \varphi}(\rho v_\varphi) + \frac{\partial}{\partial z}(\rho v_z) = 0 \tag{2.12}$$

The continuity equation may be transformed into:

$$\frac{\partial \rho}{\partial t} + \vec{v}\, \mathrm{grad}\, \rho + \rho\, \mathrm{div}\, \vec{v} = 0 \tag{2.13}$$

It is clear that the first and second terms represent the local and the convective terms of the material derivative of the density field. Thus we obtain:

$$\frac{d\rho}{dt} + \rho\, \mathrm{div}\, \vec{v} = 0 \tag{2.14}$$

For a fluid of constant density, this equation reduces to:

$$\mathrm{div}\, \vec{v} = 0 \tag{2.15}$$

whether the flow is steady or not, i.e., whether or not the flow is locally time-dependent.

The principle of conservation of mass may be formulated in either integral or differential form. Both forms express the same physical principle. When applying the integral form, it is important to remember that the enclosing surface A must be closed; it encloses a finite volume V of the space through which the fluid flows. Some parts of the boundary surface may consist of real material boundaries, such as pipe walls. Since solid walls are impervious, the normal component of the velocity at the wall must be zero, i.e., here $\vec{v} d\vec{A} = 0$. Thus a solid wall is always a stream surface. Sometimes the control volume V includes an immersed body which interrupts the continuity of the fluid as shown in Fig. 2.2. In this case, the control volume is a multiply-connected continuous region, and thus the continuity equation cannot be written in the form of Eq. (2.5). The discontinuity within the fluid mass must be excluded by introducing an additional control surface (A_2) around its boundary, while it is also necessary to introduce a cut (A_3), which makes the volume V into a single connected region bounded by a single closed surface ($A + A_2 + A_3$). In this manner, formulas written in integral form may be rendered applicable to flows involving discontinuities. In general, the integral form of a balance equation describes the relationship between certain quantities within a finite volume and across its bounding surface.

Differential balance equations, on the other hand, express the relationships for the derivatives of these quantities at a given point of the fluid. Most of the problems treated in this book make use of differential equations, but there are many cases in which application of the integral form is the more suitable.

2.2.2 The Balance of Momentum

Newton's second law of motion is perhaps the most important statement in the field of dynamics. As is well-known, this states that the rate of change of the momentum of a body equals the sum of the external forces acting on it. To apply this law to a flowing material fluid system, consider the volume V, bounded by the simple closed surface (A), made up of the identical particles of fixed identity. The momentum of an infinitesimal volume element of the fluid is $\rho \vec{v} dV$. Since the momentum is an extensive flow variable, the total momentum of the mass of fluid under consideration is the integral of the infinitesimal momentum over the material volume V. The rate of change of momentum is obviously expressed by the material derivative of this volume integral.

The external forces acting on the fluid are of two types; body forces and surface forces. Body forces arise either from action at a distance such as the gravitational or electromagnetic forces, or they occur by reason of the choice of an accelerating frame of reference, e.g., the centrifugal force or the Coriolis force. Such a body force is proportional to the mass, and may be represented by the vector $\vec{g}$ per unit mass; summed over the volume it is the volume integral of $\rho \vec{g} dV$.

The surface forces are due to whatever medium is adjacent to the bounding surface (A) for example the solid wall of a pipeline or the adjacent fluid mass around a jet. The intensity of the surface forces acting on a unit surface is represented by the stress vector $\vec{t}$, thus the force acting on an infinitesimal surface area $d\vec{A}$ is equal to $\vec{t} dA$. The total surface force acting on the fluid inside (A) is the surface integral of $\vec{t} dA$.

These mathematical terms become:

$$\frac{d}{dt} \int_V \rho \vec{v} dV = \int_V \rho \vec{g} dV + \int_{(A)} \vec{t} dA \qquad (2.16)$$

There is a further essential difference between body forces and surface forces. While $\vec{g}$ is a single-valued vector point function, the stress vector $\vec{t}$ can adopt infinitely many vectorial values at a given point for each orientation of the unit normal vector of the surface element dA. The field of the stress vectors $\vec{t}$ is not a regular vector field. The stress vector is not a vector point function. It is a function of both the position vector $\vec{r}$ and the direction of the unit normal vector $\vec{n}$, thus:

$$\vec{t} = \vec{t}(\vec{r}, t, \vec{n}) \tag{2.16a}$$

The totality of all possible corresponding values of $\vec{t}$ and $\vec{n}$ at a given point defines the state of stress. To determine the state of stress is possible without specifying every pair of stress and normal vectors.

In spite of this, we shall express surface forces in the form of a point function. This is made possible by introducing the concept of the stress tensor.

Cauchy's theorem allows us to express the stress vector for a fixed value of r as $\vec{n}$ varies. Cauchy's theorem can be stated as follows: if the stress vectors acting across three mutually perpendicular planes at a given point are known, then all stress vectors at that point can be determined. They are given by the equation:

$$\vec{t} = T(\vec{r}, t) \cdot \vec{n} \tag{2.17}$$

as linear functions of a Cartesian tensor of second order. This second-order tensor is independent of the unit normal vector of the surface. It is a tensor point function, which determines the state of stress of the fluid. T is called the stress tensor. Eq. (2.17) can be written in matrix form as:

$$\begin{bmatrix} t_x \\ t_y \\ t_z \end{bmatrix} = \begin{bmatrix} t_{xx} & t_{yx} & t_{zx} \\ t_{xy} & t_{yy} & t_{zy} \\ t_{xz} & t_{yz} & t_{zz} \end{bmatrix} \begin{bmatrix} n_x \\ n_y \\ n_z \end{bmatrix} \tag{2.18}$$

Examining this matrix equation, it can be easily recognized how the matrix of the stress tensor maps the normal vector $\vec{n}$ into the stress vector $\vec{t}$. In this sense, the stress tensor completely determines the state of stress. Knowing the components of the matrix, the stress vector can be obtained for any arbitrary surface if its (unit) normal vector $\vec{n}$ is known. The components of the stress vector in the Cartesian reference frame are given by:

$$\vec{t} = (t_{xx}n_x + t_{yx}n_y + t_{zx}n_z)\vec{i} + (t_{xy}n_x + t_{yy}n_y + t_{zy}n_z)\vec{j}$$
$$+ (t_{yz}n_x + t_{yz}n_y + t_{zz}n_z)\vec{k} \tag{2.19}$$

As it is shown in Fig. 2.3, the first letter in the index of a stress tensor element designates the reference axes, which are perpendicular to the plane in which the stress occurs. The second letter designates the direction of this component. The normal components t_{xx}, t_{yy}, and t_{zz} are the normal stresses the others, t_{xy}, t_{xz}, etc., are the shear stresses. An alternative notation is to replace t_{xx}, t_{yy}, and t_{zz}, with σ_x, σ_y, σ_z.

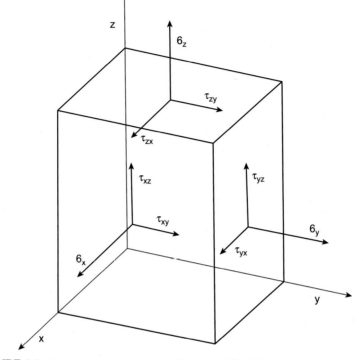

FIGURE 2.3   Stress components on an infinitesimal fluid element.

Applying Euler's transport theorem, and replacing the stress vector by the stress tensor, the balance of momentum equation can be written as:

$$\int_V \frac{\partial(\rho\vec{v})}{\partial t}\,dV + \int_{(A)} \rho\vec{v}\,(\vec{v}\,d\vec{A}) = \int_V \rho\vec{g}\,dV + \int_{(A)} T\,d\vec{A} \qquad (2.20)$$

The first term on the left-hand side of the equation expresses the rate of change of momentum within the fixed volume V. The second term represents the convective momentum flux which crosses the bounding surface of the fluid mass. On the right-hand side, the first term represents the resultant body force while the second one represents the resultant of the surface forces acting on the bounding surface.

The balance of momentum equation may be rewritten in differential form. Its surface integrals may be transformed into volume integrals by means of the divergence theorem. We first apply this to the identity, which relates to the convective momentum flux, regarding the identity:

$$\int_{(A)} \rho \vec{v}(\vec{v} d\vec{A}) = \int_{(A)} \rho(\vec{v} \circ \vec{v}) d\vec{A} \qquad (2.20a)$$

thus we obtain:

$$\int_V \left[\frac{\partial(\rho \vec{v})}{\partial t} + \text{Div}(\rho \vec{v} \circ \vec{v}) \right] dV = \int_V (\rho \vec{g} + \text{Div } T) dV \qquad (2.21)$$

Since the limits of integration are arbitrary and the integrands are continuous functions with continuous derivatives, the integral signs may be removed, yielding the differential equation:

$$\frac{\partial(\rho \vec{v})}{\partial t} + \text{Div}(\rho \vec{v} \circ \vec{v}) = \rho \vec{g} + \text{Div } T \qquad (2.22)$$

The total rate of change of the momentum within the unit volume equals the external forces acting on it. The local and the convective derivatives of the momentum can be clearly recognized on the left-hand side. Of interest is the second term on the right-hand side of the equation, which represents the resultant of the surface forces acting on a unit volume of the flowing fluid. This resultant surface force is a vector point function which depends only on the position vector and time. It is independent of either the chosen coordinate system or the shape of the unit volume. Div T forms a vector field in contrast to the stress vector $\vec{t}$, which is also a function of the unit normal vector. The divergence of the stress tensor field expresses the inhomogeneity of the state of stress of the flowing fluid.

There is another way to express completely the state of stress at an arbitrarily given point. Three stress vectors obtained on each of three mutually perpendicular planes at a given point are sufficient for this purpose. Choosing three planes perpendicular to the coordinate axes the appropriate stress vectors $\vec{t}_x$, $\vec{t}_y$ and $\vec{t}_z$ may be written in terms of its orthogonal components as:

$$\begin{aligned} \vec{t}_x &= t_{xx} \vec{i} + t_{xy} \vec{j} + t_{xz} \vec{k} \\ \vec{t}_y &= t_{yx} \vec{i} + t_{yy} \vec{j} + t_{yz} \vec{k} \\ \vec{t}_z &= t_{zx} \vec{i} + t_{zy} \vec{j} + t_{xz} \vec{k} \end{aligned} \qquad (2.23)$$

It is obvious that the divergence of the stress tensor can be replaced by the equivalent expression:

$$\text{Div } \underline{\underline{T}} = \frac{\partial \vec{t}_x}{\partial x} + \frac{\partial \vec{t}_y}{\partial y} + \frac{\partial \vec{t}_z}{\partial z} \qquad (2.24)$$

It should be noted that stress vectors $\vec{t}_x$, $\vec{t}_y$ and $\vec{t}_z$ and the scalar components of stress vector:

$$\vec{t} = t_x \vec{i} + t_y \vec{j} + t_z \vec{k}$$

are merely different quantities.

Let us examine a further physical property of the stress tensor. The momentum equation may be written as:

$$\frac{\partial(\rho \vec{v})}{\partial t} + \text{Div}(\rho \vec{v} \circ \vec{v} - T) = \rho \vec{g} \tag{2.25}$$

The terms representing the local and the convective momentum flux are easily recognized. In this form the physical meaning of the stress tensor is very obvious; the stress tensor T represents the conductive flux of the momentum. This conductive momentum flux is transported by virtue of molecular motion, i.e., the continual random interchange of molecules between adjacent fluid elements produces a transfer of momentum. This phenomenon can be described by the laws of probability and the methods of statistics; from probability distributions it is possible to determine averages or mean values for such parameters as velocity, pressure, etc. These mean values can then be related to the macroscopically measurable quantities. Molecular motion disperses the momentum. This tendency of equalization occurs macroscopically as a deterministic process. This is the reason why it is possible to give a phenomenological description of the conductive momentum flux in a continuum.

The idea of a momentum flux without accompanying macroscopic motion may be unfamiliar. Let us consider an example of this from the field of petroleum engineering. Imagine a pump connected to a pipeline. The pipeline-filling fluid is at rest. Starting the pump, the pressure increases, and the fluid flow accelerates quickly. Even if the velocity of the flowing fluid is about 1 m/s, the acceleration and the increase in pressure rising propagates as a wave throughout the pipeline with the speed of sound, some 1000 m/s. As the fluid mass starts to flow faster, the momentum is transferred by convection, at the velocity of the flow. In this example the propagation of the pressure increase and the acceleration is carried out by means of conductive momentum transfer.

Eq. (2.25) is valid for any fluid, and indeed for any continuous medium, regardless of whether mass is conserved or not. Taking into account the principle of conservation of mass, the equation of motion (Eq. 2.22) can be written in a simpler form. Expanding the derivatives of the products on the left-hand side we obtain:

$$\rho \frac{\partial \vec{v}}{\partial t} + \vec{v} \frac{\partial \rho}{\partial t} + \vec{v} \nabla(\rho \vec{v}) + \rho(\vec{v} \nabla) \vec{v} = \rho \vec{g} + \text{Div } T \tag{2.26}$$

The second and the third terms on the left-hand side vanish on account of the continuity equation. Dividing by the density, a very simple and elegant equation is obtained:

$$\frac{d\vec{v}}{dt} = \vec{g} + \frac{1}{\rho}\text{Div } T \tag{2.27}$$

This is Cauchy's equation of motion, which expresses the balance of momentum for a unit fluid mass. The scalar components of Cauchy's equation in Cartesian coordinates can be easily written as:

$$\frac{\partial v_x}{\partial t} + v_x\frac{\partial v_x}{\partial x} + v_y\frac{dv_x}{\partial y} + v_z\frac{\partial v_x}{\partial z} = g_x + \frac{1}{\rho}\left(\frac{\partial \sigma_x}{\partial x} + \frac{\partial \tau_{yx}}{\partial y} + \frac{\partial \tau_{zx}}{\partial z}\right)$$

$$\frac{\partial v_y}{\partial t} + v_x\frac{\partial v_y}{\partial x} + v_y\frac{\partial v_y}{\partial y} + v_z\frac{\partial v_y}{\partial z} = g_y + \frac{1}{\rho}\left(\frac{\partial \tau_{xy}}{\partial x} + \frac{\partial \sigma_y}{\partial y} + \frac{\partial \tau_{zy}}{\partial z}\right) \tag{2.28}$$

$$\frac{\partial v_z}{\partial t} + v_x\frac{\partial v_z}{\partial x} + v_y\frac{\partial v_z}{\partial y} + v_z\frac{\partial v_z}{\partial z} = g_z + \frac{1}{\rho}\left(\frac{\partial \tau_{xz}}{\partial x} + \frac{\partial \tau_{yz}}{\partial y} + \frac{\partial \sigma_z}{\partial z}\right)$$

The alternative form of Cauchy's equation is:

$$\frac{d\vec{v}}{dt} = \vec{g} + \frac{1}{\rho}\left(\frac{\partial \vec{t}_x}{\partial x} + \frac{\partial \vec{t}_y}{\partial y} + \frac{\partial \vec{t}_z}{\partial z}\right) \tag{2.29}$$

Its Cartesian components can be written as:

$$\frac{\partial v_x}{\partial t} + v_x\frac{\partial v_x}{\partial x} + v_y\frac{\partial v_x}{\partial y} + v_z\frac{\partial v_x}{\partial z} = g_x + \frac{1}{\rho}\left(\frac{\partial t_{xx}}{\partial x} + \frac{\partial t_{yx}}{\partial y} + \frac{\partial t_{zx}}{\partial z}\right)$$

$$\frac{\partial v_y}{\partial t} + v_x\frac{\partial v_y}{\partial x} + v_y\frac{\partial v_y}{\partial y} + v_z\frac{\partial v_y}{\partial z} = g_y + \frac{1}{\rho}\left(\frac{\partial t_{xy}}{\partial x} + \frac{\partial t_{yy}}{\partial y} + \frac{\partial t_{zy}}{\partial z}\right) \tag{2.30}$$

$$\frac{\partial v_z}{\partial t} + v_x\frac{\partial v_z}{\partial x} + v_y\frac{\partial v_z}{\partial y} + v_z\frac{\partial v_z}{\partial z} = g_z + \frac{1}{\rho}\left(\frac{\partial t_{xz}}{\partial x} + \frac{\partial t_{yz}}{\partial y} + \frac{\partial t_{zz}}{\partial z}\right)$$

This system of partial differential equations is the so-called stress equation of motion. There are many types of fluids with many different types of stress tensor. Thus we have equations of motion for inviscid fluids (Euler's equation), for linearly viscous fluids (Navier–Stokes equation), and further types of equations for the great variety of non-Newtonian fluids.

In a fluid at rest, all shear stresses vanish, and all normal stresses are equal:

$$T = \begin{bmatrix} -p & 0 & 0 \\ 0 & -p & 0 \\ 0 & 0 & -p \end{bmatrix} \tag{2.31}$$

This is the so-called hydrostatic state of stress, which can be written as:

$$T = -pI \tag{2.32}$$

It is an experimental observation that normal stresses are much greater than shear stresses, and there is a class of fluid flow problems, which can be solved rather precisely by completely neglecting shear stresses. The model of a perfect fluid is obtained in this way. Substituting Eq. (2.32) into Cauchy's equation, Euler's equation of motion is obtained:

$$\frac{\partial \vec{v}}{\partial t} + (\vec{v}\nabla)\vec{v} = \vec{g} - \frac{1}{\rho}\text{grad } p \tag{2.33}$$

Actual fluids exhibit a certain resistance to changes of shape. The viscosity of a fluid is a measure of its resistance to angular deformation. This phenomenon can be described only by taking into account the shear stresses. The first equation for the shear stresses in one-dimensional viscous flow was given by Newton:

$$\tau_{xy} = \mu \frac{dv_x}{dy} \tag{2.34}$$

Thus, the shear stress is proportional to the velocity gradient. The proportionality factor μ is called the dynamic viscosity coefficient (or just dynamic viscosity). The kinematic viscosity of a fluid is the ratio of the viscosity to the density:

$$\nu = \frac{\mu}{\rho} \tag{2.35}$$

A fluid for which the viscosity coefficient does not change with the rate of deformation is said to be a Newtonian fluid. There are certain types of fluids, especially crude oils and drilling muds, in which the viscosity varies with the rate of deformation. In this chapter, however, only Newtonian fluids are considered.

Newton's viscosity law was generalized by Stokes in tensorial form. For incompressible Newtonian fluids it can be written as:

$$T = -pI + 2\,\mu S, \tag{2.36}$$

in which:

$$S = \frac{1}{2}(\vec{v}\circ\nabla + \nabla\circ\vec{v}) \tag{2.36a}$$

Observations of viscous flows show that two types of flow may occur depending on the ratio of the inertial forces to the viscous forces in the flow. Viscous forces predominate in a body or a stream tube of small size, combined with a relatively small velocity and a large kinematic viscosity. In this case the fluid flows in well-ordered parallel layers, without any mixing. This type of flow is called laminar. Inertial forces predominate when the sizes and the velocities involved are relatively large, while kinematic viscosities are small. In most engineering applications, the latter conditions are satisfied. In such cases the disordered, fluctuating flow exhibits irregular transverse movements across the adjacent layers, with intensive mixing. The most obvious property of this type of flow is that mass, momentum, and energy are transferred across the flow at rates which are much greater than those of the molecular transport processes which occur in laminar flow. This type of flow is called turbulent. In the present chapter only laminar flows are considered.

The Navier–Stokes equation is the differential form of the momentum equation for Newtonian fluids. Cauchy's general equation of motion is:

$$\frac{d\vec{v}}{dt} = \vec{g} + \frac{1}{\rho} \text{Div } T \tag{2.37}$$

The constitutive relation of a Newtonian fluid can be substituted into Cauchy's equation.

For incompressible flow we obtain:

$$\frac{d\vec{v}}{dt} = \vec{g} - \frac{1}{\rho} \text{grad } p + \nu \Delta \vec{v} \tag{2.38}$$

This differential equation states that the time rate of the momentum of the unit mass fluid equals the sum of the body force, the pressure force, and the viscous force acting to it. The last term is the force which resists the angular deformation.

The most conspicuous feature of turbulent flow is that momentum, energy, and heat are transferred across the flow by molecular transport processes (viscous momentum transfer, diffusion), at a rate much greater than would be the case in a laminar flow. This phenomenon is the result of additional transport by the random motion of fluid particles across the flow and the concurrent mixing of the convectively transported momentum and energy between neighboring fluid elements. In steady flow, time-averaged mean values are determined. These mean values can be compared with turbulent fluctuations in a similar manner as the macroscopic quantities of laminar flow with the thermal motion of molecules. The particles which mix in a turbulent flow are much larger than molecules, thus turbulent transport is a much more intensive phenomenon than the molecular transport process. Thus the sudden increase in fluid friction or heat transfer is as suitable a method for detecting the occurrence of turbulent flow as direct visual observation. The engineer is fortunate in being able to measure such a sudden increase easily by using the usual equipment of petroleum engineering practice.

The velocity of a turbulent flow should be interpreted in terms of a mean value with a superimposed high-frequency stochastic fluctuation of irregular, changing amplitude. If instead of a hot wire anemometer, one uses a Prandtl tube for measuring the velocity, fluctuations cannot be perceived but only the time-averaged mean value. Thus the velocity of a turbulent flow can be considered as the resultant of the mean velocity and the fluctuations. The pressure of a turbulent flow can be similarly split up:

$$\vec{v}_t = \vec{v} + \vec{v}'$$ (2.39)

$$p_t = p + p'$$ (2.39a)

where $\vec{v}_t$ and p_t are the real instantaneous velocity and pressure of the flow, $\vec{v}$ and p represent their averaged values over a sufficiently long period of time, and $\vec{v}'$ and p' are the fluctuations of the velocity and pressure about the mean. Reynolds proposed the following simple relationship for determining the mean values:

$$F(\vec{r}, t) = \overline{F}_t(\vec{r}, t) = \frac{1}{t_0} \int_t^{t+t_0} F_t(\vec{r}, t)dt$$ (2.40)

in which $F_t(\vec{r}, t)$ is some arbitrarily chosen scalar or vector function. The overbar represents the time-averaged value of F_t. Assume the existence of a time interval t_0, which is large relative to the time interval of a fluctuation, but relatively small compared to the time interval over which the mean velocity was determined. Averaging over this interval yields a result which is independent of t_0, and remains unchanged during further averaging processes. Thus:

$$F(\vec{r}, t) = \overline{F}(\vec{r}, t)$$ (2.41)

i.e., the further time average of the time average remains unchanged. This mean value is not necessarily constant; all time-averaged functions can undergo slow variations, but the time-dependence of the macroscopic phenomena is of a different order of magnitude than the period of the turbulent fluctuations. Consider for example, the velocity of a fluid discharging from a tank. The velocity obtained from the Torricelli equation is the mean velocity v. The turbulent fluctuations are superimposed onto this mean. As the water level in the tank decreases, the value of v also decreases. The duration of the discharging T is considerably greater than the time interval over which t_0 is averaged, which may be the time used to measure the outflow velocity by means of a Pitot tube.

It follows from Eq. (2.41) that the time average of the fluctuations is zero:

$$\overline{F}(\vec{r}, t) = F(\vec{r}, t) - \overline{F}_t(\vec{r}, t) = 0 \tag{2.42}$$

A consequence of Eq. (2.40) is that the time average of a spatial derivative is equal to the spatial derivative of a time-averaged function, since the time-averaging operation is independent of the differentiation with respect to the coordinates:

$$\frac{\overline{\partial F_t}}{\partial x} = \frac{\partial \overline{F}_t}{\partial x} \tag{2.43}$$

The same relationship is also valid for the partial derivative w.r.t. time which relates to large scale changes in time:

$$\frac{\overline{\partial F_t}}{\partial t} = \frac{\partial \overline{F}_t}{\partial t} \tag{2.44}$$

Reynolds, using the above method of time-averaging, stated that the Navier–Stokes equation is also valid for turbulent flow provided that the instantaneous value $\vec{v}$, and p_t are substituted into it. It is necessary to carry out the averaging using Eq. (2.40) in order to replace the rather impractical values $\vec{v}_t$ and p_t are substituted into it. It is necessary to carry out the averaging using Eq. (2.40) in order to replace the rather impractical values $\vec{v}_t$ and p_t by the readily measurable values $\vec{v}$ and p, which already are suitable for boundary conditions.

Consider the equation of motion for a viscous fluid in integral form, substituting $\vec{v}_t$ and p_t into it. Further assumptions are that the density and viscosity of the fluid are constant. Thus the momentum equation is:

$$\int_V \frac{\partial(\rho \vec{v}_t)}{\partial t} dV + \int_{(A)} \rho(\vec{v}_t \circ \vec{v}_t) d\vec{A} = \int_V \rho \vec{g} dV - \int_{(A)} p_t d\vec{A}$$

$$+ \int_{(A)} \mu(\vec{v}_t \circ \nabla + \nabla \circ \vec{v}_t) d\vec{A} \tag{2.45}$$

Replacing $\vec{v}_t$ and p_t with the sum of their mean and fluctuations, and after averaging, we have:

$$\int_V \frac{\partial}{\partial t} \overline{(\rho \vec{v} + \rho \vec{v}')} dV + \int_{(A)} \rho \overline{\left[(\vec{v} + \vec{v}') \circ (\vec{v} + \vec{v}') \right]} d\vec{A}$$

$$= \int_V \rho \vec{g} dV - \int_{(A)} \overline{\left(p + p' \right)} d\vec{A}$$

$$+ \int_{(A)} \mu \overline{\left[(\vec{v} + \vec{v}') \circ \nabla + \nabla \circ (\vec{v} + \vec{v}') \right]} d\vec{A} \tag{2.46}$$

We must consider the following relationship for averaging the term representing the convective transport of momentum, since for averaging the dyadic product we have:

$$\overline{(\overrightarrow{v} + \overrightarrow{v}') \circ (\overrightarrow{v} + \overrightarrow{v}')} = \overline{\overrightarrow{v} \circ \overrightarrow{v}} = \overline{\overrightarrow{v} \circ \overrightarrow{v}'} + \overline{\overrightarrow{v}' \circ \overrightarrow{v}} + \overline{\overrightarrow{v}' \circ \overrightarrow{v}'} \qquad (2.47)$$

The average of the product of the mean velocities remains unchanged, the average of the product of the mean and the fluctuation is equal to zero, but the average of the product of the velocity fluctuations does not vanish. Thus:

$$\overline{(\overrightarrow{v} + \overrightarrow{v}') \circ (\overrightarrow{v} + \overrightarrow{v}')} = \overrightarrow{v} \circ \overrightarrow{v} + \overline{\overrightarrow{v}' \circ \overrightarrow{v}'} \qquad (2.48)$$

The average of the fluctuations vanishes in all other terms. Finally, the momentum equation can be written in the following form:

$$\int_V \frac{\partial(\rho \overrightarrow{v})}{\partial t} dV + \int_{(A)} \rho(\overrightarrow{v} \circ \overrightarrow{v}) d\overrightarrow{A} + \int_{(A)} \rho\overline{(\overrightarrow{v}' \circ \overrightarrow{v}')} d\overrightarrow{A}$$

$$= \int_V \rho\overrightarrow{g}dV - \int_{(A)} pd\overrightarrow{A} + \int_{(A)} \mu(\overrightarrow{v} \circ \nabla + \nabla \circ \overrightarrow{v}) d\overrightarrow{A} \qquad (2.49)$$

The rate of change of the momentum relative to the mean velocity field is not equal to the resultant of the external forces. The third integral on the left-hand side expresses the irreversible convective transport of momentum due to the turbulent velocity fluctuation. If this term is carried over to the right-hand side of the momentum equation, we can interpret it as an external force acting on the surface (A) with its work dissipating into the internal energy of the fluid in a manner similar to the viscous force. The dyadic product in this term can be interpreted as a second-order tensor of the apparent turbulent shear stresses:

$$\overrightarrow{T}' = -\rho\overline{(\overrightarrow{v}' \circ \overrightarrow{v}')} \qquad (2.50)$$

Its product with the vectorial surface element $d\overrightarrow{A}$ yields at the bounding surface the apparent external force caused by the momentum flux resulting from the turbulent fluctuations. It is called the apparent surface force since in reality this additional turbulent force acts on the intermixing layers of the fluid rather than on a surface.

The integral form of the momentum equation can be replaced by a differential equation. The integral equation expresses the balance of momentum for an arbitrary finite volume of fluid, in contrast, the differential equation refers to a unit volume of fluid. Using Gauss' theorem, the surface integrals may be expressed as volume integrals after which all integrals are taken over an infinitesimal control volume, i.e., the earlier finite volume shrinks to a point. By this method we obtain the equation:

$$\frac{\partial(\rho\vec{v})}{\partial t} + \text{Div}(\rho\vec{v}\circ\vec{v}) = \rho\vec{g} - \text{grad } p + \mu\Delta\vec{v} + (\mu + \kappa)\text{grad div }\vec{v}$$
$$- \text{Div}\overline{\left(\rho\vec{v}'\circ\vec{v}'\right)}$$

(2.51)

If the fluid is incompressible and no diffusion occurs, we obtain the momentum equation for a unit mass after dividing by the density. The resulting equation has the dimensions of an acceleration:

$$\frac{\partial\vec{v}}{\partial t} + (\vec{v}\nabla)\vec{v} = \vec{g} - \frac{1}{\rho}\nabla p + \nu\Delta\vec{v} - \overline{\left(\vec{v}'\circ\vec{v}'\right)}\nabla \qquad (2.52)$$

This is the so-called Reynolds equation, named after its originator Osborne Reynolds. The equations for the scalar components, in rectangular coordinates, are:

$$\frac{\partial v_x}{\partial t} + v_x\frac{\partial v_x}{\partial x} + v_y\frac{\partial v_x}{\partial y} + v_z\frac{\partial v_x}{\partial z}$$
$$= g_x - \frac{1}{\rho}\frac{\partial p}{\partial x} + \nu\left(\frac{\partial^2 v_x}{\partial x^2} + \frac{\partial^2 v_x}{\partial y^2} + \frac{\partial^2 v_x}{\partial z^2}\right) - \frac{\partial}{\partial x}\overline{\left(v_x'v_x'\right)} - \frac{\partial}{\partial y}\overline{\left(v_x'v_y'\right)} - \frac{\partial}{\partial z}\overline{\left(v_x'v_z'\right)};$$

$$\frac{\partial v_y}{\partial t} + v_x\frac{\partial v_y}{\partial x} + v_y\frac{\partial v_y}{\partial y} + v_z\frac{\partial v_y}{\partial z}$$
$$= g_y - \frac{1}{\rho}\frac{\partial p}{\partial y} + \nu\left(\frac{\partial^2 v_y}{\partial x^2} + \frac{\partial^2 v_y}{\partial y^2} + \frac{\partial^2 v_y}{\partial z^2}\right) - \frac{\partial}{\partial x}\overline{\left(v_y'v_x'\right)} - \frac{\partial}{\partial y}\overline{\left(v_y'v_y'\right)} - \frac{\partial}{\partial z}\overline{\left(v_y'v_z'\right)};$$

$$\frac{\partial v_z}{\partial t} + v_x\frac{\partial v_z}{\partial x} + v_y\frac{\partial v_z}{\partial y} + v_z\frac{\partial v_z}{\partial z}$$
$$= g - \frac{1}{\rho}\frac{\partial p}{\partial z} + \nu\left(\frac{\partial^2 v_z}{\partial x^2} + \frac{\partial^2 v_z}{\partial y^2} + \frac{\partial^2 v_z}{\partial z^2}\right) - \frac{\partial}{\partial x}\overline{\left(v_z'v_x'\right)} - \frac{\partial}{\partial y}\overline{\left(v_z'v_y'\right)} - \frac{\partial}{\partial z}\overline{\left(v_z'v_z'\right)}$$

(2.53)

It is clear from this equation, that the apparent turbulent stress tensor has six independent scalar components. There is no relationship between the apparent turbulent stresses and other kinematical quantities equivalent to Stokes's law for viscous stresses. A very good approximation, however, is provided by Prandtl's mixing-length theory for one-dimensional flow. This theory, within its range of validity, is in excellent agreement with experimental results.

Invoking the principle of the conservation of mass does not place any restriction on the general applicability of Cauchy's equation to a homogeneous one-phase flow. In multicomponent systems, mass transfer may be important, thus the equation of motion for any phase must be used in the form of Eq. (2.22).

2.2.3 The Balance of Angular Momentum

The principle of conservation of angular momentum belongs to the axioms of mechanics. In spite of its general validity, it has been largely neglected in textbooks written for petroleum engineers. The treatment of the equation for the balance of angular momentum is usually restricted to the verification of the theorem of symmetry of the stress tensor. Nevertheless the balance of angular momentum equation is of great value in solving flow problems where torques are more significant than forces, e.g., in turbomachinery as it is shown in Csanády's (1965) excellent book.

The principle of conservation of angular momentum states that the rate of change of angular momentum of a material volume of fluid V equals the resultant torque exerted by any external forces on this volume of fluid. In order to discuss this statement in more detail, consider an arbitrary material volume V bounded by the closed surface (A), which is moving with the fluid. Let $\vec{r}$ denote the position vector of an arbitrary point within the volume. At this point the infinitesimal mass element ρdV has an angular momentum, relative to the origin of the coordinate system, of $\vec{r} \times \rho\vec{v}\,dV$; the torque due the body forces is $\vec{r} \times \rho\vec{g}\,dV$, and the torque due the surface forces is $\vec{r} \times \vec{t}\,dA$. Taking the substantial derivative of the volume integral of the angular momentum over the material volume V, and integrating the torques over V and (A), we get the equation:

$$\frac{d}{dt}\int_V \vec{r} \times \rho\vec{v}\,dV = \int_V \vec{r} \times \rho\vec{g}\,dV + \int_{(A)} \vec{r} \times \vec{t}\,dA \qquad (2.54)$$

Its left-hand side may be transformed as follows:

$$\frac{d}{dt}\int_V \vec{r} \times \rho\vec{v}\,dV = \int_V \left[\frac{d}{dt}(\vec{r} \times \rho\vec{v}) + (\vec{r} \times \rho\vec{v})\mathrm{div}\,\vec{v}\right]dV$$

$$= \left[\frac{d\vec{r}}{dt} \times \rho\vec{v} + \vec{r} \times \rho\frac{d\vec{v}}{dt} + (\vec{r} \times \vec{v})\left(\frac{d\rho}{dt} + \rho\,\mathrm{div}\,\vec{v}\right)\right.$$

$$\left. \times \right]dV$$

$$(2.55)$$

The first term on the right-hand side represents the cross-product of parallel vectors (since $\vec{v} = d\vec{r}/dt$), and thus vanishes. It is easy to recognize the continuity equation in the third and fourth terms on the right-hand side, and thus their sum must also be zero. By this means:

$$\frac{d}{dt} \int_V \vec{r} \times \rho \vec{v} \, dV = \int_V \vec{r} \times \rho \frac{d\vec{v}}{dt} \, dV \qquad (2.56)$$

Apply the divergence theorem to the surface integral in Eq. (3.28) considering Eq. (3.26):

$$\int_{(A)} \vec{r} \times \vec{t} \, dA = \int_V \left[\frac{\partial}{\partial x} (\vec{r} \times \vec{t}_x) + \frac{\partial}{\partial y} (\vec{r} \times \vec{t}_y) + \frac{\partial}{\partial z} (\vec{r} \times \vec{t}_z) \right] dV$$

$$(2.57)$$

According to the derivation rule of products the following expression is obtained:

$$\int_{(A)} \vec{r} \times \vec{t} \, dA = \int_V \vec{r} \times \left(\frac{\partial \vec{t}_x}{\partial x} + \frac{\partial \vec{t}_y}{\partial y} + \frac{\partial \vec{t}_z}{\partial z} \right) dV$$

$$+ \int_V \left(\frac{\partial \vec{r}}{\partial x} \times \vec{t}_x + \frac{\partial \vec{r}}{\partial y} \times \vec{t}_y + \frac{\partial \vec{r}}{\partial z} \times \vec{t}_z \right) dV \qquad (2.58)$$

It can be recognized that:

$$\int_V \vec{r} \times \left(\frac{\partial \vec{t}_x}{\partial x} + \frac{\partial \vec{t}_y}{\partial y} + \frac{\partial \vec{t}_z}{\partial z} \right) dV = \int_V \vec{r} \times \mathrm{Div}\underline{\underline{T}} \, dV \qquad (2.59)$$

In the other hand, it is obvious that:

$$\frac{\partial \vec{r}}{\partial x} = \vec{i}; \quad \frac{\partial \vec{r}}{\partial y} = \vec{j}; \quad \frac{\partial \vec{r}}{\partial z} = \vec{k}, \qquad (2.60)$$

thus the second volume integral of Eq. (3.32) can be written:

$$\int_V \left(\frac{\partial \vec{r}}{\partial x} \times \vec{t}_x + \frac{\partial \vec{r}}{\partial y} \times \vec{t}_y + \frac{\partial \vec{r}}{\partial z} \times \vec{t}_z \right) dV$$

$$= \int_V \left(\vec{i} \times \vec{t}_x + \vec{j} \times \vec{t}_y + \vec{k} \times \vec{t}_z \right) dV \qquad (2.61)$$

The cross-products in the right-hand side can be calculated successively as:

$$\vec{i} \times \vec{t}_x = \begin{vmatrix} \vec{i} & \vec{j} & \vec{k} \\ 1 & 0 & 0 \\ t_{xx} & t_{xy} & t_{xz} \end{vmatrix} = -t_{xz}\,\vec{j} + t_{xy}\,\vec{k}$$

$$\vec{i} \times \vec{t}_y = \begin{vmatrix} \vec{i} & \vec{j} & \vec{k} \\ 0 & 1 & 0 \\ t_{yx} & t_{yy} & t_{yz} \end{vmatrix} = -t_{yz}\,\vec{i} + t_{yx}\,\vec{k} \qquad (2.62)$$

$$\vec{k} \times \vec{t}_z = \begin{vmatrix} \vec{i} & \vec{j} & \vec{k} \\ 0 & 0 & 1 \\ t_{zx} & t_{zy} & t_{zz} \end{vmatrix} = -t_{zy}\,\vec{i} + t_{zx}\,\vec{j}$$

Finally, it is obtained:

$$\vec{i} \times \vec{t}_x + \vec{j} \times \vec{t}_y + \vec{k} \times \vec{t}_z = (t_{yz} - t_{zy})\,\vec{i} + (t_{zx} - t_{xz})\,\vec{j}$$
$$+ (t_{xy} - t_{yx})\,\vec{k} \qquad (2.63)$$

Substituting (3.30), (3.33), and (3.37) into (3.38), after some rearranging we get:

$$\int_V \vec{r} \times \left(\rho\frac{d\vec{v}}{dt} - \rho\vec{g} - \mathrm{Div}\underline{\underline{T}} \right) dV = \int_V \left[(t_{yz} - t_{zy})\,\vec{i} + (t_{zx} - t_{xz})\,\vec{j} \right.$$
$$\left. + (t_{xy} - t_{yx})\,\vec{k} \right] dV \qquad (2.64)$$

Cauchy's equation of motion in the left-hand side volume integral can be recognized, thus both sides of Eq. (2.61) must be zero. In consequence of this, all scalar components of the right-hand side integral are equal to zero, that is the shear stresses satisfy the following relations.

The sum within the bracket on the left-hand side can be recognized as representing the momentum equation, which in this form equals zero. Thus the integrand on the right-hand side must also equal zero. This condition is satisfied when:

$$t_{yz} = t_{zy} \quad t_{zx} = t_{xz} \quad t_{xy} = t_{yx} \qquad (2.65)$$

This means that the stress tensor is symmetric if the flowing fluid satisfies the continuity equation, the momentum equation, and the angular momentum equation. The postulate of the symmetry of the stress tensor is equivalent to the determination of three scalar functions from three scalar equations. If the stress tensor is symmetric the momentum equation and the angular momentum equation are not independent, one of them may be chosen to solve a flow problem, while the other is replaced by the postulate of symmetry.

2.2.4 The Balance of Kinetic Energy

Let E_k denote the kinetic energy of a material volume:

$$E_k = \int_V \rho \frac{v^2}{2} dV \tag{2.66}$$

Although the kinetic energy is an extensive variable, it is not a quantity which is conserved. Thus the balance equation of kinetic energy does not express a conservation law (Batchelor, 1967). It is not an axiom of mechanics, and can be derived from the continuity and the momentum equations.

Using the transport theorem, the rate of change of E_k can be written as:

$$\frac{d}{dt} \int_V \rho \frac{v^2}{2} dV = \int_V \frac{d}{dt}\left(\rho \frac{v^2}{2}\right) dV + \int_V \rho \frac{v^2}{2} \operatorname{div} \vec{v} \, dV \tag{2.67}$$

The integrands of the right-hand side of the equation are readily obtained. Multiplying Cauchy's equation of motion by $\vec{v}$:

$$\rho \vec{v} \frac{d\vec{v}}{dt} = \rho \vec{g} \vec{v} + \vec{v} \operatorname{Div} T \tag{2.68}$$

By similarly multiplying the continuity equation by $\frac{v^2}{2}$:

$$\frac{v^2}{2} \cdot \frac{d\rho}{dt} + \frac{v^2}{2} \rho \operatorname{div} \vec{v} = 0 \tag{2.69}$$

Adding the two equations we get:

$$\rho \vec{v} \frac{d\vec{v}}{dt} + \frac{v^2}{2} \frac{d\rho}{dt} + \rho \frac{v^2}{2} \operatorname{div} \vec{v} = \rho \vec{g} \vec{v} + \vec{v} \operatorname{Div} T \tag{2.70}$$

From the chain rule it follows that:

$$\vec{v} \frac{d\vec{v}}{dt} = \frac{d}{dt}\left(\frac{v^2}{2}\right) \tag{2.71}$$

It is similarly obvious that:

$$\rho \frac{d}{dt}\left(\frac{v^2}{2}\right) + \frac{v^2}{2} \frac{d\rho}{dt} = \frac{d}{dt}\left(\rho \frac{v^2}{2}\right) \tag{2.72}$$

Thus Eq. (2.70) can be written as:

$$\frac{d}{dt}\left(\rho\frac{v^2}{2}\right) + \rho\frac{v^2}{2}\,\text{div}\,\vec{v} = \rho\vec{g}\vec{v} + \vec{v}\,\text{Div}\,T \tag{2.73}$$

Applying the identity:

$$\text{div}(T\vec{v}) = \vec{v}\,\text{Div}\,T + T\!:\vec{v}\circ\nabla \tag{2.74}$$

the following expression is obtained:

$$\frac{d}{dt}\left(\rho\frac{v^2}{2}\right) + \rho\frac{v^2}{2}\,\text{div}\,\vec{v} = \rho\vec{g}\vec{v} + \text{div}(T\vec{v}) - T\!:\vec{v}\circ\nabla \tag{2.75}$$

This equation can be integrated over an arbitrary volume V to give:

$$\int_V \frac{d}{dt}\left(\rho\frac{v^2}{2}\right)dV + \int_V \rho\frac{v^2}{2}\,\text{div}\,\vec{v}\,dV = \int_V \rho\vec{g}\vec{v}dV$$

$$+ \int_V \text{div}(T\vec{v})dV - \int_V T\!:\vec{v}\circ\nabla dV \tag{2.76}$$

Applying the transport theorem to the left-hand side and the divergence theorem to the second integral on the right-hand side of the equation, we get:

$$\frac{d}{dt}\int_V \rho\frac{v^2}{2}dV = \int_V \rho\vec{g}\vec{v}dV + \int_{(A)} \vec{v}Td\vec{A} - \int_V T\!:\vec{v}\circ\nabla dV \tag{2.77}$$

This equation states that the rate of change in kinetic energy of a moving material volume is equal to the rate at which work is being done on the volume by the body forces and the surface forces, diminished by a "dissipation" term involving the interaction of stress and deformation. This latter term must represent the rate at which work is being done in changing the volume and shape of the fluid body. A certain proportion of this power may be recoverable, but the rest must be accounted for as heat.

The local and the convective rate of change of the kinetic energy may also be separated applying the transport theorem:

$$\int_V \frac{\partial}{\partial t}\left(\rho\frac{v^2}{2}\right)dV + \int_{(A)} \rho\frac{v^2}{2}\vec{v}d\vec{A} + \int_V T\!:\vec{v}\circ\nabla dV$$

$$= \int_V \rho\vec{g}\vec{v}dV + \int_{(A)} \vec{v}TdA \tag{2.78}$$

In this equation the first term on the left-hand side represents the rate of change of kinetic energy within the fixed volume V. The next term represents the convective kinetic energy flux which crosses the bounding surface of this volume. The third term represents the loss of kinetic energy. This rate of work is transformed into internal energy; this transformation of work into heat is irreversible.

The balance of kinetic energy equation can be also written in differential form. The material derivative of the specific kinetic energy may be separated into the local and the convective parts:

$$\frac{d}{dt}\left(\rho\frac{v^2}{2}\right) = \frac{\partial}{\partial t}\left(\rho\frac{v^2}{2}\right) + \vec{v}\,\text{grad}\left(\rho\frac{v^2}{2}\right) \tag{2.79}$$

It should also be remembered that:

$$\vec{v}\,\text{grad}\left(\rho\frac{v^2}{2}\right) + \frac{v^2}{2}\text{div}\,\vec{v} = \text{div}\left(\rho\frac{v^2}{2}\vec{v}\right) \tag{2.80}$$

Substituting these into Eq. (2.75), we obtain:

$$\frac{\partial}{\partial t}\left(\rho\frac{v^2}{2}\right) + \text{div}\left(\rho\frac{v^2}{2}\vec{v} - \vec{v}T\right) = \rho\vec{g}\vec{v} - T:\vec{v}\circ\nabla \tag{2.81}$$

It is clear that the first term represents the local rate of change of the specific kinetic energy within a fixed unit volume. The divergence term includes the convective and the conductive kinetic energy fluxes. The terms on the right-hand side represent the sources and sinks of the kinetic energy. The power of the external forces gv refers to a source, the kinetic energy loss $T:\vec{v}\circ\nabla$ reflects a sink.

Note that the balance of kinetic energy equation is a scalar equation in spite of the vector and tensor variables in it. In Cartesian coordinates we can write:

$$\begin{aligned}
&\frac{\partial}{\partial t}\left(\rho\frac{v^2}{2}\right) + \frac{\partial}{\partial x}\left(\rho\frac{v^2}{2}v_x - \sigma_x v_x - \tau_{xy}v_y - \tau_{xz}v_z\right) \\
&+ \frac{\partial}{\partial y}\left(\rho\frac{v^2}{2}v_y - \tau_{yx}v_x - \sigma_y v_y - \tau_{yz}v_z\right) \\
&+ \frac{\partial}{\partial z}\left(\rho\frac{v^2}{2}v_z - \tau_{zx}v_x - \tau_{zy}v_y - \sigma_z v_z\right) = \rho\left(g_x v_x + g_y v_y + g_z v_z\right) \\
&- \left(\sigma_x\frac{\partial v_x}{\partial x} + \tau_{xy}\frac{\partial v_x}{\partial y} + \tau_{xz}\frac{\partial v_x}{\partial z} + \tau_{yx}\frac{\partial v_y}{\partial x} + \sigma_y\frac{\partial v_y}{\partial y}\right) \\
&+ \tau_{yz}\frac{\partial v_y}{\partial z} + \tau_{zx}\frac{\partial v_z}{\partial x} + \tau_{zy}\frac{\partial v_z}{\partial y} + \sigma_z\frac{\partial v_z}{\partial z}
\end{aligned} \tag{2.82}$$

An important special case is that where the body forces form a conservative field, in which:

$$\vec{g} = -\text{grad } U \tag{2.83}$$

The conservative field is steady, i.e.:

$$\frac{\partial U}{\partial t} = 0 \tag{2.84}$$

Thus the material derivative of the potential is simply:

$$\frac{dU}{dt} = \vec{v} \text{ grad } U \tag{2.85}$$

The scalar function $U(\vec{r})$ is the potential energy of a unit fluid mass. It is an extensive variable, thus the total of the potential energy of a fluid mass within a volume V is:

$$\Im = \int_V \rho U dV \tag{2.86}$$

It is easy to see that:

$$\frac{d}{dt} \int_V \rho U dV = \int_V \rho \frac{dU}{dt} dV + \int_V \left(U \frac{d\rho}{dt} + U\rho \text{div } \vec{v} \right) dV \tag{2.87}$$

Thus the rate of work done by external body forces may be expressed as the potential energy flux per unit time:

$$\int_V \rho \vec{g} \vec{v} dV = \frac{d}{dt} \int_V \rho U dV \tag{2.88}$$

Substituting into Eq. (2.59), we obtain the balance of kinetic energy equation in a conservative field:

$$\frac{d}{dt} \int_V \left(\frac{v^2}{2} + U \right) \rho dV + \int_V T : v \circ \nabla dV = \int_{(A)} \vec{v} T d\vec{A} \tag{2.89}$$

It can be seen that the rate of change of the sum of the kinetic and potential energy and the rate of conversion into internal energy equals the rate of work done by surface forces when the fluid flows in a potential field.

This is the general form of the balance of kinetic energy equation. Its special forms can be derived in accordance with how the stress tensor differs for certain type of fluids. These will be treated in detail in the sections dealing with the dynamics of perfect fluids, Newtonian fluids, and different types of non-Newtonian fluids.

2.2.5 The Principle of Conservation of Energy

Any material system is characterized by its energy content. Energy occupies a position of distinction amongst the other extensive variables. Whatever interaction occurs between a material system and its surroundings, the transfer of energy invariably accompanies the transport of other extensive variables. Certain interactions are accompanied the transport of a certain type of energy. Kinetic energy increases or decreases due to mechanical interactions; a change in internal energy accompanies thermal interactions (Truesdell and Toupin, 1960).

But the fundamental axiom is the principle of conservation of energy, which applies to closed systems only. The energy content of an open system may change depending on the action of its surroundings. Consider a system in mechanical and thermal interaction with its surroundings. In this case, the change in the total energy is the change in the sum of the kinetic and the internal energy:

$$\frac{dE}{dt} = \frac{d}{dt} \int_V \left(\frac{v^2}{2} + \varepsilon\right)\rho dV \tag{2.90}$$

where ε is the specific internal energy.

The action of the surroundings may take the form of mechanical work and/or heat.

The application of the principle of conservation of energy to this open system leads to the equation:

$$\frac{d}{dt} \int_V \left(\frac{v^2}{2} + \varepsilon\right)\rho dV = \int_V \rho \vec{g}\vec{v} dV + \int_{(A)} \vec{v} T d\vec{A} - \int_{(A)} \vec{q} d\vec{A} \tag{2.91}$$

The rate of change of the kinetic and internal energy within the material volume equals the sum of the rate of work done by external body forces, the rate of work done by surface forces, and the rate at which heat is conducted into the volume of fluid. The heat flux vector $\vec{q}$ has the dimensions (W/m^2).

Applying the transport theorem, the material derivative can be replaced by a local and a convective part:

$$\int_V \frac{\partial}{\partial t}\left(\frac{v^2}{2} + \varepsilon\right)\rho dV + \int_{(A)} \left(\frac{v^2}{2} + \varepsilon\right)\rho \vec{v} d\vec{A} = \int_V \rho \vec{g}\vec{v} dV$$
$$+ \int_{(A)} \vec{v} T d\vec{A} - \int_{(A)} \vec{q} d\vec{A} \tag{2.92}$$

In this equation, V and (A) do not represent a moving volume and surface; rather they are instantaneous configurations fixed in space. Thus, the first term gives the local rate of change of the kinetic and internal energy in the fixed volume V. The second term expresses the convective fluxes of the kinetic and internal energy due to the bulk flow across the fixed boundary surface (A).

By using the divergence theorem, the surface integrals can be replaced by volume integrals. Since the volume V is arbitrarily chosen, the equations are also valid for the integrands. In this way we obtain the energy equation in differential form. The material form of the equation can be written:

$$\rho \frac{d}{dt}\left(\frac{v^2}{2} + \varepsilon\right) = \rho \vec{g} \vec{v} + \text{div}(\vec{v}T) - \text{div}\,\vec{q} \tag{2.93}$$

The local (spatial) form is obtained as:

$$\frac{\partial}{\partial t}\left[\left(\frac{v^2}{2} + \varepsilon\right)\rho\right] + \text{div}\left[\left(\frac{v^2}{2} + \varepsilon\right)\rho\vec{v} - \vec{v}T + \vec{q}\right] = \rho \vec{g} \vec{v} \tag{2.94}$$

It can be recognized that the first term is the local rate of change of the sum of the kinetic and the internal energy content of the unit fluid volume. The second term represents the convective and the conductive fluxes of these energies. On the right hand side the power of the body forces acts as the source of the energy change.

The balance equation of the kinetic and internal energies in this form is valid for open system. It is the more general form of the first principal law of the thermodynamics.

2.2.6 The Balance of Internal Energy

Subtracting the kinetic energy equation from the total energy equation, we get the equation for the balance of the internal energy:

$$\frac{\partial}{\partial t}(\rho\varepsilon) + \text{div}(\rho\varepsilon\vec{v} + \vec{q}) = T: \vec{v} \circ \nabla \tag{2.95}$$

In integral form this can also be written as:

$$\int_V \frac{\partial}{\partial t}(\rho\varepsilon)dV + \int_{(A)} (\rho\varepsilon\vec{v} + \vec{q})d\vec{A} = \int_V T: \vec{v} \circ \nabla dV \tag{2.96}$$

It is easy to recognize the terms representing local and convective flux of the internal energy. There is also a term representing a flux of a non-mechanical power; the heat flux. The term on the right-hand side of the equation represents a source of internal energy. This is the mechanical contribution of its increase. While $-T : \vec{v} \circ \nabla$ represents a loss of mechanical energy, it produces internal energy within the volume V. If the flow is at rest $\vec{v} = 0$, thus the balance of internal energy equation becomes:

$$\frac{\partial}{\partial t}(\rho\varepsilon) + \text{div } \vec{q} = 0 \qquad (2.97)$$

Substituting:

$$\varepsilon = cT \qquad (2.98)$$

and:

$$\vec{q} = -k \text{ grad } T \qquad (2.99)$$

The well-known differential equation for the conduction of heat is obtained:

$$\frac{\partial}{\partial t}(\rho cT) = \text{div}(k \text{ grad } T) \qquad (2.100)$$

Assuming that the density, the heat capacity, and the coefficient of thermal conductivity are constant, we get:

$$\frac{\partial T}{\partial t} = \frac{k}{\rho c}\left(\frac{\partial^2 T}{\partial x^2} + \frac{\partial^2 T}{\partial y^2} + \frac{\partial^2 T}{\partial z^2}\right) \qquad (2.101)$$

For the steady state, this can be written in the simple form:

$$\frac{\partial^2 T}{\partial x^2} + \frac{\partial^2 T}{\partial y^2} + \frac{\partial^2 T}{\partial z^2} = 0 \qquad (2.102)$$

2.2.7 The Balance of Entropy

The entropy is a thermodynamic variable of states; the characteristic extensive quantity of the thermal interaction. It cannot be experienced directly; as a derived variable, entropy represents a measure of the irreversibility of a change of state.

Irreversible changes of state necessarily involve currents in the system. The increase in entropy during an irreversible process must be related to these currents.

Entropy is not a quantity which is conserved. There is no axiom to determine the rate of change of entropy during an irreversible process. Thus we can only derive the balance of entropy equation by starting from the internal energy equation:

$$\rho\frac{d\varepsilon}{dt} + \text{div } \vec{q} = T : \vec{v} \circ \nabla \qquad (2.103)$$

Separating the dissipative part of the stress tensor we get:

$$T = -pI + T_v \tag{2.104}$$

where p is the pressure, I is the unit tensor, T_v is the viscous stress tensor. Since T_v is symmetrical, the product of T_v and the asymmetrical part of $\vec{v} \circ \nabla$ must vanish. Thus:

$$\rho \frac{d\varepsilon}{dt} + \mathrm{div}\ \vec{q} = -p\ \mathrm{div}\ \vec{v} + T_v : S \tag{2.105}$$

where S is the rate of deformation tensor. The first principal law of thermodynamics can be written as:

$$d\varepsilon = Tds - pd\left(\frac{1}{\rho}\right) \tag{2.106}$$

where s is the specific entropy. Dividing by dt, we get:

$$\frac{d\varepsilon}{dt} = T\frac{ds}{dt} - p\frac{d}{dt}\left(\frac{1}{\rho}\right) \tag{2.107}$$

Substituting this into the internal energy equation, and since:

$$p\ \mathrm{div}\ \vec{v} = \rho p \frac{d}{dt}\left(\frac{1}{\rho}\right) \tag{2.108}$$

we obtain:

$$\rho T \frac{ds}{dt} = T_v : S - \mathrm{div}\ \vec{q} \tag{2.109}$$

Dividing by the temperature we have:

$$\frac{ds}{dt} = \frac{T_v : S}{T} - \frac{\mathrm{div}\ q}{T} \tag{2.110}$$

Using the identity:

$$\mathrm{div}\left(\frac{\vec{q}}{T}\right) = \frac{1}{T}\mathrm{div}\ \vec{q} - \frac{\vec{q}\ \mathrm{grad}\ T}{T^2} \tag{2.111}$$

The local form of the balance of entropy equation can be written as:

$$\frac{\partial(\rho s)}{\partial t} + \mathrm{div}\left(\rho s\vec{v} + \frac{\vec{q}}{T}\right) = \frac{T_v : S}{T} - \frac{\vec{q}\ \mathrm{grad}\ T}{T^2} \tag{2.112}$$

It is easy to recognize on the left-hand side, the terms representing the local rate of change of the entropy, its convective flux $\rho s\vec{v}$, and its conductive flux $\vec{q}/T$. On the right-hand side of the equation, terms representing the entropy sources are found. The inclusion of the conductive flux term in the entropy equation shows that entropy is always exchanged simultaneously with the internal energy.

Considering the source terms, it is obvious that the absolute temperature is always positive, while the dissipating mechanical power is also positive, i.e.:

$$T > 0; \quad T_V : S > 0$$

Since:

$$\vec{q} = -k \operatorname{grad} T \tag{2.113}$$

the right-hand side of the equation, i.e., the source of entropy, can only be positive, thus:

$$\frac{T_V : S}{T} + k\left(\frac{\operatorname{grad} T}{T}\right)^2 \geq 0 \tag{2.114}$$

The internal energy flux always produces an increase in entropy, therefore a certain fraction of the total energy content is lost. This is in accordance with the second law of thermodynamics.

2.3 MECHANICAL EQUILIBRIUM OF FLUIDS

If the velocity field is equal to zero throughout the entire field, i.e.:

$$\vec{v} \equiv 0$$

the fluid body is in mechanical equilibrium.

Substituting this into the balance of momentum equation, a simple expression is obtained:

$$0 = \int_V \rho \vec{g} dV + \int_{(A)} T d\vec{A} \tag{2.115}$$

The concept of fluidity entails that tangential stresses cannot exist at rest. Thus, the stress tensor can be obtained as:

$$T = -pI \tag{2.116}$$

It may then be shown from equilibrium considerations that the normal stress at any point has the same value in all directions. This state of stress is called hydrostatic, and can be written in matrix form as:

$$T = \begin{bmatrix} -p & 0 & 0 \\ 0 & -p & 0 \\ 0 & 0 & -p \end{bmatrix} \tag{2.117}$$

where p is the hydrostatic pressure, which at a given point is uniform in all directions.

As is well known, the stress tensor has one set of orthogonal axes — the three principal axes — for which the shear stresses vanish, and along which only the principal normal stresses σ1, σ2, and σ3 exist. Note that these three principal stresses are generally not identical; only the hydrostatic normal stresses are the same in all directions.

For a compressible fluid, the pressure is a well-defined thermodynamic variable of state which satisfies the relation:

$$Tds = d\varepsilon + pd\left(\frac{1}{\rho}\right) \tag{2.118}$$

It should be noticed that this equation defines fluid pressure as a thermodynamic variable. Therefore, in general:

$$p = p(\rho, S) \tag{2.119}$$

For an incompressible fluid a simpler relation is introduced:

$$Tds = d\varepsilon; \quad \rho = \text{const.} \tag{2.120}$$

Consequently pressure does not enter into the thermodynamic treatment of an incompressible fluid; it is an entirely different type of variable.

For an incompressible fluid the pressure is not a thermodynamic variable. It is merely a scalar variable, satisfying the momentum equation:

$$0 = \int_V \rho \vec{g} dV - \int_{(A)} pd\vec{A} \tag{2.121}$$

Applying the divergence theorem, the surface integral is changed into a volume integral, from which we get:

$$\rho \vec{g} - \text{grad} p = 0 \tag{2.122}$$

This is the law of hydrostatics; its scalar component in rectangular coordinates is written as:

$$\rho g_x = \frac{\partial p}{\partial x} \tag{2.123}$$

$$\rho g_y = \frac{\partial p}{\partial y} \tag{2.124}$$

$$\rho g_z = \frac{\partial p}{\partial z} \tag{2.125}$$

After substituting $\vec{v} \equiv 0$, the continuity equation becomes:

$$\frac{\partial \rho}{\partial t} = 0 \tag{2.126}$$

This equality expresses the condition that the density field must remain steady for the equilibrium state to be maintained.

The internal energy balance at rest can be written as:

$$\rho \frac{\partial \varepsilon}{\partial t} = -\text{div } \vec{q} \tag{2.127}$$

This determines the temperature distribution at rest.

To determine the shape of a free fluid surface, the curl of Eq. (2.126) is first taken. Since:

$$\text{curl grad } p \equiv 0 \text{ we obtain}$$

$$\rho \text{ curl } \vec{g} + \text{grad } \rho \times \vec{g} = 0 \tag{2.128}$$

Multiplying this equation by $\vec{g}$, the second term vanishes. The only term which remains is:

$$\vec{g} \text{ curl } \vec{g} = 0 \tag{2.129}$$

Expressing this in rectangular coordinates we get:

$$g_x \left(\frac{\partial g_z}{\partial y} - \frac{\partial g_y}{\partial z} \right) + g_y \left(\frac{\partial g_x}{\partial z} - \frac{\partial g_z}{\partial x} \right) + g_z \left(\frac{\partial g_y}{\partial x} - \frac{\partial g_x}{\partial y} \right) = 0 \tag{2.130}$$

This equation places a restriction on the body forces. Mechanical equilibrium can exist only if the body force field satisfies the above equation. It is clear that this condition can be satisfied by all conservative body forces, i.e., those which can be derived from a scalar potential such as:

$$\vec{g} = -\text{grad } U \tag{2.131}$$

Since curl grad $U \equiv 0$, Eq. (2.130) is satisfied.

In this case, the first term of Eq. (2.131) vanishes. The second term can thus be written:

$$\text{grad } \rho \times \text{grad } U = 0 \tag{2.132}$$

Therefore, the potential surfaces of the body force coincide with the surfaces of constant density. It can be seen that these surfaces are perpendicular to the resultant of the body forces. This is why the free surface of a liquid in a gravity field is horizontal.

For a conservative body force, the hydrostatic equation can be written as:

$$\rho \text{ grad } U = \text{grad } p \tag{2.133}$$

Therefore, the equipotential surfaces coincide with the isobaric surfaces.

The pressure distribution for a gas at rest is easily determined, since pressure and density are directly related variables:

$$p = p(\rho), \quad \rho = \rho(p) \tag{2.134}$$

A fluid in which density and pressure are directly related is called barotropic. In this case, the pressure force acting on a unit mass of gas has a potential $\wp$, thus:

$$\frac{1}{\rho}\text{grad } p = \text{grad} \int\limits_{p\overrightarrow{v}}^{p} \frac{dp}{\rho} = \text{grad } \wp \tag{2.135}$$

For a barotropic fluid in a conservative body force field, the hydrostatic equation is given by:

$$\text{grad}(U + \wp) = 0 \tag{2.136}$$

or, in another form:

$$U + \wp = \text{const.} \tag{2.137}$$

For the atmosphere in the gravity force field, assuming a uniform temperature distribution, the gravity potential is given by:

$$U = g(z - z_0) \tag{2.138}$$

where z is the vertical coordinate, and z_0 is its value on the ground surface.

The barotropic potential for the isothermic case can be expressed as:

$$\wp = \int\limits_{p_0}^{p} \frac{dp}{\rho} = \frac{p_0}{\rho_0} \int\limits_{p_0}^{p} \frac{dp}{p} = \frac{p_0}{\rho_0} \ln \frac{p}{p_0} \tag{2.139}$$

The equilibrium equation is obtained as:

$$g(z - z_0) + \frac{p_0}{\rho_0} \ln \frac{p}{p_0} = \text{const.} \tag{2.140}$$

The boundary condition $z = z_0$; $p = p_0$ shows the constant to be zero. Thus the pressure distribution along z is:

$$p(z) = p_0 \exp\frac{\rho_0 g}{p_0}(z - z_0) \tag{2.141}$$

where z_0 is any reference height and p_0 and ρ_0 are the corresponding pressure and the density.

The pressure distribution for an isentropic or a polytropic fluid can be similarly derived.

For an incompressible fluid, the hydrostatic equation is obtained in terms of the gravity field as:

$$-\rho \, \text{grad} \, U = \text{grad} \, p \tag{2.142}$$

or:

$$p + \rho U = \text{const.} \tag{2.143}$$

The pressure distribution is given by:

$$p - \rho g z = \text{const.} \tag{2.144}$$

Since at the free surface $z = 0$ and $p = p_0$, and introducing a depth-coordinate $h = -z$, we get:

$$p = p_0 + \rho g h \tag{2.145}$$

This is the most frequently used equation in hydrostatics.

References

Acheson, D.J., 1990. Elementary Fluid Dynamics. Clarendon Press, Oxford.

Batchelor, G.K., 1967. An Introduction to Fluid Dynamics. Cambridge University Press.

Csanady, T., 1965. Theory of Turbomachines. Mc Graw Hill, New York.

Truesdell, C., Toupin, R.A., 1960. The Classical Field Theories, Encyclopedia of Physics III/1. Springer, Berlin.

3.1 PROPERTIES OF POROUS MEDIA

The porous media of principal interest in our study are the rocks from underground formations that contain geothermal fluids. These rocks may be classified by their chemical composition: sandstone, limestone, dolomite, clay, etc.

Large parts of geothermal reserves of the world are contained in sandstone formations. Geothermal brine reserves of Hungary are found mainly in sandy and sandstone aquifers. Sandstone consists of grains of quartz, usually cemented together with argillaceous materials. Often hydratable materials such as montmorillonite, kaolinite, or illite clays are contained between the quartz grains in sandstone.

57
Copyright © 2017 Elsevier Inc. All rights reserved.

The most valuable geothermal reserves of the world are contained in limestone reservoirs. By common terminology, limestone reservoirs include those formations made of limestone, calcium carbonate, and those made of dolomite, the double carbonate of calcium magnesium.

Clay and shale layers are mostly the impervious boundaries of permeable reservoir rocks. The porosity of a porous medium is defined as the void volume, divided by the total (bulk) volume of the medium:

$$\phi_a = \frac{V_{pore}}{V_{bulk}} \tag{3.1}$$

where ϕ_a is the absolute porosity, V_{pore} is the pore volume (m^3), and V_{bulk} is the bulk volume (m^3). This fraction is called absolute porosity.

The so-called effective porosity has more practical importance. It is defined by as the fraction of the interconnected pore volume to the bulk volume of the porous body:

$$\phi_{eff} = \frac{V_{eff}}{V_{bulk}} \tag{3.2}$$

where ϕ_{eff} is the effective porosity, V_{eff} is the interconnecting pore volume (m^3), and V_{bulk} is the bulk volume (m^3). In the following equations, ϕ will denote the effective porosity.

Void or pore volumes are usually determined by measuring either gravimetrically or volumetrically the amount of liquid needed to saturate the dry medium. Pore volumes are also determined by gas expansion methods. Bulk volumes can be determined from measurements of the external dimensions of the medium, or from the volume of liquid displaced by immersion of the saturated medium. Porosity may be expressed as either fractions or percentages.

The average porosity of a very large porous medium, such as the Upper Pannonian aquifer, may be determined from the porosity of a number of small core samples of the reservoir rock. A simple arithmetic average will suffice when sufficient samples are available to obtain a statistical distribution of porosity in the pore samples.

Different types of porosity may be distinguished. Reservoir rocks are often classified by the types of pore space that exists in the rocks. The porosity of sandstone is usually that between sand grains, hence the type of porosity is referred to as intergranular.

Some sandstone is also fractured.

In limestone, parts of the crystals have often been dissolved into the groundwater to form a solution type of porosity. Some small isolated holes have been formed to produce a vugular type of porosity. In still others, large channels, sometimes a few meters in diameter, have been formed to produce a cavernous type of porosity. Also, limestone often contains fracture systems which make up a substantial part of the pore

TABLE 3.1 Typical Porosities

Porous Material	Porosity
Sandstone	0.08 ... 0.40
Limestone	0.01 ... 0.20
Sand	0.25 ... 0.50
Clay	0.30 ... 0.60
Shale	0.06 ... 0.35

volume. This type is referred to as fracture porosity. A typical range of values for porosity commonly measured on various types of rocks are shown in Table 3.1.

Porous media have been treated so far as rigid; an assumption which cannot always be expected to be true. The geometric quantities describing porous media, as introduced, may themselves be functions of certain dynamic quantities notable of the prevailing stresses. The simplest relationship between geometrical quantities referring to the porous medium and the stresses is obtained by assuming the porous medium is perfectly elastic according to Hooke's law. The effect of this will be that all the geometrical quantities relating to the pores are linear functions of the effective stress. Thus porosity of sedimentary rocks mostly decreases along the depth.

Considering this porosity distribution it seems to be convenient to define a local value of porosity by the equation:

$$\phi(\vec{r}, t) = \frac{dV_{eff}}{dV} \tag{3.3}$$

This $\phi(\vec{r}, t)$ is a continuous scalar function depending on the co-ordinates and time. It is obvious that these "infinitesimal" volumes must be a certain order of magnitude smaller than the observed domain, but great enough to contain a sufficient number of pores or grains to obtain a statistical limit of porosity.

Another useful method of characterizing a porous medium is that of determining the size of its pores and the pore-size distribution. While no one single dimension can describe the size or geometric shape of the holes between sand grains or crystals if limestone, it is convenient conceptually to visualize the holes as short, circular capillary tubes.

Then the pressure required forcing a non-wetting liquid such as mercury into the pore spaces can be related to the radius of the pores by:

$$p_c = \frac{2\delta \cos\alpha}{r} \tag{3.4}$$

where p_c is the capillary pressure (N/m^2), r is the radius of the pore (m), δ is the interfacial tension (N/m), and α is the contact angle (°).

The pore size distribution can be calculated from an equation of Ritter and Drake:

$$f(r) = \frac{p_c}{r} \frac{d(V - V_i)}{dp_c} \tag{3.5}$$

where f(r) is the pore-size distribution function (m^2), V is the total pore volume (m^3), V_i is the volume of the injected non-wetting fluid (m^3), and α is the contact angle (°).

From an experimental curve of capillary pressure versus. the volume of mercury injected into an evacuated rock sample, the derivative $d(V - V_i)/dp_c$ can be determined at various values of p_c, by taking the slope of the curve. Then calculating r at each value of p_c, the pore size distribution function f(r) can be calculated for each value of r.

Note that in natural sand, the sizes of the individual particles, thus the pore size, may vary over a wide range. In meshed sand, the majority of the particles have about the same diameter, thus the pores have about a uniform radius. Pore size data can be used for estimating the permeability of rock samples.

Many porous materials contain enormous surface areas per unit volume. For instance, the surface area of the grains in sandstone may be of the order of 500–5000 m^2/kg. This value may be 50,000–100,000 m^2/kg for shales. In such processes which involve absorption of materials from the fluid flowing from the medium, knowledge of the magnitude of the surface area is essential. The usual experimental method to determine specific surfaces is from nitrogen-adsorption experiments at constant temperatures. The gas-adsorption techniques involve determination of the quantity of gas necessary to form a monomolecular layer on the surface. Knowing the volume of gas in the monomolecular film, the number of gas molecules can be determined.

The permeability of a porous medium is a measure of the ease with which a fluid will flow through the medium; the higher the permeability, the higher the flow rate for a given hydraulic gradient. The permeability is a statistical average of the fluid conductivities of all the flow channels in the solid body. This average conductivity takes into account the variations in size, shape, direction, and interconnections of all the flow channels. While obviously a number of pores or flow channels must be considered in obtaining a statistically average permeability, it is often convenient for mathematical purposes to consider the permeability as the property of a point in the medium. In a homogeneous medium, the permeability at any point coincides with the average permeability. In a heterogeneous medium, the permeability varies from point to point.

3.2 DARCY'S LAW

Permeability is an experimentally defined quantity. Its theory is based on the classical experiment originally made by Darcy (1856).

A schematic drawing of Darcy's experiment is shown in Fig. 3.1. A vertical cylindrical container of diameter D is filled by uniform grain size sand. The sand column is percolated by water. If open manometer tubes are attached to the upper and lower boundaries of the sand bed of height H, the water rises to the height h_1 and h_2 in the tubes. By varying the flow rate through the bed, the corresponding Q and $h_1 - h_2$ values are measured. The result is a linear relationship:

$$Q = K \frac{D^2 \pi}{4} \frac{h_1 - h_2}{H} \qquad (3.6)$$

where Q is the flow rate (m^3/s), K is the hydraulic conductivity (m/s), h_1 is the piezometric level at point 1 (m), h_2 is the piezometric level at point 2 (m), H is the height of the sand bed (m), and D is the diameter of the vessel (m).

This relationship is the so-called Darcy's law; analogous to the Hagen−Poiseuille law. Here Q is the flow rate of the percolated water, the hydraulic conductivity of dimension (m/s) depending on the properties

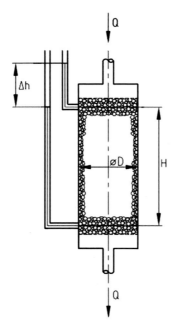

FIGURE 3.1 Darcy's experiment.

of the fluid and the porous medium. A constant of the type K is not very satisfactory because we would like to separate the influence of the porous medium from that of the fluid.

Nutting (1930) has already generalized the original form of Darcy's law for sand-filled vessels in an arbitrary position:

$$q_A = \frac{Q}{\frac{D^2 \pi}{4}} = \frac{k}{\mu} \frac{p_1 - p_2 + \rho g(h_1 - h_2)}{H} \tag{3.7}$$

where q_A is the seepage velocity (m/s), k is the permeability (m^2), p_1 is pressure at point 1 (N/m^2), p_2 is pressure at point 2 (N/m^2), μ is the dynamic viscosity (kg/(s·m)), ρ is the density (kg/m^3), h_1 is the piezometric level at point 1 (m), h_2 is the piezometric level at point 2 (m), H is the length of the sand bed (m), and D is the diameter (m).

The hydraulic conductivity K is separated into permeability k and viscosity μ. The ratio of the flow rate and total cross-section of the granular bed is called seepage velocity. This generalized form is valid for any bed in arbitrary position as it is sketched in Fig. 3.2.

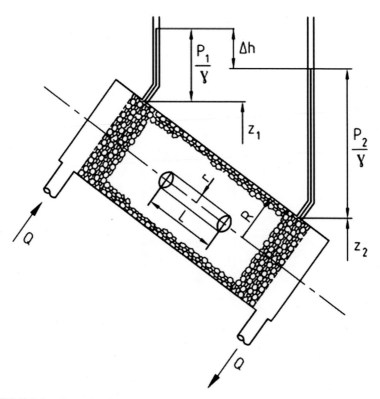

FIGURE 3.2   Generalized Darcy's experiment.

The hydraulic gradient J can be easily recognized in Eq. (3.7):

$$J = \frac{P_1 - P_2}{\rho g H} + \frac{h_1 - h_2}{H} \tag{3.8}$$

thus we obtain:

$$q_A = \frac{\rho g k}{\mu} J \tag{3.9}$$

In this scalar form, Darcy's law is still restricted in its application, as it is appropriate to a finite bed of very particular geometrical shape. For greater generalization, it should be expressed as a differential equation by letting H become infinitesimal. Naturally, in this process q_A is replaced by a vector $\vec{q}$ which might be called the local seepage velocity vector. Ferrandon (1948) generalized Darcy's law in the form:

$$\vec{q} = -\frac{\rho k}{\mu} \text{grad}(U + \Pi) \tag{3.10}$$

where U is the potential of the body forces, and Π is the well-known barotropic potential:

$$\Pi = \int_{P_0}^{P} \frac{dp}{\rho} \tag{3.11}$$

The body force is mostly the gravity force, thus the potential U is:

$$U = gh \tag{3.12}$$

The barotropic potential π for an incompressible fluid can be written in the simple form:

$$\Pi = \frac{P}{\rho} \tag{3.13}$$

Therefore, the differential form of Darcy's law in a gravity field for incompressible fluid is obtained as:

$$\vec{q} = -\frac{\rho k}{\mu} \text{grad}\left(gh + \frac{P}{\rho}\right) \tag{3.14}$$

or

$$\vec{q} = -\frac{gk}{\nu} \text{grad}\left(h + \frac{P}{\rho g}\right) \tag{3.15}$$

Hydraulic conductivity K, or permeability k can be determined experimentally based on Darcy's law. In laboratory, K or k is measured by means of an instrument called a permeameter. In a permeameter, the flow pattern is one of steady or unsteady one-dimensional flow through a small, cylindrical porous medium sample. The so-called constant head permeameter is used most commonly to determine permeability.

The porous rock sample is placed between two porous plates that provide almost no resistance between them. A constant head difference Δh is applied across the tested sample, producing a flow rate Q. When incompressible liquid is used, hydraulic conductivity and permeability can be determined from Darcy's law:

$$K = \frac{QL}{\frac{D^2\pi}{4}\cdot\Delta h} \tag{3.16}$$

where K is the hydraulic conductivity (m/s), Q is the flow rate (kg/s), L is the length of the specimen (m), D is the diameter (m), and Δh is the difference of the piezometric height (m), and:

$$k = \frac{\mu\cdot QL}{\frac{D^2\pi}{4}\cdot\rho g\Delta h} \tag{3.17}$$

where k is the permeability (m^2).

To obtain more reliable results, several tests are performed under different heads and flow rates. Various units are used in practice for hydraulic conductivity and permeability. The SI unit for K is m/s. Hydrologists prefer the unit m/day. Permeability k is measured in SI system m^2. Reservoir engineers widely use the unit darcy, defined by:

$$1\ \text{Darcy} = \frac{1\frac{cm^3}{s}\ 1\ \text{cm}\ 1\ \text{centipoise}}{1\ cm^2\ 1\ \text{bar}}$$

Thus a porous medium is said to have a permeability of one darcy if a single phase fluid of 1 cp viscosity that completely fills the void space of the sample will flow through it at a rate of 1 cm^3/s, per 1 cm^2 area, and 1 cm length under a pressure 1 bar. In many cases, the darcy is a rather large unit, so that the millidarcy (mD) is frequently used. Approximately:

$$1\ m^2 = 10^{12}\ \text{darcy} = 10^{15}\ \text{millidarcy}$$

Some typical values for permeability of porous media can be found in Table 3.2.

3.3 THE COMPLEX CONTINUUM MODEL

In the following, the fundamental equations describing the flow of a fluid through a porous medium are developed. These are the familiar differential equations, which express the balance laws of mass, momentum, energy, etc. Accordingly, in the derived equations, both the fluid and the porous solid are considered as continua, each filling the entire space.

TABLE 3.2 Permeability of Different Materials

Permeability	Pervious			Semi-Pervious			Impervious					
Unconsolidated Sand & Gravel	Well Sorted Gravel		Well Sorted Sand or Sand & Gravel		Very Fine Sand, Silt, Loess, Loam								
Unconsolidated Clay & Organic					Peat		Layered Clay		Unweathered Clay				
Consolidated Rocks	Highly Fractured Rocks				Oil Reservoir Rocks		Fresh Sandstone		Fresh Limestone, Dolomite			Fresh Granite	
κ (cm^2)	10^{-3}	10^{-4}	10^{-5}	10^{-6}	10^{-7}	10^{-8}	10^{-9}	10^{-10}	10^{-11}	10^{-12}	10^{-13}	10^{-14}	10^{-15}
κ (millidarcy)	10^{+8}	10^{+7}	10^{+6}	10^{+5}	10,000	1,000	100	10	1	0.1	0.01	0.001	0.0001

Modified from Bear, J., 1972. Dynamics of Fluids of porous Meida. Elsevier, New York.

In the continuum model, the actual material is replaced by continuous functions of physical quantities. The actual physical processes are represented by the evolution of these continuous functions in space and time. The value of such a function at a certain point is obtained as an average value in the infinitesimal volume dV surrounding the point. The essential mathematical simplification of the continuum model is that this average quantity is assigned in the limit to the point itself.

This "infinitesimal" element of the saturated porous medium must contain a sufficient number of pore channels to statistically allow the continuum interpretation. The solid-fluid phase boundaries within the volume element are neglected. The saturated porous medium is replaced by a complex continuum, in which continuous distribution functions of physical quantities are doubled referring to the fluid and the solid. Thus, at any point of the complex continuum two densities (ρ_F, ρ_S), two velocities ($\vec{v}_F$, $\vec{v}_S$), and two stress tensors $\underline{\underline{T}}_F$, $\underline{\underline{T}}_S$ are present simultaneously, in accordance to the principle of equipresence.

The infinitesimal volume element of the saturated porous medium may be separated to a fluid volume, determined by:

$$dV_F = \phi dV \tag{3.18}$$

and a solid volume, determined by:

$$dV_S = (1 - \phi)dV \tag{3.19}$$

Because the velocity fields are doubled, the material derivatives of any physical quantities must be doubled:

$$\frac{d}{dt_F} = \frac{\partial}{\partial t} + (\vec{v}_F \nabla) \tag{3.20}$$

and

$$\frac{d}{dt_S} = \frac{\partial}{\partial t} + (\vec{v}_S \nabla) \tag{3.21}$$

These notations are used in the following section.

3.4 THE PRINCIPLE OF CONSERVATION OF MASS

The principle of conservation of mass states that the mass of a body is constant during its motion. This can be stated in the rate form, as the time rate of change of the mass of a body is zero. It is obvious that this statement must be expressed mathematically for a material system.

Consider the volume $V(\vec{r}, t)$ filled by the complex continuum, bounded by the closed surface (A). The mass of an infinitesimal volume element:

$$dm = \phi \rho_F dV + (1 - \phi) \rho_S dV \tag{3.22}$$

The total mass of the volume V can be expressed as:

$$M = \int_V \phi \rho_F dV + \int_V (1 - \phi) \rho_S dV \tag{3.23}$$

Note, we must take two material derivatives for the two volume integrals, expressing the conservation of mass:

$$\frac{d}{dt_F} \int_V \rho_F \phi dV + \frac{d}{dt_S} \int_V (1 - \phi) \rho_S dV = 0 \tag{3.24}$$

Applying the transport theorem, the material derivative can be replaced by a local and convective term:

$$\int_V \frac{\partial(\phi \rho_F)}{\partial t} dV + \int_V \frac{\partial[(1 - \phi)\rho_S]}{\partial t} dV + \int_{(A)} \phi \rho_F \vec{v}_F d\vec{A} + \int_{(A)} (1 - \phi)\rho_S \vec{v}_S d\vec{A}$$
$$= 0 \tag{3.25}$$

The mass balance equation may be written for the fluid and solid phase separately:

$$\int_V \frac{\partial(\phi \rho_F)}{\partial t} dV + \int_{(A)} \phi \rho_F \vec{v}_F d\vec{A} = \int_V \xi_F dV \tag{3.26}$$

$$\int_V \frac{\partial[(1-\phi)\rho_S]}{\partial t} dV + \int_{(A)} (1-\phi)\, \rho_S \vec{v}_S d\vec{A} = \int_V \xi_S dV \qquad (3.27)$$

where ξ_F and ξ_S is the rate at which mass of the fluid is produced within the unit volume of the system by chemical reactions. Similarly ξ_S is the same, referring to the solid phase. It is obvious that:

$$\int_V \xi_F dV + \int_V \xi_S dV = 0 \qquad (3.28)$$

that is, the whole mass of the system is constant. The integral form of the mass balance equation may be written in differential form applying the divergence theorem:

$$\frac{\partial(\phi\rho_F)}{\partial t} + \text{div}(\phi\rho_F \vec{v}_F) = \xi_F \qquad (3.29)$$

$$\frac{\partial[(1-\phi)\rho_S]}{\partial t} + \text{div}[(1-\phi)\rho_S \vec{v}_S] = \xi_S \qquad (3.30)$$

If the interphase mass transfer may be neglected, the equation are obtained as:

$$\frac{\partial(\phi\rho_F)}{\partial t} + \text{div}(\phi\rho_F \vec{v}_F) = 0 \qquad (3.31)$$

$$\frac{\partial[(1-\phi)\rho_S]}{\partial t} + \text{div}[(1-\phi)\rho_S \vec{v}_S] = 0 \qquad (3.32)$$

For a homogeneous, isotropic, and non-deformable solid matrix:

$$\frac{\partial\phi}{\partial t} = 0; \quad \vec{v}_s \equiv 0,$$

Thus we get:

$$\phi\frac{\partial\rho_F}{\partial t} + \text{div}(\rho_F\phi \vec{v}_F) = 0 \qquad (3.33)$$

In this equation the product:

$$\vec{q} = \phi\vec{v}_F \qquad (3.34)$$

is the so-called local seepage velocity, which is a vector point function:

$$\vec{q} = \vec{q}(\vec{r}, t) \qquad (3.35)$$

The continuity equation for this case is obtained as:

$$\phi\frac{\partial\rho_F}{\partial t} + \text{div}(\rho_F \cdot \vec{p}) = 0 \qquad (3.36)$$

For an incompressible fluid, if $\rho_F = $ const, we get:

$$\operatorname{div} \vec{q} = 0 \tag{3.37}$$

It can be written by its orthogonal components as:

$$\frac{\partial q_x}{\partial x} + \frac{\partial q_y}{\partial y} + \frac{\partial q_z}{\partial z} = 0 \tag{3.38}$$

At any instant, there is at every point in the flow domain a local seepage velocity vector with a definite direction. The instantaneous curves (which at any point tangent to the direction of the local seepage velocity at that point) are called seepage streamlines of the flow. The mathematical expression defining a seepage streamline is therefore:

$$\vec{q} \times d\vec{r} = 0 \tag{3.39}$$

or written by orthogonal coordinates:

$$\frac{dx}{q_x} = \frac{dy}{q_y} = \frac{dz}{q_z} \tag{3.40}$$

In steady flow, i.e. one in which flow variables remain invariant with time, streamlines are constant. In unsteady flow, we can speak only of an instantaneous picture of the streamlines, as the picture varies continuously.

3.5 THE BALANCE EQUATION OF MOMENTUM

Newton's second law states: the rate of change of the momentum of a body equals the sum of the external forces acting on it. To apply this law to a porous medium saturated by fluid, consider the volume V of this medium, bounded by the simple closed surface (A) constituted of the same particles of fixed identity. The momentum of an infinitesimal volume element of the fluid-saturated porous medium is:

$$[\rho_F \phi \vec{v}_F + \rho_S (1 - \phi) \vec{v}_S] \, dV \tag{3.41}$$

Since the momentum is an extensive flow variable, the total momentum of the considered fluid-filled porous body is:

$$\int_V [\rho_F \phi \vec{v}_F + \rho_S (1 - \phi) \vec{v}_S] dV \tag{3.42}$$

The rate of change is obviously expressed by the material derivative of this volume integral, considering that different material derivatives are obtained for the fluid and solid phase:

$$\frac{d}{dt_F} \int_V \phi \rho_F \vec{v}_F dV + \frac{d}{dt_S} \int_V (1 - \phi) \rho_S \vec{v}_S dV \tag{3.43}$$

The external forces acting on the body are body forces and surface forces. The body force due to gravity can be expressed as:

$$\int_V \left[\phi \rho_F \vec{g}_F + (1 - \phi) \rho_S \vec{g}_S \right] dV \tag{3.44}$$

There are many physical phenomena involving interaction among different processes; i.e. mass transfer caused by temperature gradient, or electro-osmotic motion. Thus the resultant body forces $\vec{g}_F$ and $\vec{g}_S$ may be different, but usually the gravity acceleration $\vec{g} = \vec{g}_F = \vec{g}_S$ is the only dominant body force.

The surface forces are due to whatever medium is adjacent to the bounding surface (A). In the complex continuum, the state of stress is determined by two stress tensors $\underline{\underline{T_F}}$ and $\underline{\underline{T_S}}$ referring to the fluid and solid phases.

The result of the surface forces acting on the bounding surface (A) can be expressed as:

$$\iint_{(A)} \left[\phi \underline{\underline{T_F}} + (1 - \phi) \underline{\underline{T_S}} \right] d\vec{A} \tag{3.45}$$

where $\underline{\underline{T_F}}$ and $\underline{\underline{T_S}}$ are the stress tensors referring to the fluid and solid phase. Finally, the balance equation of momentum is obtained as:

$$\frac{d}{dt_F} \int_V \phi \rho_F dV + \frac{d}{dt_S} \int_V (1 - \phi) \rho_S \vec{v}_S dV$$

$$= \int_V \left[\phi \rho_F \vec{g}_F + (1 - \phi) \rho_S \vec{g}_S \right] dV + \int_{(A)} \left[\phi \underline{\underline{T_F}} + (1 - \phi) \underline{\underline{T_S}} \right] d\vec{A} \tag{3.46}$$

The left hand side of this equation expresses the rate of change of momentum for the fluid and solid phases respectively, while on the right hand side the first integral is the resultant body force, the second one is the result of the surface forces acting on the bounding surface.

The balance equation of momentum may be rewritten in differential form. Its surface integral can be transformed into volume integral by means of the divergence theorem. Applying the transport theorem to the left hand side terms, considering the arbitrary nature of the limits of integrations we get:

$$\phi \rho_F \frac{d\vec{v}_F}{dt_F} + (1 - \phi) \rho_S \frac{d\vec{v}_S}{dt_S} = \phi \rho_F \vec{g}_F + (1 - \phi)$$

$$\times \left\{ \rho_S \vec{g}_S + \text{Div}[\phi T_F + (1 - \phi)] T_S \right\} \tag{3.47}$$

This equation is valid for any porous body saturated by fluid. The acting forces, the rheological properties of the fluid and the solid material, are arbitrary. Therefore, it is convenient to replace this complicated physical system by some fictitious simpler one, because otherwise the mathematical treatment (i.e. the formulation of the boundary conditions, the method of solution) is practically impossible. Certain restrictions and simplifying assumptions will obtain an equation system which is convenient to solve engineering problems.

Expressing the rate of change of the momentum of the fluid phase, we obtain:

$$\phi \rho_F \frac{d\vec{v}_F}{dt_F} = \phi \rho_F \vec{g}_F + \text{Div}\left(\phi \underline{T_F}\right)$$
$$+ \left\{ (1 - \phi)\rho_S \left(\vec{g}_S - \frac{d\vec{v}_S}{dt_S} \right) + \text{Div}\left[(1 - \phi)\underline{T_S} \right] \right\}$$

(3.48)

The analogy with the momentum equation of a single phase of a multicomponent system can be easily recognized. The last term of the right hand side of the equation in the curly bracket can be considered to be the interphase momentum transfer as the fluid and solid phase interacts on the interphase boundary surfaces, i.e. on the pore channel walls. This term clearly shows that any mechanical process of the solid matrix gives size to flow of the pore fluid. A well-known example the flow induced by the consolidation of sedimentary rocks.

At this point in the development of the equation of motion, we introduce an assumption that the terms in the square bracket expressed by the variables of the solid phase are replaced by semi-empirical constitutive relations.

In saturated porous media, a force is exerted by the flowing fluid on the solid matrix, acting on the solid-fluid interphase surface. It is obvious that a reaction force of the same magnitude and an opposite direction exists, according to Newton's third law. The force acting on the unit volume of the porous medium:

$$\vec{f}_s = \vec{f}_u + \vec{f}_D$$

(3.49)

therefore, it is the sum of an uplifting force and a drag. The uplifting force can be expressed easily:

$$\vec{f}_u = -(1 - \phi)\rho_F \vec{g}$$

(3.50)

The drag force acting on a particle is composed of a skin friction drag and a form drag. Assuming laminar flow the drag is equal to:

$$\vec{f}_{D1} = \alpha^2 \cdot \mu \cdot \delta \vec{v}_F$$

(3.51)

where α is the particle shape factor, μ is the dynamic viscosity ($N \, s/m^2$), and δ is the mean diameter of the particle (m). The number of the particles inside the unit volume can be expressed as:

$$N = \frac{\beta(1 - \phi)}{\delta^3} \tag{3.52}$$

where β is a coefficient of proportionality of dimension (m^3). Thus the drag acting on the unit volume of the porous media is obtained as:

$$\overrightarrow{f}_D = N \cdot \overrightarrow{f}_{D1} = (1 - \phi) \cdot \beta \alpha^2 \mu \frac{\overrightarrow{v}}{\delta^2} \tag{3.53}$$

Substituting the local seepage velocity we get:

$$\overrightarrow{f}_D = \frac{(1 - \phi) \cdot \beta \alpha^2 \mu}{\phi \delta^2} \, \overrightarrow{q} \tag{3.54}$$

The reaction forces of these can be written as:

$$\overrightarrow{f}_F = -\overrightarrow{f}_s = (1 - \phi) \cdot \rho_F \overrightarrow{g} - \frac{(1 - \phi)\alpha^2 \beta \mu}{\phi \delta^2} \, \overrightarrow{q} \tag{3.55}$$

Thus the momentum equation of the fluid phase is:

$$\phi \rho_F \frac{d\overrightarrow{v}_F}{dt_F} = \phi \rho_F \overrightarrow{g} + \mathrm{Div}(\phi \underline{\underline{T}}_F) + (1 - \phi)\rho_F \overrightarrow{g} - \frac{(1 - \phi)\alpha^2 \beta \mu \overrightarrow{q}}{\phi \delta^2} \tag{3.56}$$

Let's consider a homogeneous isotropic porous medium in which a steady laminar seepage flow is developed. The porosity distribution is naturally uniform:

$$\phi = \mathrm{const.} \tag{3.57}$$

For steady flow:

$$\frac{\partial \overrightarrow{v}_F}{\partial t} = 0 \tag{3.58}$$

while the convective momentum flux may also be neglected, because of the very small values of the velocity.

The normal stresses in the fluid are much greater (10^4–10^5 times) than the shear components, thus we may assume that:

$$\mathrm{Div} \cdot T_F = -\mathrm{grad} p \tag{3.59}$$

Considering the above assumptions the momentum equation is obtained as:

$$0 = \rho_F \overrightarrow{g} - \mathrm{grad} p - \frac{(1 - \phi)\alpha^2 \beta \mu \overrightarrow{q}}{\delta^2 \phi^2} \tag{3.60}$$

Assuming barotropic flow it is obvious that:

$$-\frac{1}{\rho_F}\text{grad}_p = \text{grad}\,\Pi \tag{3.61}$$

where Π is the barotropic potential (m^2/s^2).
For incompressible fluids:

$$\Pi = \frac{p}{\rho_F} \tag{3.62}$$

The field of the body forces may be also assumed having a potential:

$$\vec{g} = -\text{grad}U \tag{3.63}$$

Thus:

$$-\text{grad}(U + \Pi) = \frac{(1-\phi)\alpha^2\beta\mu}{\phi^2\delta^2\rho}\,\vec{q} \tag{3.64}$$

Then the local seepage velocity can be expressed as:

$$\vec{q} = -\frac{\phi\delta^2}{(1-\phi)\alpha^2\beta}\frac{\rho}{\mu}\text{grad}(U + \Pi) \tag{3.65}$$

The comparison of this equation with the empirical Darcy's law:

$$\vec{q} = -k\frac{\rho}{\mu}\text{grad}\left(gh + \frac{p}{\rho}\right) \tag{3.66}$$

leads to the recognition that:

$$k = \frac{\phi^2\delta^2}{(1-\phi)\alpha^2\beta} \tag{3.67}$$

Therefore Darcy's law can be obtained as a consequence of the momentum equation. Thus Darcy's law is used as an equivalent expression, naturally for steady laminar flow of a barotropic Newtonian fluid flowing through porous media only. Notwithstanding, it must be kept in mind that Darcy's law is not an equation of motion; it cannot describe the flow within an individual pore channel. Strictly speaking, Darcy's law represents the statistical macroscopic equivalent of the Hagen–Poiseuille equation.

Note that the experiment of Darcy does not show what happens if the permeability and viscosity are not constant. It is quite familiar the permeability density and viscosity to be taken into the gradient:

$$\vec{q} = -\text{grad}\left[\frac{k}{\mu}(\rho gh + p)\right] \tag{3.68}$$

This is allowed only for constant permeability, density, and viscosity as the preceding show.

On the other hand, there is an indirect experimental indication of the correctness of Eq. (3.66). If Darcy's law is extended to immiscible multiple phase flow, it leads to the relative permeability concept. Relative permeability is actually variable through the porous medium during a flow experiment and therefore a distinction between Eq. (3.63) and Eq. (3.68) is important. The fact is that the empirically verified relative permeability equations are originated in Eq. (3.66) and not in Eq. (3.68).

Darcy's law together with continuity equation and equation of state is sufficient to determine the flow pattern in a porous medium for given boundary conditions. The equation system contains five unknown functions (q_x, q_y, q_z, p, ρ), thus the five equations can be solved. Two-dimensional problems can be especially easily treated by applying complex variable functions. Such specific applications are exercises in mathematics rather than fluid mechanics. Therefore, certain specific cases will be discussed regarding the practical importance.

3.6 THE BALANCE EQUATION OF INTERNAL ENERGY

Any material system is characterized by its energy content. Energy occupies a position of distinction amongst the other extensive variables. Whatever interaction occurs between a material system and its surroundings, the transfer of energy invariably accompanies the transport of other extensive variables. Certain interactions are accompanied the transport of a certain type of energy. Kinetic energy increases or decreases due to mechanical interactions; a change in internal energy accompanies thermal interactions. The balance equation of internal energy can be written for a fluid saturated porous material as:

$$\int_V \frac{\partial}{\partial t}\left[(1-\phi)\rho_S c_S T_S + \phi \rho_F c_F T_F\right]dV + \int_A \left[(\rho_1 - \phi)\rho_S c_S T_S \vec{v_s} + \phi \rho_F c_F T_F \vec{v_F}\right]d\vec{A}$$

$$+ \int_A \left[(1-\phi)k_S \mathrm{grad} T_S + \phi k_F \mathrm{grad} T_F\right]d\vec{A}$$

$$= \int_V \left[(1-\phi)\underline{\underline{T}}_S \cdot \vec{v_S}^\circ \nabla + \phi \underline{\underline{T}}_F \vec{v_F}^\circ \nabla\right]d\vec{V} + \int_V \left[(1-\phi)h_S + \phi h_F\right]dV \qquad (3.69)$$

The first integral expresses the local rate of change of the internal energies of the rock and the fluid. The first surface integral is the convective transferred internal energy by macroscopic motion. The second surface integral expresses the transferred internal energy by heat conduction. On the right hand side, the volume integral expresses the heat generated by friction in the solid and the fluid phase. The second volume integral is the heat source generated by radioactive decay, chemical reactions, and phase change. These are the volumetric sources of the internal energy.

The Eq. (3.69) can be simplified remarkably due to some approximate assumptions. It is reasonable to take the velocity of the rock matrix to zero, thus there is neither convective transfer nor mechanical energy dissipation in the solid phase. The temperatures of the rock and the fluid can be considered to equal. An experimental fact is that the dissipation of the mechanical energy is negligible relative to the convective and conductive fluxes of the internal energy. Thus we obtain a remarkable simple equation for the internal energy balance:

$$\int_V \frac{\partial}{\partial t}\left[(1-\phi)\rho_S c_S + \phi\rho_F c_F\right]T dV + \int_A \left\{\phi\rho_F c_F \overrightarrow{v}_F + [(1-\phi)k_S + \phi k_F]\right\}T d\overrightarrow{A}$$

$$= \int_V [(1-\phi)h_S + \phi h_F]dV$$

$$(3.70)$$

It can be written in differential form applying the divergence theorem:

$$\frac{\partial}{\partial t}[(1-\phi)\rho_S c_S + \phi\rho_F c_F]T + \text{div}\left(\phi\rho_F c_F \overrightarrow{v}_F\right) + \text{div}[(1-\phi)k_S + \phi k_F]\text{grad}T$$

$$= [(1-\phi)h_S + \phi h_F]$$

$$(3.71)$$

It is an important particular case, when there is no remarkable convective heat flux, there are no sources of the internal energy and the thermal state of the system is steady:

$$\text{div}[(1-\phi)k_S + \phi k_F]\text{grad}T = 0 \qquad (3.72)$$

Characterizing the given formation by the average values of the porosity and heat conduction we can define an overall heat conductivity:

$$k = (1-\phi)k_S + \phi k_F \qquad (3.73)$$

Similarly we can obtain an overall density:

$$\rho = (1-\phi)\rho_S + \phi\rho_F \qquad (3.74)$$

and heat capacity:

$$c = \frac{(1-\phi)\rho_S c_S + \phi\rho_F c_F}{(1-\phi)\rho_S + \phi\rho_F} \qquad (3.75)$$

Applying these notations, the equation of heat conduction is obtained in the form:

$$\frac{\partial}{\partial t}(\rho cT) = \text{div}(k\text{grad}T) \tag{3.76}$$

We assume that the density, heat capacity, and heat conductivity are constant. Thus we get:

$$\frac{\partial T}{\partial t} = \frac{k}{\rho c}\left(\frac{\partial^2 T}{\partial x^2} + \frac{\partial^2 T}{\partial y^2} + \frac{\partial^2 T}{\partial z^2}\right) \tag{3.77}$$

For steady state, these can be written in the simple form:

$$\frac{\partial^2 T}{\partial x^2} + \frac{\partial^2 T}{\partial y^2} + \frac{\partial^2 T}{\partial z^2} = 0 \tag{3.78}$$

Heat conduction problems have paramount importance in the investigation of geothermal systems. This is the subject of the next chapter.

References

Bear, J., 1972. Dynamics of Fluids of porous Meida. Elsevier, New York.
Darcy, H., 1856. Les Fontaines Publiques de la Ville Dijon. Dalmont, Paris.
Ferrandon, J., 1948. Genie Civil 125, 24.
Nutting, P.G., 1930. Bulletins of American Association of Petroleum Geology 14, 1337.

4.1 DIFFERENTIAL EQUATION OF HEAT CONDUCTION

The most important interaction between a geothermal reservoir and its surroundings is heat transfer. Heat transfer may be carried out by conduction, convection, or radiation. The internal energy supply of a geothermal reservoir is mostly heat conduction across the impervious homogeneous bedrock beneath it. Heat conduction is the transfer of internal energy by microscopic diffusion and collisions of molecules, atoms, and electrons. Internal energy propagates as rapidly vibrating atoms and molecules interact with neighboring particles. Macroscopically, heat conduction is the propagation of internal energy without macroscopic motion in opposite direction of the temperature gradient. This process can be described by the differential equation of heat

Copyright © 2017 Elsevier Inc. All rights reserved.

conduction (Roshenov and Hartnett, 1973). The rate of change of the internal energy inside a unit volume material equals the heat fluxes crossing its boundaries:

$$\frac{\partial(\rho cT)}{\partial t} = -\text{div}\, \vec{q} \tag{4.1}$$

where ρ is the density, c is the specific heat capacity, and $\vec{q}$ is the heat flux vector (W/m²). The heat flux vector satisfies Fourier's law:

$$\vec{q} = -k\, \text{grad}\, T, \tag{4.2}$$

where k is the heat conductivity coefficient. The ability with which a certain material conducts heat is quantified by the expression of thermal conductivity. Diamond has the highest thermal conductivity of about 2.300 W/m K. Metals are usually good conductors of heat. Metallic bonds have mobile, free electrons, which are able to transfer internal energy with high intensity. Steel has conductivity with an average of 50 W/m K. In weak conductors and insulators, the heat flux is carried almost entirely by phonon vibration.

One can easily recognize that heat propagates from the hotter region to cooler as the negative sign shows (Rybach and Muffler, 1981). Substituting Eq. (4.2) into Eq. (4.1), we get an orthogonal coordinate system:

$$\frac{\partial(\rho cT)}{\partial t} = k\left(\frac{\partial^2 T}{\partial x^2} + \frac{\partial^2 T}{\partial y^2} + \frac{\partial^2 T}{\partial z^2}\right) \tag{4.3}$$

It is an acceptable approximation to consider the material as homogeneous and isotropic. Thus ρ, c, and k are constant. In this case, it can be written that:

$$\frac{\partial T}{\partial t} = \frac{k}{\rho c}\left(\frac{\partial^2 T}{\partial x^2} + \frac{\partial^2 T}{\partial y^2} + \frac{\partial^2 T}{\partial z^2}\right) \tag{4.4}$$

The group of coefficients $k/\rho c$ is called thermal diffusivity.
Some typical rock parameters are tabulated below (Table 4.1.)

4.2 STEADY ONE-DIMENSIONAL HEAT CONDUCTION

Consider a horizontally infinite layer of constant thickness H. beneath the surface.

The differential Eq. (4.4) can be written in the simplest form in the case of a steady. one-dimensional heat conduction as:

$$\frac{d^2 T}{dz^2} = 0 \tag{4.5}$$

TABLE 4.1   Thermal Properties of Different Rocks

Type of Rock	Density (kg/m³)	Specific Heat (kJ/kg K)	Heat Conductivity (W/m K)
Basalt	2900	0.879	1.7–2.5
Dolomite	2600	0.882	5.0
Granite	2670	0.840	2.5–3.8
Limestone	2600	0.880	1.7–3.3
Marl	2100	0.885	0.8–2.1
Sandstone	2400	0.870	2.0–4.2
Tuff	2650	1.255	1.2–2.1
Water	1000	4.187	0.6

Since the coordinate axis z is directed downward. the terrestrial heat flow is directed upward. Thus:

$$q_z = -k\frac{dT}{dz} \tag{4.6}$$

The solution of Eq. (4.5), after integrating twice, is:

$$T = az + b \tag{4.7}$$

The surface temperature is $z = 0$; $T = T_0$. This boundary condition provides the value of the integration constant b. Substituting into Eq. (4.7), we get $b = T_0$. The second boundary condition based on Eq. (4.6) is heat conduction in sinking and filling sedimentary basins:

$$\frac{dT}{dz} = -\frac{q_z}{k} \tag{4.8}$$

Thus we get:

$$a = -\frac{q_z}{k} \tag{4.9}$$

Finally, it is obtained:

$$T = T_0 - \frac{q_z}{k}z \tag{4.10}$$

Note that q_z is negative, thus the temperature increases along the depth. The heat flux can be expressed as:

$$q_z = -\frac{k}{H}(T_1 - T_0) \tag{4.11}$$

Consider now three parallel layers with thicknesses of H_1, H_2, and H_3, and with heat conductivities of k_1, k_2, and k_3. Each layer can be considered to have serially fitted thermal resistances, through which the same heat flows upward. Thus we get:

$$q = \frac{k_1}{H_1}(T_1 - T_0) = \frac{k_2}{H_2}(T_2 - T_1) = \frac{k_3}{H_3}(T_3 - T_2) \qquad (4.12)$$

where $q = |q_z|$ the absolute value of the heat flux. Expressing the temperature differences from Eq. (4.12), it is obtained:

$$T_1 - T_0 = \frac{qH_1}{k_1}$$

$$T_2 - T_1 = \frac{qH_2}{k_2} \qquad (4.13)$$

$$T_3 - T_2 = q\frac{H_3}{k_3}$$

Summing the expressions T_1 and T_2 drop out and we get for the heat flux:

$$q = \frac{T_3 - T_0}{\frac{H_1}{k_1} + \frac{H_2}{k_2} + \frac{H_3}{k_3}} \qquad (4.14)$$

It is easy to generalize this result for a heat flux through layer n. Their thicknesses and heat conductivities must be known:

$$q = \frac{T_n - T_0}{\sum_{i=1}^{n} \frac{H_i}{k_i}} \qquad (4.15)$$

It can be recognized that the temperature distribution is a fraction-stroke linear function. The derivative:

$$\gamma = \frac{dT}{dz} = \frac{a}{k} \qquad (4.16)$$

varies from layer to layer, depending primarily on the heat conductivity of a certain formation. If the thicknesses of the different sedimentary layers are close to each other, and the heat conductivities varies quite periodically as the Pannonian sediments in the Carpathian Basin, the temperature distribution can be approximated by a single straight line as it can be seen in Fig. 4.1. These measured data are obtained in several Hungarian deep boreholes across sedimentary formations. An exception is the Tótkomlós well, in which the boundary between the sedimentary formations and the bedrock can be well recognized where there is a sudden change of the slope of the temperature distribution.

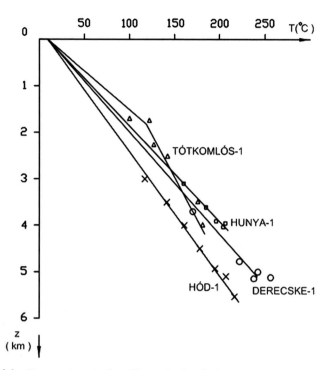

FIGURE 4.1   Temperatures in deep Hungarian boreholes.

4.3 STEADY AXISYMMETRIC HEAT CONDUCTION

Steady two-dimensional axisymmetric heat conduction is the most common phenomenon of thermal interaction between a well and its surroundings. Heat is transferred by radial heat conduction through the walls of tubing and casing, the cement sheet, and the surrounding rock. Consider the unit length piece of a pipe of inner and outer radii R_1 and R_2 as it can be seen in Fig. 4.2. The temperatures at these cylindrical surfaces are T_1 and T_2. There is no temperature variation along the pipe axis, which is the z-axis of the coordinate system. An annular area is obtained at the xy plane. The differential equation of heat conduction in this case can be written as:

$$\frac{\partial^2 T}{\partial x^2} + \frac{\partial^2 T}{\partial y^2} = 0 \tag{4.17}$$

It is suitable to change the orthogonal coordinate system to a cylindrical one according to the geometry of the system. Since:

$$r = \sqrt{x^2 + y^2} \tag{4.18}$$

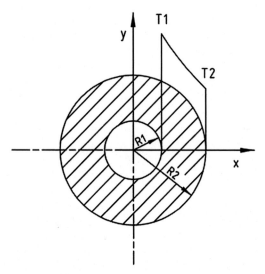

FIGURE 4.2   Axisymmetric heat conduction.

we can get that:

$$\frac{\partial r}{\partial x} = \frac{x}{r}, \text{ and } \frac{\partial r}{\partial y} = \frac{y}{r} \qquad (4.19)$$

Applying the chain rule:

$$\frac{\partial T}{\partial x} = \frac{\partial T}{\partial r} \cdot \frac{\partial r}{\partial x} = \frac{\partial T}{\partial r} \cdot \frac{x}{r} = \frac{dT}{dr} \cdot \frac{x}{r} \qquad (4.20)$$

and:

$$\frac{\partial T}{\partial y} = \frac{\partial T}{\partial r} \cdot \frac{\partial r}{\partial y} = \frac{\partial T}{\partial r} \cdot \frac{y}{r} = \frac{dT}{dr} \cdot \frac{y}{r} \qquad (4.21)$$

The second derivative is obtained as:

$$\frac{\partial^2 T}{\partial x^2} = \frac{d^2 T}{dr^2} \cdot \frac{x^2}{r} + \frac{dT}{dr} \frac{r - \frac{x^2}{r}}{r^2} = \left(\frac{d^2 T}{dr^2} - \frac{1}{r} \frac{dT}{dr} \right) \frac{x^2}{r^2} + \frac{1}{r} \frac{dT}{dr} \qquad (4.22)$$

and:

$$\frac{\partial^2 T}{\partial y^2} = \frac{d^2 T}{dr^2} \cdot \frac{y^2}{r^2} + \frac{dT}{dr} \frac{r - \frac{y^2}{r}}{r^2} = \left(\frac{d^2 T}{dr^2} - \frac{1}{r} \frac{dT}{dr} \right) \frac{y^2}{r^2} + \frac{1}{r} \frac{dT}{dr} \qquad (4.23)$$

Adding Eqs. (4.22) and (4.23), we get the Laplace equation in cylindrical form:

$$\frac{d^2 T}{dr^2} + \frac{1}{r} \frac{dT}{dr} = 0 \qquad (4.24)$$

In order to get an easier integration, it can be written as:

$$\frac{d}{dr}\left(r\frac{dT}{dr}\right) = 0 \tag{4.25}$$

Its first integral is:

$$\frac{dT}{dr} = \frac{A}{r}, \tag{4.26}$$

then the second integration leads to the expression:

$$T = A\ \ln r + B \tag{4.27}$$

The two constants of integration can be determined satisfying the boundary conditions:

If $r = R_1$, then $T = T_1$.
If $r = R_2$, then $T = T_2$.

After substitution we get:

$$A = \frac{T_1 - T_2}{\ln\frac{R_2}{R_1}} \tag{4.28}$$

and:

$$B = T_1 - \frac{T_1 - T_2}{\ln\frac{R_1}{R_2}} \tag{4.29}$$

Thus, the final result is:

$$T = T_1 - \frac{T_1 - T_2}{\ln\frac{R_1}{R_2}}\ln\frac{r}{R_1} \tag{4.30}$$

Since $T_1 > T_2$, the temperature decreases logarithmically across the pipe wall.

The overall heat flux that outflows at the unit length section of the pipe is:

$$Q = 2\pi r q \tag{4.31}$$

where q is the heat flux through a unit surface:

$$q = -k\frac{dT}{dr} = -k\frac{A}{r} \tag{4.32}$$

Thus we get for the overall heat flux:

$$Q = 2\pi k \frac{T_1 - T_2}{\ln\frac{R_2}{R_1}} \tag{4.33}$$

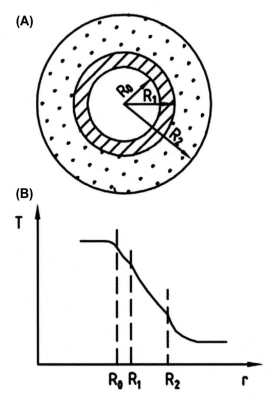

FIGURE 4.3   Axisymmetric heat conduction through two-layered pipe wall.

Consider now a complex system as it is shown in Fig. 4.3. It consists of two coaxial annuli of different thermal conductivities k_1 and k_2. At the internal radius R_0, the temperature is T_0, while the outermost surface has a radius of R_2 and the temperature T_2. If $T_0 > T_2$, the heat flux is directed outward. It is obvious that the overall heat fluxes are the same for both annuli:

$$Q = 2\pi k_1 \frac{T_0 - T_1}{\ln \frac{R_1}{R_0}} \tag{4.34}$$

and:

$$Q = 2\pi k_2 \frac{T_1 - T_2}{\ln \frac{R_2}{R_1}} \tag{4.35}$$

Expressing the temperature differences, we get:

$$T_0 - T_1 = \frac{Q}{2\pi k_1} \ln \frac{R_1}{R_0} \tag{4.36}$$

and:

$$T_1 - T_2 = \frac{Q}{2\pi k_2} \ln \frac{R_2}{R_1} \tag{4.37}$$

Adding Eqs. (4.30) and (4.37), it is obtained for the overall heat flux:

$$Q = 2\pi \frac{T_0 - T_2}{\frac{1}{k_1} \ln \frac{R_1}{R_0} + \frac{1}{k_2} \ln \frac{R_2}{R_1}} \tag{4.38}$$

Generalizing this result for a more complex system containing an annulus, we get:

$$Q = 2\pi \frac{T_0 - T_n}{\sum_{i=1}^{n} \frac{1}{k_i} \ln \frac{R_i}{R_{i-1}}} \tag{4.39}$$

This expression can be applied for complex well completion having an arbitrary number of casings and cemented annulus.

4.4 TRANSIENT AXISYMMETRIC HEAT CONDUCTION

The differential equation of heat conduction for a transient two-dimensional axisymmetric temperature field can be written as:

$$\frac{\rho c}{k} \frac{\partial T}{\partial t} = \frac{1}{r} \frac{\partial}{\partial r} \left(r \frac{\partial T}{\partial r} \right) \tag{4.40}$$

Let's introduce the auxiliary variable:

$$s = \frac{\rho c}{k} \frac{r^2}{4t} \tag{4.41}$$

Its derivatives are:

$$\frac{\partial s}{\partial t} = -\frac{\rho c}{k} \frac{r^2}{4t^2} \tag{4.42}$$

and:

$$\frac{\partial s}{\partial r} = \frac{\rho c}{k} \frac{r}{2t} \tag{4.43}$$

Applying the chain rule, Eq. (4.40) can be written as:

$$\frac{\rho c}{k} \frac{\partial T}{\partial s} \frac{\partial s}{\partial t} = \frac{1}{r} \frac{\partial}{\partial s} \cdot \frac{\partial s}{\partial r} \left(r \frac{\partial T}{\partial s} \cdot \frac{\partial s}{\partial r} \right) \tag{4.44}$$

Substituting the derivatives (4.42) and (4.43), we get:

$$-\frac{\rho c}{k}\cdot\frac{\rho c}{k}\frac{r^2}{4t^2}\frac{\partial T}{\partial s} = \frac{1}{r}\frac{\rho c}{k}\cdot\frac{r}{2t}\frac{\partial}{\partial s}\left(r\frac{2\rho c}{k}\frac{r}{2t}\frac{\partial T}{\partial s}\right) \tag{4.45}$$

After simplifying it is obtained, considering that T depends on s only:

$$-s\frac{dT}{ds} = \frac{d}{ds}\left(s\frac{dT}{ds}\right) \tag{4.46}$$

After some manipulation, first we get:

$$-s\frac{dT}{ds} = \frac{d}{ds}\left(s\frac{dT}{ds} + s\frac{d^2T}{ds^2}\right) \tag{4.47}$$

then it follows that:

$$s\frac{d^2T}{ds^2} = -(s+1)\frac{dT}{ds} \tag{4.48}$$

Rearranging Eq. (4.48), we get on expression suitable for the integration easily:

$$\frac{\frac{d^2T}{ds^2}}{\frac{dT}{ds}} = -1 - \frac{1}{s} \tag{4.49}$$

The integration obtains the following result:

$$\ln\frac{dT}{ds} = -s - \ln s + \ln C \tag{4.50}$$

where C is the constant of integration. After some manipulation we get:

$$\ln\frac{dT}{ds} = \ln e^{-s} - \ln s + \ln C, \tag{4.51}$$

Finally, the differential equation is obtained:

$$\frac{dT}{ds} = C\frac{e^{-s}}{s} \tag{4.52}$$

For determination the constant of the integration C, it is considered that the overall heat flux is:

$$Q = 2\pi rq = -2\pi rk\cdot\frac{dT}{ds} \tag{4.53}$$

Expressing r dT/dr from this equation we get:

$$r\frac{dT}{dr} = \frac{Q}{2\pi k} = r\frac{dT}{ds}\cdot\frac{\partial s}{\partial r} = \frac{\rho c}{k}\frac{r^2}{2t}\frac{dT}{ds} = 2s\frac{dT}{ds} \tag{4.54}$$

Thus:

$$\frac{Q}{2\pi k} = 2C \cdot e^{-s} \tag{4.55}$$

We can take the boundary condition that in the symmetry axis of the cylinder $r = 0$, and the value of the auxiliary variable $s = 0$, thus:

$$\frac{Q}{2\pi k} = 2C \tag{4.56}$$

from which the value of the constant is obtained:

$$C = \frac{Q}{4\pi k} \tag{4.57}$$

Thus we must integrate the differential equation:

$$\frac{dT}{ds} = \frac{Q}{4\pi k} \cdot \frac{e^{-s}}{s} \tag{4.58}$$

Now we can take the initial condition $t = 0$, $T = T_0$, to which it belongs $s = \infty$. Separating the variables we get:

$$\int_{T_0}^{T} dT = \frac{Q}{4\pi k} \int_{s=\infty}^{x} \frac{e^{-s}}{s} ds \tag{4.59}$$

Thus the temperature at an arbitrary point r, and an arbitrary time t is:

$$T(r,t) = T_0 - \frac{Q}{4\pi k} \int_{x}^{\infty} \frac{e^{-s}}{s} ds, \tag{4.60}$$

where x is:

$$x = \frac{\rho c}{k} \frac{r^2}{4t} = s \tag{4.61}$$

The result obtained in the form of Eq. (4.60) contains the integral:

$$ei(x) = \int_{x}^{\infty} \frac{e^{-s}}{s} ds \tag{4.62}$$

which is not an elementary function. It is the so-called exponential integral function.

The values of $ei(x)$ are given in tables. For small x, they are represented in Fig. 4.4. It can be seen that $ei(x)$ can be approximated by a logarithmic function in the interval $x < 0.01$, as:

$$ei(x) = -\ln x - 0.5772 \tag{4.63}$$

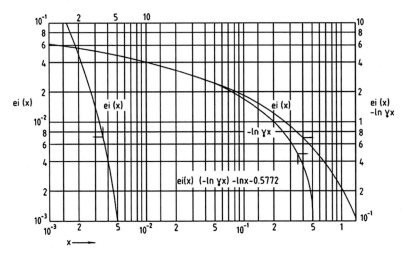

FIGURE 4.4   Approximation of ei(x) by ln function.

where the value of 0.5772 is the so-called Euler—Mascheroni constant. This approximation is especially suitable for slow transient heat conduction problems as the development of the heated region around the production well.

4.5 HEAT CONDUCTION WITH HEAT GENERATION

Heat generation in rock originates in the natural radioactivity of rocks. All rocks contain small amounts of radioactive elements of which only four isotopes contribute to heat generation: two uranium isotopes, U^{238} and U^{235}; thorium, Th^{232}; and the relatively rare potassium isotope, K^{40}. Their half lives have an order of magnitude of 10^9 years. Thus the volume heat source distribution of the crust can be considered to steady. The strength of volume heat source of rocks can be approximated by the formula of Rybach (1976):

$$h = 10^{-5}\rho(9.52 \cdot c_U + 3.48 \cdot c_{Th} + 2.56 \cdot c_K) \qquad (4.64)$$

The concentrations c_U and c_{Th} are given in ppm (parts per million). c_K is given as a percentage. In this case, h is obtained in $\mu W/m^3$. These values depends primarily on the type of rock, and certain regularity can be observed. Some typical values are summarized in Table 4.2.

The heat source strength h changes along the depth. As experience shows, this dependence is exponential:

$$h = h_o e^{-\frac{z}{H}} \qquad (4.65)$$

TABLE 4.2   Heat Source Strength Values

Type of Rock	Heat Source Strength ($\mu W/m^3$)
Basalt	0.31
Dolomite	0.36
Limestone	0.78
Sandstone	0.85
Diorite	1.08
Dacite	1.49
Marl	1.80
Granite	2.45

in which h_o is the source strength at the surface, while at the depth $z = H$ and $h = \frac{h_o}{e}$. This characteristic depth depends on certain geologic and tectonic parameters. Its typical range is 7.5–15 km.

In this case, the differential equation is a steady, one-dimensional and can be written as:

$$k\frac{d^2T}{dz^2} + h_o e^{-\frac{z}{H}} = 0 \tag{4.66}$$

Integrating twice we get:

$$T = -\frac{H^2 h_o}{k} e^{-\frac{z}{H}} + Az + B = 0 \tag{4.67}$$

Boundary conditions are taken into account to determine the constants of integration A and B. One condition is that at the surface $z = 0$, the temperature $T = T_0$. The other boundary condition is that the heat flux obtained at the surface:

$$q_o = -k\left(\frac{dT}{dz}\right)_{z=0} \tag{4.68}$$

after substitution the temperature distribution along the depth is obtained as:

$$T = T_0 + \frac{q_o - H \cdot h_o}{k} + \frac{H^2 \cdot h_o}{k}\left(1 - e^{-\frac{z}{H}}\right) \tag{4.69}$$

This solution is valid for hot dry rocks without porosity in deep, thick formations. In this case, the heat conductivity depends on the temperature as:

$$k = \frac{k_o}{1 + \kappa T} \tag{4.70}$$

The parameter $\kappa(1/°C)$ depends on the type of the rock. Its order of magnitude is 10^{-3}. k_0 is the heat conductivity at the surface.

The differential equation of heat conduction in this case is:

$$\frac{k_o}{1 + \kappa T} \cdot \frac{dT^2}{dz^2} + k_o \cdot e^{-\frac{z}{H}} = 0 \tag{4.71}$$

Its solution is obtained as:

$$T = \frac{1}{\kappa}\left\{(1 + \kappa T_o)\exp\left[H^2 h_o\left(1 - e^{-\frac{z}{H}}\right) - Hh_o z + q_o z\right] - 1\right\} \tag{4.72}$$

This type of temperature distribution is proven in very deep boreholes; for example, one of the deepest (5672 m) is Hungarian borehole HOD-1.

4.6 HEAT CONDUCTION IN AND FILLING SINKING SEDIMENTARY BASINS

It is a well-known observation that the geothermal gradient is lower than the average in rapidly sinking and filling sedimentary basins. Such a region is the Gulf Coast in the United States of America, or Makó–Hódmezővásárhely trench in the Pannonian Basin, Hungary. As the crystalline bedrock sinks, the thickness of the sedimentary layers grows permanently. The upper surface of the sediment layer is filled with fresh sediments having a temperature of the annual mean value. During the development of the basin, a given sediment layer warms up as it sinks from the surface toward the depth. This phenomenon can be described by a transient heat conduction equation:

$$\frac{\partial T}{\partial t} + U\frac{\partial T}{\partial z} = \frac{k}{\rho c}\frac{\partial^2 T}{\partial z^2} \tag{4.73}$$

The conduction is obtained by a steady process in a coordinate system, which originates at the surface while the subsequent sediment layers sink with a constant velocity U. The coordinate axis z is directed downward. At the point $z = 0$, the temperature is obviously constant $T = T_0$, and at a given depth z, each sediment layer passing through this depth must have the same temperature. In this steady temperature, only the fixed convective derivative component can change. We can get a second-order linear differential equation:

$$U\frac{\partial T}{\partial z} = \frac{k}{\rho c}\frac{\partial^2 T}{\partial z^2} \tag{4.74}$$

Rearranging the equation we get:

$$\frac{\frac{\partial^2 T}{\partial z^2}}{\frac{\partial T}{\partial z}} = \frac{\rho c U}{k} \tag{4.75}$$

The first integration leads to the expression:

$$\ln\left(\frac{dT}{dz}\right) = \frac{\rho c U}{k}z + \ln C_1 \tag{4.76}$$

After some manipulation, the following equation is obtained:

$$\frac{dT}{dz} = C_1 \cdot e^{\frac{\rho c U}{k}z} \tag{4.77}$$

The second integration gives the expression:

$$T = \frac{C_1 \cdot k}{\rho c U} \cdot e^{\frac{\rho c U}{k}z} + C_2 \tag{4.78}$$

To get the constants of integration, the following boundary conditions are taken:

If $z = 0$, $k\frac{dT}{dz} = q_0$
If $z = 0$, $T = T_0$

where q_0 is the terrestrial heat flux value at the surface. Substituting into Eqs. (4.77) and (4.78), for the constants we obtain:

$$C_1 = \frac{q_0}{k} \quad \text{and} \quad C_2 = T_0 - \frac{q_0}{\rho c U}$$

Thus the temperature distribution along the depth is:

$$T = T_0 + \frac{q}{\rho c U}\left(e^{\frac{\rho c U}{k}} - 1\right) \tag{4.79}$$

The annual mean temperature on the surface T_0, and the surface heat flux q_0, can be determined easily. The velocity U can be calculated by the thickness and age of the given sediment layer.

It can be recognized that the geothermal gradient is not constant; it changes with the depth exponentially as follows:

$$\frac{dT}{dz} = \frac{q_0}{k} \cdot e^{\frac{\rho c U}{k}z} \tag{4.80}$$

The terrestrial heat flow changes exponentially; it grows along the depth as follows:

$$q = q_0 \cdot e^{\frac{\rho c U}{k}z} \tag{4.81}$$

This simple model has many approximations. It is obvious that the sinking velocity U is not constant. The material parameters; the density ρ, the heat conductivity k, and the specific heat capacity c, can change stochastically. The Eqs. (4.79–4.81) provide a qualitative description of this process.

References

Roshenov, W.N., Hartnett, J.P., 1973. Handbook of Heat Transfer. Mc' Graw Hill, New York.

Rybach, L., 1976. Radioactive heat production: a physical property determined by the chemistry of rocks. In: Sterns, R.G.J. (Ed.), The Physics and Chemistry of Minerals and Rocks. Wiley and Sons, pp. 309–318.

Rybach, L., Muffler, L.J.P., 1981. Geothermal Systems. Wiley, New York.

Natural State of Undisturbed Geothermal Reservoirs

5.1 GEOTHERMAL RESERVOIRS IN HYDROSTATIC STATE

The natural state of an undisturbed geothermal reservoir is hydrostatic equilibrium. Hydrostatics in porous media is governed by the same principal laws as those in homogeneous fluids. However, there are certain differences between them. The rock matrix is not a homogeneous formation. There are thick under-consolidated clayey layers in it, where the pore fluid must sustain the overburden load. Thus over-pressured reservoirs are developed as in Fábiánsebestyén—Nagyszénás or the Gulf Coast.

In this section, pure hydrostatic reservoirs, such as the Upper Pannonian sedimentary aquifer, are considered. The most important

Copyright © 2017 Elsevier Inc. All rights reserved.

differences between a vessel and the saturated porous media are in their size, temperature, and density distribution. Even in a large vessel with a vertical size of only a few meters, the thickness of a sedimentary basin can be 2000–3000 m. Thus, the fluid temperature in the vessel as well as its density is uniform; the porous rock has an increasing temperature distribution along the depth and the density decreases accordingly. A further difference is that the proximity of the walls to the interior of pore fluid introduces certain modifications.

Equilibrium exists in the gravitational field, thus the hydrostatic equation is valid as:

$$\frac{dp}{dz} = \rho g \tag{5.1}$$

where the vertical coordinate z is directed downward.

It is well known that the natural temperature distribution along the depth is linear:

$$T = T_0 + \gamma z, \tag{5.2}$$

where T_0 is the temperature at the surface, and γ is the geothermal gradient. The density of the water depends on its temperature. It is an acceptable approximation to describe this relationship by a quadratic equation:

$$\rho = \rho_0 \left[1 - A(T - T_0) - B(T - T_0)^2 \right] \tag{5.3}$$

where ρ_0 is the density of the water at the reference temperature T_0, and A and B are experimentally obtained constant coefficients. Substituting $T - T_0$ from Eq. (5.2), we get the density distribution along the depth:

$$\rho = \rho_0 \left(1 - A\gamma z - B\gamma^2 z^2 \right) \tag{5.4}$$

The hydrostatic equation after substitution of Eq. (5.4) leads to the expression of the pressure gradient:

$$\frac{dp}{dz} = \rho_0 g \left(1 - A\gamma z - B\gamma^2 z^2 \right) \tag{5.5}$$

After integration, satisfying the boundary condition at $z = h$; $p = p_0$ the pressure distribution along the depth is obtained as:

$$p = p_0 + \rho_0 g \left[z - h + A\gamma \frac{(z-h)^2}{2} - B\gamma^2 \frac{(z-h)^3}{3} \right] \tag{5.6}$$

Here p_0 is the atmospheric pressure; h is the depth of the groundwater surface.

It is a slight approximation to consider h = 0, because the depth of the reservoir is far deeper than the depth of the groundwater surface, h < z. In this case, the pressure distribution is simply:

$$p = p_0 + p_0 g \left[z - A\gamma \frac{z^2}{z} - B\gamma^2 \frac{z^3}{3} \right] \tag{5.7}$$

The constants A and B are evaluated by means of a 3D phase diagram, where the three thermodynamic quantities being measured are pressure (p), volume (v) and temperature (T). It is obtained that:

$$A = 1.712 \cdot 10^{-4} \, {}^\circ C^{-1}$$

$$B = 3.232 \cdot 10^{-6} \, {}^\circ C^{-2}$$

The reservoir pressure can be slightly different because of the amount of total dissolved solids in the water, or the natural fluid motion in the reservoir. Substantial differences can arise by the restricted compaction leading to high overpressure in the reservoir.

Natural recharge or thermal convection can disturb the pure hydrostatic state of geothermal reservoirs.

As an example, we can determine the pressure in a reservoir at the depth of 2000 m. Further data in this example is: $\rho_0 = 999 \text{ kg/m}^3$, $\gamma = 0.05°C/m$, $p_0 = 101{,}325 \text{ N/m}^2$, $g = 9.81 \text{ m/s}^2$. Applying Eq. (5.7) we get:

$$p = 101325 + 999 \cdot 9.81 \left(2000 - \frac{1.712 \cdot 10^{-4} \cdot 5 \cdot 10^{-2}}{2} \cdot 4 \cdot 10^6 \right.$$

$$\left. - \frac{3.232 \cdot 10^{-6} \cdot 25 \cdot 10^{-4}}{3} \cdot 8 \cdot 10^9 \right)$$

$$= 19322790 \frac{N}{m^2} \cong 193.22 \text{ bar}$$

5.2 CONSOLIDATION OF A SEDIMENTARY AQUIFER

Consolidation is the process by which the volume of granular materials decreases due to the overburden loading. In a sedimentary basin, the upper layer of the depositing sediment has a framework of particles with high porosity while the pore space is filled with water.

The initial porosity of the top sediment layer depends on its lithology. For sands, it is about 40%, for clays it can be more than 60%. As more sediment is deposited above this layer, the effect of the increased loading leads to the increase of the particle-to-particle stress. The increasing stress in the solid matrix results in the reduction of the initial porosity. The porosity decrease is carried out due the more dense packing of the grains.

As the reduction of the pore space takes place, the pore water is forced to leave the consolidating sediment layer. The flow resistance counteracts the displacement of pore water, even permeable rock matrix, but it cannot hamper the drainage of the layer. In this case, so-called normal compaction takes place, and the porosity decreases exponentially. If impermeable layers block the drainage of the pore water, the pore pressure increases, attaining the lithostatic pressure while the porosity decreases scarcely. Thus, so-called under-compacted zones are formed with high overpressure. A measured porosity—depth profile along a borehole shows the exponential porosity decrease in the normal compaction domain. The local porosity maximum can be recognized in the under-compacted, over-pressured region.

Before the mathematical model of consolidation can be interpreted, it is necessary to briefly explain the concept of stresses in saturated porous media.

As overburden loading acts on a saturated porous media, it is carried by both the solid grains and the pore water. Terzaghi (1925) proposed the distinction of three different kinds of stresses in saturated porous materials. These stresses can be obtained from the static equations. The surface is horizontal; the coordinate axis z is directed to downward. At a certain depth z, the weight of the column above a horizontal surface of unit area obtains the so-called total stress:

$$\sigma_z = [(1 - \phi)\rho_s + \phi\rho_F]gz \qquad (5.8)$$

The effective stress is the average of the grain-to-grain forces carried by the solid matrix. The Archimedean lifting force must be regarded as the fluid and the solid phases are interacting. This force is directed in opposite to the weight as:

$$\sigma_z' = [(1 - \phi)\rho_s - (1 - \phi)\rho_F]gz \qquad (5.9)$$

The effective stress is the force that keeps the ensemble of the solid particles rigid. It determines the strength of a saturated porous material.

The pore pressure of the fluid is considered to the neutral stress. Applying the hydrostatic equation, the reaction force of the Archimedean lifting force, which acts on the fluid at the surfaces of the grains, must be regarded:

$$\phi\rho_f g + (1 - \phi)\rho_f g - \frac{dp}{dz} = 0 \qquad (5.10)$$

Thus it is obtained the well-known simple expression:

$$p = p_0 + \rho_f gz \qquad (5.11)$$

Note that z is not necessarily the same for the solid and the fluid phase. The top of the sediment layer can be covered by the water under the free-fluid surface.

Comparing the equations obtained for the total of the effective and natural stresses, it can be recognized that the effective stress is the difference of the total and the neutral stresses:

$$\sigma_z' = \sigma_z - p \tag{5.12}$$

In order to describe the process of consolidation, some approximate assumptions are made. It is assumed that the pore space of the rock is perfectly saturated by water. Both the water and the rock grains are incompressible.

The volume decrease results from the porosity reduction only. The seepage flow of the expelled pore water can be described by the Darcy's law. The development of consolidation in time depends on the displacement process of the pore water.

Thus the continuity equation can be written as:

$$\frac{\partial(\rho\phi)}{\partial t} + \text{div}(\rho\phi\vec{v}) = 0 \tag{5.13}$$

It can be decomposed as the derivative of the product:

$$\rho\frac{\partial\rho}{\partial t} + \phi\frac{\partial\rho}{\partial t} + \phi\vec{v}\text{grad}\rho + \rho\text{div}(\phi\vec{v}) = 0 \tag{5.14}$$

The second and the third term variable due the incompressibility is:

$$\phi\left(\frac{\partial\rho}{\partial t} + \vec{v}\text{grad}\rho\right) = \phi\frac{d\rho}{dt} = 0 \tag{5.15}$$

It is obtained that:

$$\frac{\partial\phi}{\partial t} + \text{div}\,\vec{q} = 0 \tag{5.16}$$

where $\vec{q} = \phi\vec{v}$ is the seepage velocity. Since the Darcy's law is:

$$\vec{q} = -\frac{K}{\mu}\text{grad}p \tag{5.17}$$

after substitution, the result is:

$$\frac{\partial\phi}{\partial t} - \frac{K}{\mu}\text{div}\,\text{grad}p = 0 \tag{5.18}$$

The other familiar form is:

$$\frac{\partial\phi}{\partial t} = \frac{K}{\mu}\left(\frac{\partial^2 p}{\partial x^2} + \frac{\partial^2 p}{\partial y^2} + \frac{\partial^2 p}{\partial z^2}\right) \tag{5.19}$$

In Terzaghi's original theory, it is assumed that the only possible mode of deformation is in the vertical direction, so the volume change $\frac{\Delta V}{V}$ can be identified with the vertical strain ε_{zz}, and assuming a linearly elastic Hookeian material, this vertical strain can be expressed by the vertical effective stress as:

$$\varepsilon_{zz} = -m_v \sigma'_{zz} \tag{5.20}$$

where m_v is the vertical compressibility. It is a material property that can be determined in an oedometer test. Because the effective stress is the difference of the total stress and the pore pressure, it is obtained that:

$$\frac{\partial \varepsilon}{\partial t} = \frac{\partial \phi}{\partial t} = -m_v \left(\frac{\partial \sigma_z}{\partial t} - \frac{\partial p}{\partial t} \right) \tag{5.21}$$

Substituting this expression into Eq. (5.20), it follows that:

$$\frac{\partial p}{\partial t} = \frac{\partial \sigma'_z}{\partial t} + c_v \left(\frac{\partial^2 p}{\partial x^2} + \frac{\partial^2 p}{\partial y^2} + \frac{\partial^2 p}{\partial z^2} \right) \tag{5.22}$$

The coefficient c_v is the consolidation coefficient:

$$c_v = \frac{K}{\mu m_v} = \frac{k}{m_v \rho_f g} \tag{5.23}$$

where k is the hydraulic conductivity:

$$k = \frac{K \rho_f g}{\mu} \tag{5.24}$$

The simplest application of the consolidation theory is the vertical consolidation of a clay sample: it is a one-dimensional compression test or an oedmater test. The sample is loaded at a certain instant ($t = 0$) by a given load $q = $ const. Because the load is constant, the pore pressure can vary in the vertical direction only. In this case, the differential equation of consolidation is obtained as:

$$\frac{\partial p}{\partial t} = c_v \frac{\partial^2 p}{\partial z^2} \tag{5.25}$$

This equation is analogous with the one-dimensional transient heat conduction equation. The solution is also similar. Initially, the pore pressure sustains the load q, and the effective stress remains zero:

$$t = 0 \quad \text{and} \quad p = q$$

The lower boundary of the sample holder is impermeable. At the top of the sample, the porous plate transferring the load is very permeable, thus the pore pressure at the top is zero:

$$z = 0 \quad \text{and}$$

$$z = H \quad \text{and} \quad p = 0$$

The solution of the differential equation is found in the form of a product of two particular solutions. One of them depends on the time t only, while the other is the function of depth z only:

$$p = X(z) \cdot Y(t) \tag{5.26}$$

Substituting it into Eq. (5.25), it is obtained:

$$X \cdot \frac{dY}{dt} = c_v^2 Y \cdot \frac{d^2X}{dz^2} \tag{5.27}$$

After a slight modification it is:

$$\frac{1}{Y}\frac{dY}{dt} = \frac{c_v^2}{X}Y \cdot \frac{d^2X}{dz^2} \tag{5.28}$$

It can be recognized that the left-hand side of the equation depends only on t; the right-hand side depends only on z. It is possible only if both sides equal the same constant α. Thus two ordinary differential equations can be solved. The final result is than product in the following form:

$$\frac{p}{q} = \frac{4}{\pi}\sum_{j=1}^{\infty}\frac{(-1)^{j-1}}{2j-1}\cos\left[(2j-1)\frac{\pi}{2}\frac{z}{H}\right]\cdot e^{-(2j-1)^2\frac{\pi^2 c_v}{4H^2}\cdot t} \tag{5.29}$$

The solution is represented as a series of curves.

Along the certain curve, the parameter $\frac{c_v t}{H^2}$ is constant. The consolidation process is started from the $\frac{p}{q} = 1$ value of the abscissa, and practically is ended where $\frac{c_v t}{H^2} = 2$.

At this time the overpressure is eliminated, the excess water is displaced, and the consolidation is finished.

5.3 OVER-PRESSURED GEOTHERMAL RESERVOIRS

The fundamental difference between a hydrostatic and an over-pressured reservoir is that in an over-pressured reservoir, water cannot be communicated with the surrounding saturated rocks. Over-pressured reservoirs develop mainly in rapidly filling tertiary sedimentary basins such as the Gulf Coast, the Niger Delta, and certain parts of Pannonian Basin. As it is known, sediments are deposited at the delta front of rivers. The fresh delta sediments have a high porosity, even more than 40% completely saturated with water. The loose packed grains compact under the influence of gravity and the overburden created by the deposition of even more overlaying fresh sediments. The compaction of a sedimentary layer occurs as the porosity decreases, expelling the water from the pore space. If the displacement of the pore water is unhindered, the pressure distribution along the depth is hydrostatic. There are some mechanisms providing a seal, which hinders the displacement of the pore water. The most common seals are clay and shale layers. Clay or shale sequences have very low permeability, but they are not perfect seals. The favorable combination of low permeability and large thickness is necessary to retain the dewatering of the compacting sedimentary layer.

Consider a thick, compacting clay or silt layer. It can be recognized that the boundary zone of the layer can be compacting and dewatering, more so than its inner domain. The expelled pore water must flow along a longer path from the inner part of the layer. Its pressure loss is necessarily greater. It requires higher-pressure energy to maintain the flow against the greater pressure loss. The boundary zone of the formation is almost perfectly compacted while the inner region remains under-compacted, containing higher-pressure pore water. As the compaction develops, the permeability decreases. Thus an almost perfectly sealed domain develops at the boundary zone of the layer. The closing pore channels don't expel the water. The substantial overpressure remains in the inner part of the layer until it attains the lithostatic pressure, due to the overburden load. Because the fluid has low compressibility, it supports the majority of the additional overburden load and retards the further compaction.

However, overpressure can also be present within the pore space by other fluid expansion mechanisms such as aquathermal expansion, hydrocarbon maturation, transformation of smectite to illite, and clay dehydration. The depth interval over which this transition occurs depends on the local geology and the state of stress of the crust.

5.4 RECOVERABLE FLUID MASS BY ELASTIC EXPANSION

Consider a closed reservoir of volume V. It contains a porous rock of homogeneous porosity ϕ. The pore volume is filled with a fluid of homogeneous density ρ. The thermal state of the fluid is determined unambiguously by its pressure and temperature. The reservoir has an outflow opening; the production well. The pressure at the outflow opening can be controlled arbitrarily.

The balance equation of mass can be written in integral form for the reservoir fluid:

$$\int_V \frac{\partial(\rho\phi)}{\partial t} dV + \int_{(A)} \phi\rho\vec{v}d\vec{A} = 0 \qquad (5.30)$$

That is the sum of rate of change of pore fluid mass within the volume V, and the mass flux $\phi\rho\vec{v}$ across the bounding surface of V is zero. Since ρ and ϕ has uniform distribution, the integrand of the volume integral is homogeneous within V. This may be multiplied instead of integration. The bounding surface of the closed reservoir is impermeable everywhere except at the outflow opening, where integral obtains the withdrawing mass flow rate $\dot{m}$. Thus we get the equation:

$$V\frac{\partial(\rho\phi)}{\partial t} + \dot{m} = 0 \tag{5.31}$$

Since the deformation of the rock matrix is negligible, and the porosity is constant, it can be written:

$$V\phi\frac{\partial\rho}{\partial t} + \dot{m} = 0 \tag{5.32}$$

Since the density distribution is uniform in the reservoir, its convective derivatives are zero. Thus the local and the material derivatives are equal:

$$\frac{\partial\rho}{\partial t} = \frac{d\rho}{dt} \tag{5.33}$$

Tapping the reservoir, its pressure decreases inducing the elastic expansion of the reservoir fluid mass. It is known through experimental experience that temperature is unchanged in the case of relatively small pressure change. Thus the elastic expansion of the reservoir fluid can be considered to be isothermal. Applying the chain rule we get:

$$V\phi\frac{d\rho}{dp}\cdot\frac{dp}{dt} + \dot{m} = 0 \tag{5.34}$$

where:

$$\beta = \frac{1}{\rho}\frac{d\rho}{dp} \tag{5.35}$$

is the isothermal compressibility coefficients of the fluid. Thus the outflowing mass flow rate can be expressed as:

$$\dot{m} = -V\phi\beta\rho\frac{dp}{dt} \tag{5.36}$$

Separating the variables we get the differential equation:

$$\dot{m}dt = -V\phi\beta\rho dp \tag{5.37}$$

If the production rate is constant, the integration obtains:

$$M = \dot{m}(t_2 - t_1) = V\phi\beta\rho(p_1 - p_2) \tag{5.38}$$

The subscript 1 refers to the beginning of the production, while 2 refers to the end of it. M is the cumulated recoverable fluid mass within the time interval $t_2 - t_1$, while the reservoir pressure decreases from p_1 to p_2. The proportional factor, the so-called storativity, is:

$$S = V \phi \beta \rho. \tag{5.39}$$

The recoverable fluid mass from a finite reservoir depends on the reservoir parameters V and ϕ, the fluid properties ρ and β, and the reservoir pressure decrease $p_1 - p_2$. If the lower pressure limit comes close to the atmospheric pressure, the production will be depleted. In this case artificial lift methods must be applied. The $p_1 - p_2$ pressure difference can be increased almost arbitrarily applying submersible centrifugal pumps. Thus the recoverable fluid mass can be proliferated.

As an illustrative example, we can determine the recoverable water mass from the Upper Pannonian sandstone aquifer by elastic expansion.

The horizontal extension of the aquifer is 40,000 km^2, its average thickness is 200 m. The porosity is 20%, the average temperature is 80°C. The density of water 970 kg/m^3, its isothermal compressibility is $4.68 \cdot 10^{-8}$ m^2/N. The pressure decrease during production is $3 \cdot 10^5$ N/m^2.

The volume of the aquifer is:

$$V = A \cdot h = 4 \cdot 10^{10} \text{ m}^2 \cdot 200 \text{ m} = 8 \cdot 10^{12} \text{ m}^3$$

Substituting into Eq. (5.38) we get:

$$M = 8 \cdot 10^{12} \text{m}^3 0.2 \cdot 4.68 \cdot 10^{-8} \frac{\text{m}^2}{\text{N}} \cdot 970 \frac{\text{kg}}{\text{m}^3} \cdot 3 \cdot 10^5 \frac{\text{N}}{\text{m}^2} = 2.179 \cdot 10^{13} \text{ kg}$$

It is known that the useable part of the geothermal energy belongs to the temperature interval over the adjacent temperature at the surface:

$$E = M \cdot c (T - T_0) \tag{5.40}$$

where c is the heat capacity of the water, T is its temperature, and T_0 is the annual mean temperature at the surface. The specific heat capacity of water is 4187 J/kgK, the temperature $T = 80°C$, and $T_0 = 10.5°C$; after substitution we get:

$$E = 2,179 \cdot 10^{13} \text{ kg} \cdot 4,187 \cdot 10^3 \frac{\text{J}}{\text{kgK}} (180 - 10.5) \text{K} = 6,341 \cdot 10^{18} \text{ J}$$

Using a more familiar unit, it is 6341 PJ. Note that the present geothermal energy utilization in Hungary is approximately 20 PJ/year. This simple lumped-parameter model is only accurate to an order of magnitude, for the recoverable part of the geothermal resource found in the Upper Pannonian sandstone aquifer and fed by naturally upflowing wells. It is obvious that the recovery of a given geothermal reservoir needs a more sophisticated simulation model.

5.5 THERMAL CONVECTION CURRENTS IN POROUS MEDIA

The flow of fluid, which is caused solely by density differences brought about by temperature gradients, is called free or thermal convection. A fluid body may be in mechanical equilibrium without being in thermal equilibrium. The hydrostatic equation:

$$\rho \vec{g} = \text{grad} p \qquad (5.41)$$

may be satisfied even though the temperature distribution is not uniform. There are certain temperature gradients that allow the mechanical equilibrium to be maintained. Other temperature distributions may induce movements in the fluid. Similar to the bulk fluid bodies, this phenomenon also occurs in saturated porous media. If the saturated porous medium is heated from below, convection currents will arise in it. The reason is that the density of the hotter fluid at the bottom is lower than the cooler fluid at the top region. This density difference may cause convection currents as the hotter fluid tends to rise.

Naturally, these convection currents cannot develop in a saturated porous medium as easily as in a bulk fluid. The energy to maintain the thermal convection is taken from the terrestrial heat flow.

The aim of the following qualitative analysis is to determine the conditions necessary for the occurrence of the thermal convection currents in a porous aquifer.

For simplicity, the reservoir is considered to be a homogenous, isotropic, fluid-saturated, porous medium, bounded by two parallel horizontal planes. An orthogonal coordinate system is chosen: the x and y axes are in the horizontal upper boundary surface, while the z axis is directed vertically downward. Analogously with regard to convection in bulk fluid bodies, the thermal convection cells will occur in the form of irregular hexagonal or pentagonal columns with vertical axes. These may be considered a series of two-dimensional flow patterns. In this case, it is convenient to investigate the two-dimensional flow in the x-z plane. The number of the unknown functions is five: the x and z components of the seepage velocity, the pressure p, the temperature T, and the density ρ. There are five equations for determination of the five unknowns: the two scalar components of the Darcy's law, the continuity equation, the balance equation of the internal energy, and the equation of the state. The Darcy equations are:

$$q_x = -\frac{gK}{\upsilon}\frac{\partial H}{\partial x} \qquad (5.42)$$

and

$$q_z = -\frac{gk}{\upsilon}\left(\frac{\Delta\rho_f}{\rho_f} - \frac{\partial H}{\partial z}\right) \tag{5.43}$$

where K is the permeability.

The continuity can be written for incompressible fluid as:

$$\frac{\partial q_x}{\partial x} + \frac{\partial q_z}{\partial z} = 0 \tag{5.44}$$

The balance equation of the internal energy for the unit volume of fluid-saturated porous rock is:

$$[(1-\phi)\rho_k c_k + \phi\rho_f c_f]\frac{\partial T}{\partial t} + \rho_f c_f\left(q_x\frac{\partial T}{\partial x} + q_z\frac{\partial T}{\partial z}\right)$$
$$= [(1-\phi)k_k + \phi k_f]\cdot\left(\frac{\partial^2 T}{\partial x^2} + q_z\frac{\partial^2 T}{\partial z^2}\right) \tag{5.45}$$

Finally, the equation of state, neglecting the pressure dependence of the density, is:

$$\frac{\Delta\rho_f}{\rho_f} = \beta\Delta T \tag{5.46}$$

One of the boundary conditions is that the velocity is zero along the bounding surfaces. The temperature along both the lower and the upper boundary surfaces is constant. The temperature distribution along the depth is linear:

$$T = T_0 + \gamma z \tag{5.47}$$

where γ is the geothermal gradient. The temperature change in the x-direction is neglected. The solution of the equation system found in the following form:

$$q_x = -A\cdot e^{\frac{t}{t_\infty}}\sin\frac{\pi\cdot x}{1}\sin\frac{\pi\cdot z}{1} \tag{5.48}$$

$$q_z = -A\cdot e^{\frac{t}{t_\infty}}\cos\frac{\pi\cdot x}{1}\sin\frac{\pi\cdot z}{1} \tag{5.49}$$

$$\Delta T = B\cdot e^{\frac{t}{t_\infty}}\cos\frac{\pi\cdot x}{1}\cos\frac{\pi\cdot z}{1} \tag{5.50}$$

A and B are temporarily unknown coefficients, t_∞ is the duration of the development of the convective cell, and l is the distance between the horizontal boundary surfaces. Substituting into the Darcy Eqs. (5.42) and (5.43) we obtained the expression:

$$\frac{\partial H}{\partial x} = \frac{A\cdot\upsilon}{Kg}\cdot e^{\frac{t}{t_\infty}}\sin\frac{\pi\cdot x}{1}\sin\frac{\pi\cdot z}{1} \tag{5.51}$$

$$\frac{\partial H}{\partial z} = \left(\frac{A \cdot \upsilon}{Kg} + B\beta\right) \cdot e^{\frac{t}{t\infty}} \cos\frac{\pi \cdot x}{1} \cos\frac{\pi \cdot z}{1} \qquad (5.52)$$

The piezo metric head H must be satisfied by the Laplace or the Poisson equations. Thus it obviously can be derived twice. Since the above equations are compatible only if the following expression is fulfilled:

$$A = \frac{gkB\beta}{2\upsilon} \qquad (5.53)$$

Substituting the Eqs. (5.51)–(5.53) into the internal energy balance Eq. (5.45), regarding the Eq. (5.46) after derivations, it is obtained:

$$\frac{1}{t_\infty} = -\frac{2\pi^2}{1^2}\frac{(1-\phi)k_k + \phi k_f}{(1-\phi)\rho_k c_k + \phi \rho_f c_f} + \frac{gK\gamma B \cdot \rho_f c_f}{2\upsilon} \qquad (5.54)$$

There is the possibility of the occurrence of convective cells only if the right-hand side of Eq. (5.54) is positive, that is:

$$\frac{gK\gamma B \cdot \rho_f c_f}{2\upsilon} \rangle \frac{2\pi^2}{1^2}\frac{(1-\phi)k_k + \phi k_f}{(1-\phi)\rho_k c_k + \phi \rho_f c_f} \qquad (5.55)$$

A certain yields geothermal gradient belongs to the fulfillment of the condition Eq. (5.55), if:

$$\gamma \rangle \gamma_y = \frac{4\pi^2}{1^2}\frac{(1-\phi)k_k + \phi k_f}{(1-\phi)\rho_k c_k + \phi \rho_f c_f} \cdot \frac{\upsilon}{gK\beta \rho_f c_f} \qquad (5.56)$$

Some important consequences at Eq. (5.56) can be recognized directly.

As large the vertical thickness of the reservoir is, a small geothermal gradient is efficient for the thermal convection. Rocks of high porosity and permeability provide good conditions to form convection cells. It is obvious that poor heat conductivity of the rock matrix actuates the convective currents of the pore fluid. A high coefficient of thermal expansion of the pore fluid leads to an increase in the Archimedean lifting force, which actuates the thermal convection. Finally, low viscosity advantageously influences the capability of thermal convection.

These conclusions are in agreement with the basic laws of physics. Thus, the obtained expression for the calculated geothermal gradient seems to be qualitatively acceptable. It must be noted that many simplifications and approximate assumptions were made in the process of calculating the geothermal gradient. Thus, Eq. (5.56) has qualitative value only.

Reference

Terzaghi, K., 1925. Erdbaumechanik auf boden physikalischer grundlage. Deuticke, Leipzig.

CHAPTER

6

Two-Dimensional Steady Flow Through Porous Media

OUTLINE

6.1 BASIC EQUATION

Physically, all flow systems extend in three dimensions. However, in many problems, the features of pore fluid motion are essentially two-dimensional, with the motion being substantially the same in parallel planes. For these problems, we need to concern ourselves with two-dimensional flow only, and thereby we are able to considerably reduce the work necessary to obtain a solution. Fortunately, in reservoir engineering, many problems fall into this category.

Flow and Heat Transfer in Geothermal Systems
http://dx.doi.org/10.1016/B978-0-12-800277-3.00006-2

Copyright © 2017 Elsevier Inc. All rights reserved.

In this chapter incompressible steady flows are only considered. In this case, the continuity equation can be written as:

$$\frac{\partial q_x}{\partial x} + \frac{\partial q_y}{\partial y} = 0 \tag{6.1}$$

For steady flow, Darcy's law is obtained in the form:

$$\vec{q} = -k\,\mathrm{grad}H \tag{6.2}$$

This is equivalent to the existence of a velocity potential:

$$\phi = -k\left(\frac{p}{\rho g} + z\right) + C \tag{6.3}$$

where C is an arbitrary constant. Thus:

$$q_x = \frac{\partial \phi}{\partial x}; \quad q_y = \frac{\partial \phi}{\partial y} \tag{6.4}$$

Substituting these expressions into the equation of continuity, we obtain the Laplace equation:

$$\Delta\phi = \frac{\partial^2 \phi}{\partial x^2} + \frac{\partial^2 \phi}{\partial y^2} = 0 \tag{6.5}$$

This indicates that for conditions of incompressible, steady-state laminar flow, the pore fluid motion can be completely determined by solving one equation, subject to the boundary conditions of the flow domain.

The continuity equation becomes identified with the stream function Ψ, where the velocity components can be expressed as:

$$q_x = \frac{\partial \psi}{\partial y}; \quad q_y = -\frac{\partial \psi}{\partial x} \tag{6.6}$$

It can be easily recognized that the velocity potential ϕ and the stream function Ψ can be related as:

$$\frac{\partial \phi}{\partial x} = \frac{\partial \psi}{\partial y} \tag{6.7}$$

$$\frac{\partial \phi}{\partial y} = -\frac{\partial \psi}{\partial x} \tag{6.8}$$

These two extremely important equations are called the Cauchy—Riemann equations. If ϕ and Ψ have continuous derivatives of the first order in a given region A, the necessary and sufficient condition that:

$$W = \phi + i\psi \tag{6.9}$$

is an analytic function of:

$$z = x + iy \tag{6.10}$$

in A, the Cauchy—Riemann equations are satisfied.

Of the many important properties of analytic functions $W(z)$, we note the following:

Both real (ϕ) and imaginary (Ψ) parts of an analytic function satisfy Laplace's equation in two dimensions. It was shown that:

$$\frac{\partial^2 \phi}{\partial x^2} + \frac{\partial^2 \phi}{\partial y^2} = 0 \tag{6.11}$$

Since the existence of the velocity potential implies that the flow is irrotational, it can be written that:

$$\frac{\partial q_y}{\partial x} - \frac{\partial q_x}{\partial y} = 0 \tag{6.12}$$

Substituting the derivatives of the stream function into an equation where rotation does not occur, we obtain:

$$\frac{\partial^2 \psi}{\partial x^2} + \frac{\partial^2 \psi}{\partial y^2} = 0 \tag{6.13}$$

Any function that satisfies this equation is called a harmonic function. Two harmonic functions such as ϕ and Ψ that are related so that $W = \phi + i\Psi$ is an analytic function, are called conjugate harmonic functions, forming a complex potential W.

The following families of curves are the equipotential lines:

$$\phi = \phi(x, y) = \text{const} \tag{6.14}$$

and the next are the streamlines:

$$\Psi = \Psi(x, y) = \text{const.} \tag{6.15}$$

Equipotential lines and streamlines intersect each other at right angles. To prove this, we obtain the slope of each family of curves, respectively. Along a potential line:

$$d\phi = \frac{\partial \phi}{\partial x}\, dx + \frac{\partial \phi}{\partial y}\, dy = 0 \tag{6.16}$$

from which:

$$\frac{dy}{dx_\phi} = -\frac{\dfrac{\partial \phi}{\partial x}}{\dfrac{\partial \phi}{\partial y}} = -\frac{q_x}{q_y} \tag{6.17}$$

Similarly for the streamlines:

$$d\psi = \frac{\partial \psi}{\partial x} \, dx + \frac{\partial \psi}{\partial y} \, dy = 0 \tag{6.18}$$

$$\frac{dy}{dx_\psi} = -\frac{\dfrac{\partial \psi}{\partial x}}{\dfrac{\partial \psi}{\partial y}} = \frac{q_y}{q_x} \tag{6.19}$$

Thus, we obtain:

$$\frac{dy}{dx_\phi} = -\frac{1}{\dfrac{dy}{dx_\psi}} \tag{6.20}$$

If a function $W(z)$ is not analytic at some point z_0, then z_0 is said to be singular point or a singularity of the function. For example, if:

$$W = \ln(z) \tag{6.21}$$

then the function is analytic at every point except the point $z = 0$, where it is discontinuous; hence $z = 0$ is a singular point. The flow induced by singularities has a particular importance in two-dimensional flow theory.

Since the velocity field of any steady, two-dimensional potential flow satisfies the Cauchy–Riemann equations, any analytic, single-valued complex variable function $W(z)$ must represent such a flow in the z-plane.

The derivative of an analytic complex function is independent of the manner in which Δz approaches zero:

$$\frac{dW}{dz} = \lim_{\Delta z \to 0} \frac{\Delta W}{\Delta z} + \frac{\partial \phi}{\partial x} + i \frac{\partial \psi}{i \partial y} \tag{6.22}$$

It can be recognized that:

$$\frac{dW}{cz} = q_x - i q_y \tag{6.23}$$

Thus, the conjugate velocity equals the derivative of the complex potential, and the absolute value of the velocity at any point is simply:

$$q = \left| \frac{dW}{dz} \right| \tag{6.24}$$

The lines along which the real part of W is constant are called potential lines:

$$\mathrm{Re}(W) = \phi = \mathrm{const} \tag{6.25}$$

The lines along which the imaginary part of W is constant are the streamlines

$$\ln(W) = \Psi = \text{const} \tag{6.26}$$

Streamlines and potential lines form an orthogonal net. If one chooses the increment of Ψ to be constant, the discharge between two streamlines is obtained as the difference of their constants; consider Fig. 6.1. Let's choose two arbitrary points in the x_y plane. The streamlines passing through the points 1 and 2 are determined by the constant values of the stream functions Ψ_1 and Ψ_2. The tangent of the arbitrary curve 1–2 at some point is denoted by α. It can be recognized easily that:

$$\frac{dx}{ds} = \cos\alpha \tag{6.27}$$

$$\frac{dy}{ds} = \sin\alpha \tag{6.28}$$

The normal component of the velocity, which is perpendicular to the curve, can be expressed as

$$q_n = q_y \cos\alpha - q_x \sin\alpha \tag{6.29}$$

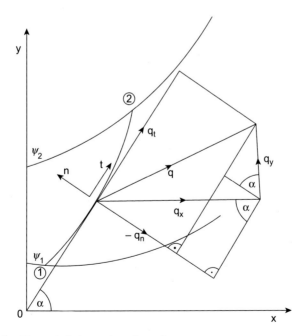

FIGURE 6.1 Flow rate between two streamlines.

The discharge across the arc 1–2 is:

$$Q = \int_1^2 q_n ds \tag{6.30}$$

After substitution we get:

$$Q = -\int_1^2 \left(q_x \frac{dy}{ds} - q_y \frac{dx}{ds} \right) = \int_1^2 \left(\frac{\partial \psi}{\partial x} dx + \frac{\partial \psi}{\partial y} dy \right) = \psi_2 - \psi_1 \tag{6.31}$$

6.2 INTEGRATION OF THE CONJUGATE VELOCITY FIELD

Cauchy's integral theorem is well known in the complex variable theory; if C is a closed curve interior to a region R within and on which $\overline{q}(z)$ is analytic, then:

$$\int_C \overline{q}(z) \, dz = 0 \tag{6.32}$$

An important corollary of Cauchy's integral theorem is: if $\overline{q}(z)$ is analytic within an on a region R, then the value of the line integral between any two points within R is independent the path of integration.

If the conjugate of the velocity field is analytic within and on a closed contour C of a simply connected region R, and if a point a is interior to C, then:

$$\overline{q}(a) = \frac{1}{2\pi i} \int_C \frac{\overline{q}(z) dz}{z - a} \tag{6.33}$$

This equation is Cauchy's integral formula. It shows that the value of the conjugate velocity is an analytic function within a region that is completely determined through the region in terms of its values on the boundary.

In order to integrate the velocity field around singularities, consider Fig. 6.2, applying its notation taking the line integral between two arbitrary points A and B.

$$\int_A^B \overline{q} dz = \int_A^B (q_x dx + q_y dy) + i \int_A^B (q_x dy - q_y dx) \tag{6.34}$$

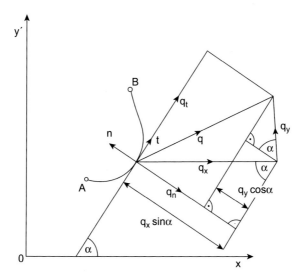

FIGURE 6.2   Line integral of the conjugate velocity field.

It can be recognized easily that the real part of the integral is the integral of the tangential velocity component by the arc ds (note ds $= |dz|$). Calculating the imaginary part we obtain that:

$$\int_A^B \left(q_x \frac{dy}{ds} - q_y \frac{dx}{ds} \right) ds = \int_A^B (q_x \sin\alpha - q_y \cos\alpha)\, ds = -\int_A^B q_n ds, \quad (6.35)$$

as it is shown in Fig. 6.3.

$$-q_n = q_x \sin\alpha - q_y \cos\alpha \quad (6.36)$$

The unit vectors $\vec{t}$ and $\vec{n}$ forms a right-handed coordinate system. Because of this condition, the normal component of the seepage velocity has a negative sign. Thus the line integral of the conjugate velocity field can be written as:

$$\int_A^B \bar{q}dz = \int_A^B (q_t - iq_n)ds \quad (6.37)$$

Extending the integration to a closed curve we obtain:

$$\int_{CF} \bar{q}dz = \int_C q_t ds - i \int_C q_n ds \quad (6.38)$$

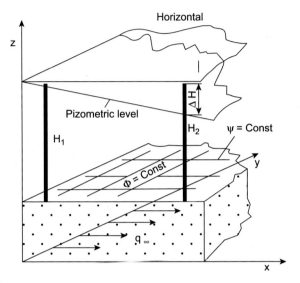

FIGURE 6.3   Parallel flow.

The first integral at the right hand side is the circulation of the velocity field:

$$\Gamma = \int_C q_t ds \tag{6.39}$$

This, as it is known, means that the circulation around C is equal to the flux of, through the area bounded by C. Thus if the integral on the closed contour C is non-vanishing, obtaining a real quantity, there must be vortex-type singularities in the region bounded by the curve.

If the imaginary part of the complex integral of the velocity field on a closed contour doesn't vanish, then there is a surplus through this closed curve. This surplus is positive if the outflowing fluid is greater than the inflow, thus:

$$Q = - \int_C q_n ds > 0 \tag{6.40}$$

which is called source, while $q < 0$ is a sink.

6.3 EXAMPLES OF ANALYTIC FUNCTIONS REPRESENTING TWO-DIMENSIONAL POTENTIAL FLOWS

Each analytic function W(z) represents a potential flow in the z-plane. The simplest example is the linear function:

$$W = Az = Ax + iAy \qquad (6.41)$$

where A is a real constant. Since:

$$\frac{dW}{dz} = A \qquad (6.42)$$

The Eq. (6.41) complex potential corresponds to a uniform flow with velocity A in the x-direction. Let's note this uniform velocity by $q_{\infty x}$, thus, the complex potential of the uniform parallel flow is:

$$W = q_{\infty x} z \qquad (6.43)$$

The potential lines and streamlines are obtained as the real and imaginary parts of the complex potential:

$$\Phi = q_{\infty x} x = \text{const} \qquad (6.44)$$

$$\Psi = q_{\infty x} y = \text{const} \qquad (6.45)$$

Thus the orthogonal straight flow net is obtained as it is shown in Fig. 6.3.

Since the velocity has a real component q_x only, it is obtained from the velocity potential.

$$q_x = \frac{\partial \phi}{\partial x} \qquad (6.46)$$

The velocity potential depends linearly on the piezometric head:

$$\Phi = -KH + C \qquad (6.47)$$

where K is the hydraulic conductivity, while C is an arbitrary additive constant. Thus the velocity of a homogeneous parallel flow is:

$$q_x = -\frac{\partial (KH)}{\partial x} \qquad (6.48)$$

The velocity is determined by the slope of the piezometric surface (see Fig. 6.3.).

The effect of different kind of singularities will be discussed in the following.

Consider the analytic function:

$$W = C \cdot \ln z \qquad (6.49)$$

in which C is an arbitrary real quantity. It is obvious that at $z = 0$, the function has a pole, thus $z = 0$ is a singular point. The conjugate complex velocity is obtained as:

$$\bar{q}(z) = \frac{dW}{dz} = \frac{C}{z} \tag{6.50}$$

It can be recognized that the point $z = 0$ is also a singular point of the velocity field. To determine the kind of singularity, we take the integral of the velocity field on a closed contour around $z = 0$. Let's choose for this purpose a circle of radius R and with the center at $z = 0$. On this circle the line integral of the velocity field can be determined by:

$$\int\limits_{(R)} \bar{q}dz = \int\limits_{(R)} \frac{C}{R}(\cos\varphi - i\sin\varphi)iR(\cos\varphi + i\sin\varphi)d\varphi \tag{6.51}$$

The real part of the integral will vanish, thus we get:

$$\int\limits_{(R)} \bar{q}dz = i2\pi C = iQ \tag{6.52}$$

This is a purely imaginary quantity, thus the singularity at the point $z = 0$ must be a source of intensity Q.

Thus, a two-dimensional source of discharge (per unit length in the direction perpendicular to the xy-plane) placed at the origin can be represented by the complex potential:

$$W = \frac{Q}{2\pi}\ln z \tag{6.53}$$

Its conjugate velocity field is:

$$\bar{q} = \frac{Q}{2\pi z} \tag{6.54}$$

The potential lines and streamlines are obtained as the real- and imaginary part of the complex potential W:

$$\phi + i\psi = \frac{Q}{2\pi}\ln(r \cdot e^{i\psi}) = \frac{Q}{2\pi}\ln r + i\psi\frac{Q}{2\pi} \tag{6.55}$$

Thus the potential lines are concentric circles determined by the expression:

$$\frac{Q}{2\pi}\ln r = \text{const} \tag{6.56}$$

where the radius of each circle can be calculated by the equation:

$$r = e^{\frac{2\pi n}{Q}}, \quad n = 0, \pm1, \pm2, \ldots \tag{6.57}$$

The streamlines form a family of straight lines passing through the origin, determined by the equation:

$$\frac{Q}{2\pi}\varphi = \text{const} \tag{6.58}$$

The flow net is shown in Fig. 6.4.

It is obvious that the complex potential of a sink is given with a negative sign, because of $Q < 0$:

$$W = -\frac{Q}{2\pi}\ln z \tag{6.59}$$

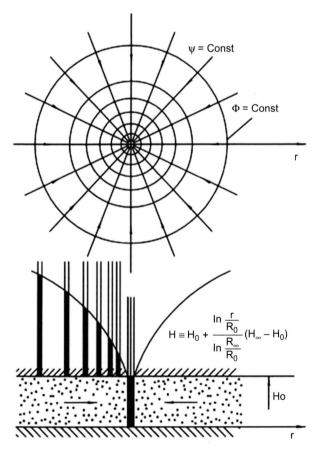

FIGURE 6.4 Streamlines and potential lines of a source.

The velocity field is:

$$\bar{q} = -\frac{Q}{2\pi z} \tag{6.60}$$

Its potential lines are concentric circles with the common center at the origin, while the streamlines are radial straight lines. The only difference is the direction; both the velocity and streamlines are directed into the origin.

Although sources and sinks are simply mathematical conveniences, they are of great value in that when combined with other simple flow patterns, they can closely reproduce many complicated natural flow patterns. The power of the method of sources and sinks stems from the linearity of the Laplace equation. For if we add together a number of complex potentials, each of which satisfies Laplace's equation, the sum of these will satisfy it, too. This allows the application of the principle of superposition.

The complex potential of a sink can represent the flow pattern around a production well in a confined reservoir of constant thickness. The velocity potential can be expressed by the piezometric head:

$$\Phi = -KH + \text{const} \tag{6.61}$$

At the wall of the borehole, the potential:

$$-KH_0 + \text{const.} = -\frac{Q}{2\pi}\ln R_0 \tag{6.62}$$

while at the infinity (practically at sufficiently large distance from the well), it can be written:

$$-KH_\infty + \text{const.} = -\frac{Q}{2\pi}\ln R_\infty \tag{6.63}$$

If the thickness of the aquifer or the reservoir is b, the discharge of the well is obtained as:

$$Q_w = 2\pi bK \frac{H_\infty - H_0}{\ln\dfrac{R_\infty}{R_0}} \tag{6.64}$$

The piezometric level around the well depends on the distance logarithmically:

$$H = H_0 + (H_\infty - H_0)\frac{\ln\dfrac{r}{R_0}}{\ln\dfrac{R_\infty}{R_0}} \tag{6.65}$$

The in situ value of hydraulic conductivity for a reservoir of thickness b is obtained as:

$$K = \frac{Q_w}{2\pi b} \frac{\ln\dfrac{R_2}{R_1}}{H_2 - H_1} \tag{6.66}$$

where H_2 and H_1 are the piezometric levels in two observation wells placed in distances R_2 and R_1 from the origin, where the production well is placed.

6.4 METHOD OF SUPERPOSITION

As an example of the method of superposition, we shall consider a uniform flow of velocity $q_{\infty x}$ past a well in a layer of thickness b. For convenience, the well will be located at the origin, as it is shown in Fig. 6.5. Noting that the complex potential for the well is:

$$W = -\frac{Q_w}{2\pi b}\ln z, \tag{6.67}$$

and the complex potential of the uniform flow is:

$$W = q_{\infty x}z \tag{6.68}$$

For the combined flow, the superposition obtains the complex potential is:

$$W = q_{\infty x}z - \frac{Q_w}{2\pi b}\ln z \tag{6.69}$$

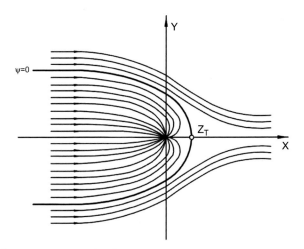

FIGURE 6.5 Source in a parallel flow.

The conjugate velocity field is obtained as:

$$\bar{q} = q_{\infty x} - \frac{Q_w}{2\pi b z} \tag{6.70}$$

The potential and stream functions can be written as:

$$\phi = q_{\infty x} x + \frac{Q_w}{2\pi b} \ln r \tag{6.71}$$

$$\psi = q_{\infty x} y + \phi \frac{Q_w}{2\pi b} \tag{6.72}$$

This may be expressed by Cartesian coordinates:

$$\phi = q_{\infty x} x + \frac{Q_w}{4\pi b} \ln \frac{x^2 + y^2}{r_w^2} \tag{6.73}$$

$$\psi = q_{\infty x} y + \frac{Q_w}{2\pi b} \text{arc tg} \frac{y}{x} \tag{6.74}$$

Behind the well, a stagnation point S occurs at which the velocity is obviously equal to zero:

$$\bar{q}_x = q_{\infty x} - \frac{Q_w}{2\pi b z_s} = 0 \tag{6.75}$$

Expressing z_s, we get:

$$z_s = \frac{Q_w}{2\pi b q_{\infty x}} \tag{6.76}$$

It can be recognized that the stagnation point falls to the x-axis, as large the capacity of the sink (the discharge of the well), and as far the stagnation point from the origin. The so-called zero-streamline passes through the stagnation point. Its equation can be written as:

$$\psi_0 = q_{\infty x} y + \frac{Q_x}{2\pi b} \text{arctg} \frac{y}{x} = 0 \tag{6.77}$$

from which we obtain the implicit formula:

$$y = \frac{Q_w}{2\pi b q_{\infty x}} \text{arctg} \frac{y}{x} \tag{6.78}$$

The streamlines of this flow pattern are shown in Fig. 6.5.

The complex potential of uniform flow can be replaced by the piezometric head. Thus we get:

$$y = -\frac{Q_w x}{2\pi b K H} \text{arctg} \frac{y}{x} \tag{6.79}$$

The next example of the method of superposition the flow pattern around a so-called doublet is a production and injection well of equal strength. It is a rather familiar configuration in geothermal fields. Consider a source and a sink of equal strengths Q and $-Q$, spaced at $-a$ and a respectively, from the origin on the x-axis (see Figs. 6.6 and 6.7.),

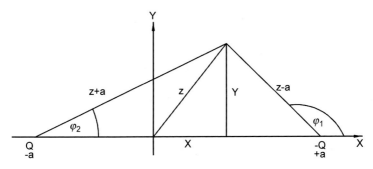

FIGURE 6.6 Source and sink on the plane of the complex variable z.

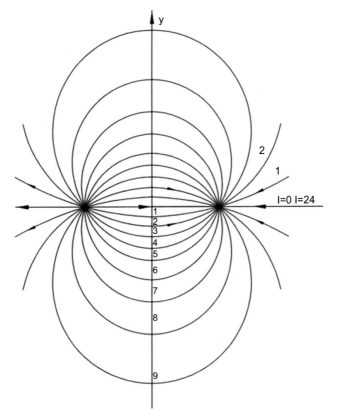

FIGURE 6.7 Streamlines of a doublet.

The resulting complex potential is obviously:

$$W = \frac{Q}{2\pi b}\ln(z+a) - \frac{Q}{2\pi b}\ln(z-a) \tag{6.80}$$

It can also be written in the form:

$$W = \frac{Q}{2\pi b}\ln\frac{z+a}{z-a} \tag{6.81}$$

Applying the notations:

$$z + a = r_2 e^{i\psi_e} \tag{6.82}$$

and

$$z - a = r_1 e^{i\varphi_1} \tag{6.83}$$

we obtain for the potential and stream functions:

$$\phi = \frac{Q}{2\pi b}\ln\frac{r_2}{r_1} \tag{6.84}$$

and

$$\psi = \frac{Q}{2\pi b}(\varphi_2 - \varphi_1) \tag{6.85}$$

To find the equation of the equipotential lines, we note that if $\phi = $ constant, r_2/r_1 must be similarly constant. Hence, setting:

$$\left(\frac{r_2}{r_1}\right)^2 = \frac{(x+a)^2 + y^2}{(x-a)^2 + y^2} = C \tag{6.86}$$

After some manipulation, the equation of the potential lines is found to be:

$$\left(x + \frac{C+1}{1-C}a\right)^2 + y^2 = \frac{4a^2 C}{(1-C)^2} \tag{6.87}$$

This equation shows that the equipotential lines are a family of circles with radii $2a\sqrt{C}/(1-C)$, and with centers at:

$$x = \frac{C+1}{1-C}a; \quad y = 0 \tag{6.88}$$

To obtain the equation of the streamlines we can take the imaginary part of the complex potential:

$$\text{tg}\frac{2\pi b\psi}{Q} = \text{tg}(\varphi_2 - \varphi_1) = C \tag{6.89}$$

whence, noting that:

$$\varphi_2 = \text{arctg}\,\frac{y}{x+a} \tag{6.90}$$

and

$$\varphi_1 = \text{arctg}\,\frac{y}{x-a} \tag{6.91}$$

we finally obtain:

$$x^2 + \left(y - \frac{a}{C}\right)^2 = a^2\left(1 + \frac{1}{C^2}\right) \tag{6.92}$$

This equation shows that the streamlines are also another family of circles with radii:

$$R = a\sqrt{1 + \frac{1}{C^2}} \tag{6.93}$$

and with centers at $x = 0$ and $y = a/C$.

A more general case is when the flow rate of the production well Q_{pw}, and the flow rate of the reinjection well Q_{rw}, are different. The complex potential can be written as:

$$W = \frac{Q_{rw}}{2\pi b}\ln(z+a) - \frac{Q_{pw}}{2\pi b}(z-a) \tag{6.94}$$

The conjugate of the velocity field is obtained:

$$\bar{q} = \frac{Q_{rw}}{2\pi b}\frac{1}{z+a} - \frac{Q_{pw}}{2\pi b}\frac{1}{z-a} \tag{6.95}$$

A stagnation point occurs where the velocity equals zero:

$$0 = \frac{Q_{rw}}{2\pi b}\frac{1}{z_s+a} - \frac{Q_{rw}}{2\pi b}\frac{1}{z_s-a} \tag{6.96}$$

The complex coordinate of the stagnation point is obtained as:

$$z_s = \frac{Q_{rw} + Q_{pw}}{Q_{rw} - Q_{pw}}a \tag{6.97}$$

Three different flow patterns can be recognized.

If the discharge of the production well is equal to the reinjected flow rate:

$$Q_{pw} = Q_{rw} \tag{6.98}$$

then there is no solution for Eq. (6.96), therefore the stagnation point cannot occur.

If the reinjected flow rate is greater than the discharge of the production well:

$$Q_{rw} > Q_{pw} \qquad (6.99)$$

then the stagnation point falls to the positive half of the x-coordinate axis, further from the origin than the production well:

$$zs > a \qquad (6.100)$$

If the discharge of the production well is greater than the reinjected flow rate:

$$Q_{pw} > Q_{rw} \qquad (6.101)$$

then the stagnation point occurs on the negative half of the x-coordinate axis, further from the origin than the recharge well:

$$-zs < -a \qquad (6.102)$$

The flow patterns of these cases are shown in Fig. 6.8.

Consider now a semi-infinite, confined aquifer between two parallel horizontal planes, bounded by a straight impermeable contour at one side. Let a well be placed at the point $(z - a)$ as it is shown in Fig. 6.9, where a is the distance between the well and the impervious boundary of the aquifer. In this case, the so-called method of images can be employed to find the complex potential. The impervious boundary line may be considered to be a streamline. This streamline can occur due to some singularity; an added fictitious sink of equal capacity induces it. The

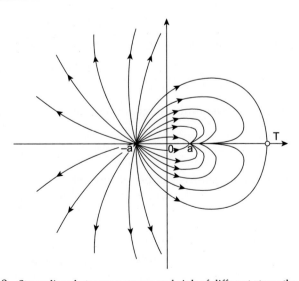

FIGURE 6.8   Streamlines between a source and sink of different strengths.

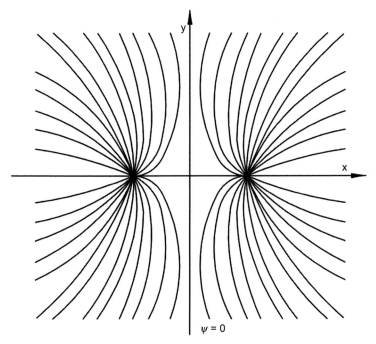

$\psi = 0$

FIGURE 6.9   Method of images: an impervious boundary is replaced by a source.

material sink, and its image symmetrically to the boundary forms a flow pattern with the complex potential:

$$W = -\frac{Q}{2\pi b}\ln(z-a) - \frac{Q}{2\pi b}\ln(z+a) \qquad (6.103)$$

The conjugate of the velocity field can be written as:

$$\bar{q} = -\frac{Q}{2\pi b}\left(\frac{1}{z-a} + \frac{1}{z+a}\right) \qquad (6.104)$$

It is obvious that a stagnation point occurs at the origin; the solution of the equation:

$$\frac{1}{z_s - a} + \frac{1}{z_s + a} = 0 \qquad (6.105)$$

is $z_s = 0$.

The stream function can be separated easily introducing the notations:

$$z + a = r_2 e^{i\varphi_2} \qquad (6.106)$$

and

$$z - a = r_1 e^{i\varphi_1} \tag{6.107}$$

After substitution we get:

$$W = -\frac{Q}{2\pi b}\ln(z + a)(z - a) = -\frac{Q}{2\pi b}[\ln(r_2 r_1) + i(\varphi_2 + \varphi_1)\ln e] \tag{6.108}$$

Thus the equation of the streamlines is:

$$\psi = -\frac{Q(\varphi_1 + \varphi_2)}{2\pi b} = \text{const.} \tag{6.109}$$

The zero-streamline coincides with the y-coordinate axis. The streamline pattern is shown in Fig. 6.10. Since the velocity has no component perpendicular to the streamlines, this zero-streamline forms the boundary of drainage area of the well. The image-sink has no additional importance, because it falls out of the drainage area. Its role is that the fictitious

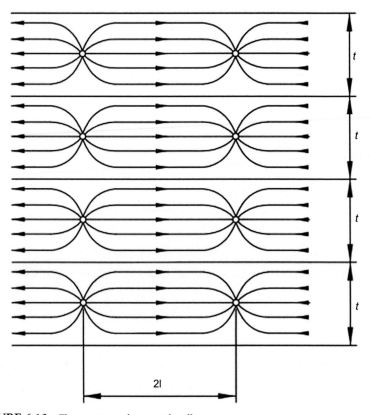

FIGURE 6.10   Flow pattern of a row of wells.

hydrodynamic singularity may replace the impervious boundary of the drainage area equivalently.

Consider, finally, the configuration of production and reinjection wells as it is shown in Fig. 6.10. The discharge of production wells and the capacity of reinjection wells is uniform. The wells are placed along two parallel rows with constant spacing. Let the number of wells be m along each straight line. Thus, the complex potential of the system can be written as:

$$W = \frac{Q}{2\pi b} \sum_{v=-m}^{m} [\ln(z - a + iвt) - \ln(z + a + iвt)] \qquad (6.110)$$

The velocity distribution is obtained:

$$\bar{q} = \frac{Q}{2\pi b} \sum_{v=-m}^{m} \left[\frac{1}{z - a + iвt} - \frac{1}{z + a + iвt} \right] \qquad (6.111)$$

The streamline pattern is shown in Fig. 6.10.

As the number of wells increases, the boundary lines between their drainage areas tends to be straight parallel lines midway between two wells.

6.5 THE HELES–SHAW FLOW

A possible way to describe the flow through fractured media is replacing the family of fractures by a simple equivalent fracture. The flow occurring between two narrow, parallel planes is the so-called Hele–Shaw flow.

Consider a steady, two-dimensional laminar flow of an incompressible fluid between two parallel planes. The Navier–Stokes equation is valid for this flow. It can be written as:

$$\frac{\partial \vec{v}}{\partial t} + (\vec{v}\nabla)\vec{v} = \vec{g} - \frac{1}{\rho}\nabla p + v\Delta\vec{v} \qquad (6.112)$$

Since the flow is steady, the local derivative $\partial\vec{v}/\partial t = 0$. In a two-dimensional flow, the velocity component $v_z = 0$. The only body force is the gravity, thus:

$$g_x = -\frac{\partial(gh)}{\partial x} \qquad g_y = -\frac{\partial(gh)}{\partial y} \qquad (6.113)$$

Since the velocity at the bounding plane walls must be zero, the change of the velocity components v_x and v_y with respect to z will be much greater than the velocity changes with respect to x and y. Hence, in comparison to their respective derivatives in the z-direction we may assume that:

$$\frac{\partial v_x}{\partial z} >> \frac{\partial v_x}{\partial x}; \quad \frac{\partial v_x}{\partial z} >> \frac{\partial v_x}{\partial y}$$

$$\frac{\partial v_y}{\partial t} >> \frac{\partial v_y}{\partial x}; \quad \frac{\partial v_y}{\partial z} >> \frac{\partial v_y}{\partial y}$$

(6.114)

Thus, the derivatives $\partial v_x/\partial x$; $\partial v_x/\partial y$; $\partial v_y/\partial x$; and $\partial v_y/\partial y$ can be neglected together as the second derivatives with respect x and y.

$$\frac{dv_x}{dz} = \frac{g}{\nu} \cdot \frac{\partial H}{\partial x} z + C_1$$

(6.115)

$$\frac{dv_y}{dz} = \frac{g}{\nu} \cdot \frac{\partial H}{\partial y} z + C_2$$

(6.116)

Integrating once more we get:

$$v_x = \frac{g}{\nu} \frac{\partial H}{\partial x} \frac{z_2}{2} + C_1 z + C_3$$

(6.117)

$$v_y = \frac{g}{\nu} \frac{\partial H}{\partial y} \frac{z^2}{2} + c_2 z + c_4$$

(6.118)

The following boundary conditions can be taken to determine the four constants of integration:

$$z = \delta; \quad v_x = v_y = 0; \quad z = -\delta; \quad v_x = v_y = 0$$

Substituting these values we get:

$$0 = \frac{g}{\nu} \cdot \frac{\partial H}{\partial x} \frac{\delta^2}{2} + c_1 \delta + c_3$$

(6.119)

$$0 = \frac{g}{\nu} \cdot \frac{\partial H}{\partial x} \frac{\delta^2}{2} + c_1 \delta + c_3$$

(6.120)

$$0 = \frac{g}{\nu} \cdot \frac{\partial H}{\partial x} \frac{\delta^2}{2} + c_2 \delta + c_4$$

(6.121)

$$0 = \frac{g}{\nu} \cdot \frac{\partial H}{\partial x} \frac{\delta^2}{2} + c_2 \delta + c_4$$

(6.122)

The c_1 and c_2 constants of integration are zero, while:

$$c_3 = c_4 = -\frac{g}{\nu} \cdot \frac{\partial H}{\partial x} \frac{\delta^2}{2}$$

(6.123)

Finally, the two velocity components are obtained as:

$$v_x = \frac{g}{\nu} \frac{\partial H}{\partial x} \frac{z^2 - \delta^2}{2} \tag{6.124}$$

and

$$v_y = \frac{g}{\nu} \frac{\partial H}{\partial y} \frac{z^2 - \delta^2}{2} \tag{6.125}$$

which show a parabolic velocity distribution, as it is shown in Fig. 6.2. The Eqs. (6.124) and (6.125) can be written as:

$$v_x = \frac{\partial}{\partial x} \left(\frac{gH}{\nu} \frac{z^2 - \delta^2}{2} \right) \tag{6.126}$$

and

$$v_y = \frac{\partial}{\partial y} \left(\frac{gH}{\nu} \frac{z^2 - \delta^2}{2} \right) \tag{6.127}$$

It can be recognized that velocities v_x and v_y can be interpreted as the velocities of a potential flow of which velocity potential is:

$$\phi = \frac{gH}{\nu} \frac{z^2 - \delta^2}{2} \tag{6.128}$$

Strictly speaking, the potential function ϕ defines a family of potential functions of the parameter z. Each parallel plane z = const. has its own potential function. Nevertheless, this flow is a quasi-potential flow. The streamline patterns obtained each parallel plane are congruent, but the size of the velocities are different along them accordingly to the Eqs. (6.126) and (6.127). The equation of the streamlines can be written as:

$$\frac{dy}{dx} = \frac{v_y}{v_x} = \frac{\dfrac{\partial H}{\partial y}}{\dfrac{\partial H}{\partial x}} \tag{6.129}$$

This equation doesn't contain the variable z, thus the streamline pattern are really congruent in all planes.

Based on the Eqs. (6.124) and (6.125), the cross-sectional integral mean velocities of both v_x and v_y between $-\delta$ and $+\delta$ the two planes can be obtained:

$$c_x = \frac{1}{2\delta} \int\limits_{-\delta}^{+\delta} v_x dz = \frac{g\delta^2}{3\nu} \frac{\partial H}{\partial x} \tag{6.130}$$

and

$$c_y = \frac{1}{2\delta} \int\limits_{-\delta}^{+\delta} v_y dz = \frac{g\delta^2}{3\nu} \frac{\partial H}{\partial y} \qquad (6.131)$$

The mean velocities c_x and c_y can be obtained as the derivatives of the potential function:

$$\Phi = \frac{g\delta^2}{3\nu} \qquad (6.132)$$

Since the continuity is also valid for the mean velocities, it can be written:

$$\frac{\delta c_x}{\delta x} + \frac{\delta c_y}{\delta y} = 0 \qquad (6.133)$$

thus, a stream function can be defined as:

$$c_x = \frac{\delta \Psi}{\delta y} \quad \text{and} \quad c_y = -\frac{\delta \Psi}{\delta x} \qquad (6.134)$$

These equations lead to the well-known Cauchy—Riemann equations:

$$\frac{\delta \Phi}{\delta x} + \frac{\delta \Phi}{\delta y} = 0 \qquad (6.135)$$

$$\frac{\delta \Psi}{\delta y} - \frac{\delta \Psi}{\delta x} = 0 \qquad (6.136)$$

The velocity of the Cauchy—Riemann equations guarantee the existence of a complex potential function:

$$W(z) = \Phi(x, y) + i\Psi(x, y) \qquad (6.137)$$

This is the reason that the use of the method of complex potentials is also feasible to describe the flows between parallel planes. Recognizing this relationship, Hele-Shaw (1897) devised an apparatus whereby two-dimensional flows through porous media could be investigated experimentally for structures with complex boundaries.

Reference

Hele-Shaw, H.S., 1897. Experiments on the surface resistance in pipes and on ships. Institute of Naval Architects, Transactions 39, 145—146.

Flow Through Producing Wells

7.1 FLOW TOWARD THE WELL IN A POROUS RESERVOIR

Along a streamline from the contour of the drainage area to the well-head, two hydraulic sub-systems can be discerned with merely different flow behavior. One of them is a seepage flow in the reservoir toward the wall. This flow satisfies Darcy's law; the dependence of the pressure loss on the flow rate is linear. The other sub-system is the upward turbulent flow through the tubing. Its pressure loss depends on the flow rate parabolic. Nevertheless, these flows are interdependent; any change in the flow variables of one has a certain reaction on the other flow. These flows are serially connected; their flow rates are the same while their pressure losses are added. Notwithstanding, the different nature of the two flows necessitates to separately determine both of them, and the obtained solution can be fitted by suitable boundary conditions. First, the reservoir flow is studied.

Copyright © 2017 Elsevier Inc. All rights reserved.

It is known that the permeable layers of the porous sedimentary geothermal reservoirs of the Pannonian Basin have a relatively large, mainly a few kilometers, horizontal extension, but a small vertical thickness, in most cases a few meters only. Accordingly to the natural conditions, it is reasonable to assume a two-dimensional Darcy flow in these reservoirs toward the wells.

Darcy's law is the generalization of an extended set of experiments carried out on flow through granulated columns.

The granulated column is considered to homogeneous and isotropic, having a permeability of K. The fluid is incompressible, Newtonian; its dynamic viscosity is μ. The flow is steady and laminar. In this case the flow rate depends on the pressure gradient linearly:

$$Q = -\frac{A \cdot K}{\mu} \cdot \frac{dp}{dx} \qquad (7.1)$$

Since the fluid flows in the direction of the decreasing pressure, the sign of the right-hand side of the equation must be negative. The cross-section of the column A is perpendicular to the velocity.

Ferrandon (1948) generalized Darcy's equation as:

$$\vec{q} = -\frac{K}{\mu} \text{grad } p, \qquad (7.2)$$

where q is the so-called seepage velocity, the flow rate across a unit cross-section. The flow toward wells is considered two-dimensional and axi-symmetrical. Thus we get:

$$q_r = -\frac{K}{\mu} \frac{dp}{dr} \qquad (7.3)$$

consider the simplest model of a tapped reservoir as it is depicted in Fig. 7.1.

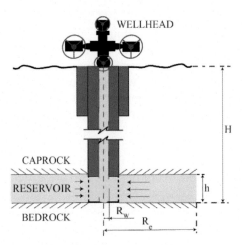

FIGURE 7.1   Simple axi-symmetrical reservoir model.

The outer boundary of the reservoir is open. In accordance to the continuity equation, the produced flow rate through the well is the same as the cross-flowing flow rate through the outer boundary of the drained area. The continuity equation for the case of non-deformable porous matrix is:

$$\phi \frac{\partial \rho}{\partial t} + \mathrm{div}(\rho \vec{q}) = 0 \tag{7.4}$$

For incompressible fluids, it can be written as:

$$\mathrm{div}\, \vec{q} = 0 \tag{7.5}$$

Substituting Darcy's law:

$$\mathrm{div}\left(-\frac{K}{\mu}\mathrm{grad}p\right) = 0 \tag{7.6}$$

In a cylindrical coordinate system we get:

$$\frac{1}{r}\frac{d}{dr}\left(r\frac{dp}{dr}\right) = 0 \tag{7.6a}$$

Integrating twice, we obtain:

$$p = C_1 \ln r + C_2 \tag{7.7}$$

In order to get the constants, the following boundary conditions are chosen:

$$r = R_e \quad \text{and} \quad p = p_w \tag{7.8}$$

$$r = R_e \quad \text{and} \quad p = p_c \tag{7.9}$$

After substitution we get:

$$C_1 = \frac{P_e - P_w}{\ln\frac{R_e}{R_w}} \quad \text{and} \quad C_2 = p_w + \frac{P_e - P_w}{\ln\frac{R_e}{R_w}}\ln R_w \tag{7.10}$$

Finally, we obtain:

$$p = p_w + \frac{P_e - P_w}{\ln\frac{R_e}{R_w}} \cdot \ln\frac{r}{R_w} \tag{7.11}$$

The pressure distribution can be expressed by the flow rate:

$$Q = 2\pi r q \tag{7.12}$$

while:

$$q = -\frac{K}{\mu}\frac{dp}{dr} \tag{7.13}$$

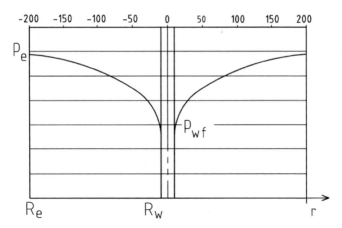

FIGURE 7.2 Pressure distribution along the radius.

Thus we get:

$$p = p_w + \frac{\mu Q}{2\pi hK} \ln \frac{r}{R_w} \tag{7.14}$$

It can be seen in Fig. 7.2 that the pressure distribution along the radius is a logarithmic function.

As the flow toward the well attains the borehole, the pressure drop is more considerable than it would be obtained by the logarithmic distribution. The reason for the increased pressure drop is a thin damaged region around the borehole with reduced permeability. When the well has been drilled, the suspended particles of the drilling mud penetrate to the pore channels of the formation across the surface of the borehole. These particles can partially plug the pore channels, substantially reducing the permeability of a relative thin region like a skin on the borehole wall. This weak permeability zone leads to an additional pressure drop, which can be calculated by the equation:

$$\Delta p_{skin} = \frac{\mu Q}{2\pi hK} s \tag{7.15}$$

in which s is the so-called mechanical skin factor, which is a dimensionless quantity. The two pressure drops obtained by Eqs. (7.14) and (7.15) are added as the fluid flows through serially connected elements. Thus the total pressure drop is obtained as:

$$p_e - p_{wf} = \frac{\mu Q}{2\pi hK} \left(\ln \frac{R_e}{R_w} + s \right) \tag{7.16}$$

Expressing the flow rate, we obtain:

$$Q = \frac{2\pi h K}{\mu} \cdot \frac{P_e - P_{we}}{\ln\frac{R_e}{R_w} + s} \tag{7.17}$$

The reservoir and the production well are interacting elements in serial connection from a hydraulic point of view. Accordingly, the flow rate of the reservoir flow toward the well and the flow rate of the upflowing fluid in the production well must be the same. On the other hand, pressure drops obtained in the reservoir and in the well must be added. These are the joint conditions of the two sub-systems; the reservoir and the well.

Consider now a geothermal reservoir, in which the undisturbed reservoir pressure is 200 bar $= 200 \cdot 10^5$ N/m^2, the radius of the drainage area is 500 m, the radius of the inflow cross-section of the well is 0.08 m, the reservoir thickness is 16 m, and the permeability is 500 md $= 0.5 \cdot 10^{-12}$ m^2; the flow rate is 0.02 m^3/s, and dynamic viscosity is $0.3 \cdot 10^{-3}$ N/m^2. The skin factor may be neglected. Calculate the flowing bottom-hole pressure.

Using Eq. (7.16), after some arrangement:

$$P_{wf} = P_e - \frac{\mu Q}{2\pi h K} \ln \frac{R_e}{R_w}$$

Substituting the above data we get:

$$P_{wf} = 200 \cdot 10^5 \ \frac{N}{m^2} - \frac{0.3 \cdot 10^{-3} \ \frac{Ns}{m^2} \cdot 0.02 \ \frac{m^3}{s}}{2 \cdot 3.14 \cdot 16 \ m \ \ 0.5 \cdot 10^{-12} \ m^2} \ln \frac{500}{0.08}$$

$$= 18,956,699 \ \frac{N}{m^2} \cong 189.56 \text{ bar}$$

7.2 THE FLUID UPFLOW THROUGH THE WELL

In the initial period of geothermal energy production, the natural flow of thermal water resulting from elastic expansion of the reservoir fluid was almost exclusive. In this case, the flow in the well starts by pumping until the whole tubing was filled by hot water. Because of this, the bottom-hole pressure becomes lower than the pressure of the undisturbed reservoir. The pressure decrease induces the elastic expansion of the reservoir fluid, inflowing the well and rising through the tubing to the surface. After this artificial starting, the well flow becomes self-supporting, while the separation of the dissolved gases helps also to maintain the flow. A particularly simple special case involves only the flow of homogeneous liquid phase in the well. A steady flow is considered. The flow is turbulent, and the velocity and the density can be taken as constants. The mechanical energy equation can be written between the bottom-hole (1) and the wellhead (2):

$$\rho\frac{c_1^2}{2} + p_1 + z_1\rho g = \rho\frac{c_2^2}{2} + p_2 + z_2\rho g + \Delta p_{12}' \qquad (7.18)$$

Since the fluid is incompressible and the diameter of the tubing is constant, the velocities are the same: $c_1 = c_2 = c$. The reference level $z_2 = 0$ is at the surface, thus the bottom-hole depth $z_1 = -H$. The pressure at the bottom-hole $p_1 = p_{wf}$, and at the wellhead $p_2 = p_{wh}$. The pressure loss $\Delta p_{12}'$ can be determined by the Weisbach equation:

$$\Delta p_{12}' = \lambda\frac{H}{D}\rho\frac{c^2}{2} \qquad (7.19)$$

where λ is the friction factor. The friction factor can be calculated depending on the Reynolds number, and the relative roughness of the tubing.

The Reynolds number is obtained as:

$$Re = \frac{cD\rho}{\mu}, \qquad (7.20)$$

while the relative roughness depends on the pipe material and the technology. It can be estimated applying the diagram of Colebrook (Fig. 7.3).

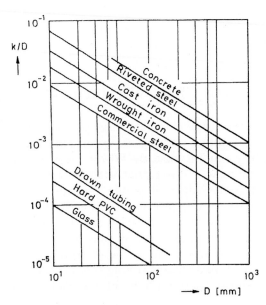

FIGURE 7.3   Relative roughness of different type pipes.

Knowing the Reynolds number, and the relative roughness k/D, the friction factor can be determined by Colebrook's equation:

$$\frac{1}{\sqrt{\lambda}} = -2\lg\left(\frac{2.51}{Re\sqrt{\lambda}} + \frac{1}{3.715 \cdot \frac{D}{k}}\right) \tag{7.21}$$

The hydraulically smooth pipe, and the wholly rough pipe are involved in Eq. (7.4) as a special case. For hydraulically smooth pipe, the second term, while wholly rough pipes, the first term, disappears in the bracket.

Since the equation of the friction factor is obtained in implicit form, the λ can be calculated by iteration. The convergence of the iteration is rather fast, particularly when the starting value of λ is chosen using the diagram of Fig. 7.4. In such a case, two, or at the most three, iteration steps are sufficient.

Finally, after some arrangement, the mechanical energy equation can be written as:

$$p_{wf} = p_{wh} + \rho g H + \lambda \frac{H}{D} \rho \frac{c^2}{2} \tag{7.22}$$

It can be recognized that bottom-hole pressure p_{wf} holds dynamical equilibrium with the wellhead pressure p_{wh}, the hydrostatic pressure of the water column of height H, and the pressure loss due to fluid friction in

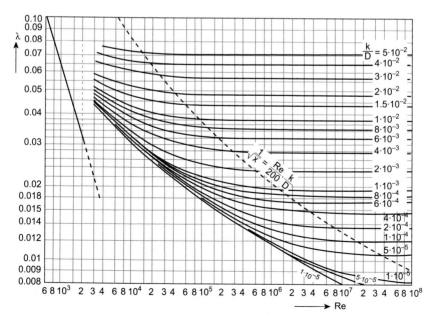

FIGURE 7.4 Moody's friction factor diagram for commercial pipes.

the tubing. Note that the wellhead pressure includes the pressure losses of the surface facilities, too.

On the other hand, the pressure of the reservoir fluid at the inflow cross-section of the well is:

$$P_{wf} = P_e - \frac{\mu Q}{2\pi h K}\left(\ln\frac{R_e}{R_w} + s\right) \tag{7.23}$$

The comparison of these equations leads to the expression for the undisturbed pressure of the reservoir p_e:

$$P_e = \rho g H + P_{wh} + \frac{\mu Q}{2\pi h K}\left(\ln\frac{R_e}{R_w} + s\right) + \lambda\frac{H}{D}\rho\frac{c^2}{2} \tag{7.24}$$

Since:

$$Q = c\frac{D^2\pi}{4} \tag{7.25}$$

this pressure equilibrium can be expressed as:

$$P_e = \rho g H + P_{wh} + \frac{\mu Q}{2\pi h K}\left(\ln\frac{R_e}{R_w} + s\right) + \lambda\frac{H}{D^5}\frac{8}{\pi^2}\rho Q^2 \tag{7.26}$$

The undisturbed pressure of the reservoir holds the equilibrium with the hydrostatic pressure of the water column of height H, the wellhead pressure, and indirectly the pressure drop of the surface facilities; filters, heat exchangers, etc., provide the pressure losses of the reservoir fluid flow toward the well and the upflowing fluid in the tubing. It is noteworthy that the hydraulic resistance of the flow in the reservoir depends on the flow rate linearly, while the dependence of the pressure loss of the upflowing fluid in accordance to its turbulent nature is parabolic. Note the sensibility to the tubing diameter; there is an inverse proportion to the fifth power of the diameter.

It is remarkable to consider the pressure difference maintaining the upflow through the well. The undisturbed pressure of the reservoir at the depth of H is obtained as:

$$P_e = P_0 + \rho_0 g\left(H - A\gamma\frac{H^2}{2} - B\gamma^2\frac{H^3}{3}\right) \tag{7.27}$$

This pressure is greater than the hydrostatic pressure of the hot water column in the well. On the other hand, the reservoir pressure at the entrance cross-section of the well and the flowing bottom-hole pressure are equal. Thus, an excess pressure develops in the tubing as it is shown in Fig. 7.5.

Closing the flowing well during a short time, pressure waves propagate forward and backwards through the tubing until the fluid friction dissipates their mechanical energy. As the waves are decayed the fluid

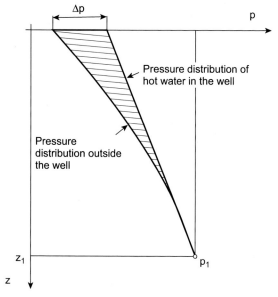

FIGURE 7.5 Excess pressure in the tubing.

comes to equilibrium. This hydrostatic state can be expressed analogously to the communicating vessels; for the hot water filled tubing and the adjacent aquifer:

$$p_{wh} + \rho g H = p_0 + \rho_0 g \left(H - A\gamma \frac{H^2}{2} - B\gamma^2 \frac{H^3}{3} \right) \qquad (7.28)$$

Considering that:

$$\rho = \rho_0 \left(1 - A\gamma H - B\gamma^2 H^2 \right), \qquad (7.29)$$

substituting it into Eq. (7.18), we obtain the excess pressure at the closed wellhead:

$$p_{wh} - p_0 = \rho_0 g \left(\frac{A\gamma H^2}{2} + \frac{2B}{3}\gamma^2 H^3 \right) \qquad (7.30)$$

This excess pressure maintains the flow through the production well. This is the so-called thermal lift.

Equalizing Eq. (7.26) and Eq. (7.27), we obtain the following expression:

$$p_0 + \rho_0 g \left(H - A\gamma \frac{H^2}{2} - 2B\gamma^2 \frac{H^3}{3} \right) = \rho g H + p_{wh} + \frac{\mu Q}{2\pi h K} \ln \left(\frac{R_e}{R_w} + s \right)$$

$$+ \lambda \frac{H}{D^5} \frac{8}{\pi^2} \rho Q^2$$

$$(7.31)$$

Considering Eq. (7.29), we can get a simple quadratic equation for the flow rate:

$$\lambda \frac{H}{D^5} \frac{8}{\pi^2} \rho Q^2 + \frac{\mu}{2\pi Kh}\left(\ln\frac{R_e}{R_w} + s\right)Q + P_{wh} - P_0$$
$$- \rho_0 gH\left(A\gamma\frac{H}{2} + 2B\gamma^2\frac{H^2}{3}\right) = 0 \tag{7.32}$$

Since all parameters of this equation are known, the flow rate of a naturally flowing well is obtained as the solution of an equation type of:

$$aQ^2 + bQ + c = 0 \tag{7.33}$$

The performance curve of a production well is the graphical representation of the relationship between the wellhead pressure and the flow rate. After some rearrangements of Eq. (7.9), this function is obtained as:

$$P_{wh} = P_e - \rho gH - \frac{\mu Q}{2\pi hK}\left(\ln\frac{R_e}{R_w} + s\right) - \lambda\frac{H}{D^5}\rho Q^2 \tag{7.34}$$

The first and the second term on the right-hand side don't depend on the flow rate. Their sum is the static component of the wellhead pressure. This static wellhead pressure exists in a closed well as the mechanical equilibrium is developed in it. The pressure drop of the formation depends on the flow rate linearly, while the pressure loss in the well is a parabolic function of the flow rate. This is shown in Fig. 7.6. The calculated performance curve describes correctly the behavior of the upflow

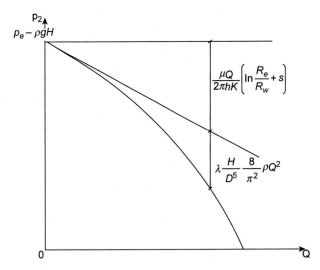

FIGURE 7.6   Calculated performance curve.

through the well in the simplest case of the homogeneous liquid flow. The actual performance curve of a production well can be determined experimentally by in situ measurements of the flow rate and the wellhead pressure in several performance states.

Knowing the performance curve of the well, and the pressure–loss curve of the surface facilities, the performance point of the system can be determined. The hydraulic load of the surface pipeline is also a parabolic function of the flow rate:

$$p_{wh} = p_{st} + \lambda_s \frac{L_s}{D_s^5} \frac{8}{\pi^2} \rho Q^2, \tag{7.35}$$

where p_{st} is the static pressure to overcome the height difference between the outflowing and inflowing cross-section of the surface pipeline, the second term is the pressure loss of it. The values λ_s, L_s, and D_s belong the surface piping. The intersection of the two curves represents the common performance point, where the flow rates and the pressures are equal. Thus, according to the Eqs. (7.31) and (7.32) we get:

$$p_e - \rho g H - \frac{\mu}{2\pi h K}\left(\ln\frac{R_e}{R_w} + s\right)Q - \lambda \frac{H}{D^5}\frac{8}{\pi^2}\rho Q^2 = p_{st} + \lambda_s \frac{L_s}{D_s}\frac{8}{\pi^2}\rho Q^2 \tag{7.36}$$

It can be outlined clearly that finally the undisturbed reservoir pressure provides the necessary mechanical energy to maintain the flow in the reservoir, through the well and in the surface elements in the system.

Following a streamline from the boundary of the undisturbed region of the reservoir to the wellhead, the mechanical energy equation can be written. Referring to the unit volume of the flowing fluid, the equation is obtained in a dimension of pressure. Since the flow is steady, the sum of mechanical energy and its dissipated parts, the pressure loss is constant. The graphical representation of this equation is the so-called energy diagram. The length of the streamline is the abscissa, while the energy terms occur in the ordinate.

The streamline begins at the contour of the drainage area at the point $r = R_e$. Assuming a horizontal reservoir, the total mechanical energy occurs in the form of pressure p_e. As the fluid flows toward the well, the pressure decreases along the streamline logarithmically as it is expressed by Eq. (7.9). At the entrance section of the well, the mechanical energy is equal to the flowing bottom-hole pressure p_{wf}. The pressure loss in the reservoir is dissipated to heat. In the tubing, the upflowing fluid has a frictional pressure loss, which increases linearly along the streamline. This part of mechanical energy dissipates also into heat. The fluid sizes in the tubing its potential energy $\rho g z$ increases linearly along the vertical streamline. At the wellhead, the mechanical energy is the sum of the

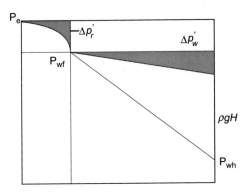

FIGURE 7.7 Energy diagram.

wellhead pressure p_{wh}, and the potential energy ρ_{gH}, which is the hydrostatic pressure of the hot water column in the tubing.

The energy diagram of this system is shown in Fig. 7.7. This diagram characterizes the energy relations of a naturally outflowing well.

Consider a hot water production well. Its depth is 2000 m, the internal diameter of the tubing is 160 mm, and the relative roughness $D/k = 1000$. The flow rate is 0.02 m³/s. The density of water 958 kg/m³, and the kinematic viscosity is $0.3 \cdot 10^{-6}$ m²/s. Calculate the pressure loss in the tubing between the bottom-hole and the wellhead.

The cross-sectional average velocity is:

$$c = \frac{Q}{\frac{D^2\pi}{4}} = \frac{0.02 \; \frac{m^3}{3} \cdot 4}{0.16^2 \; m^2 \cdot 3.14} = 0.994 \; \frac{m}{s}$$

The Reynolds number is obtained as:

$$Re = \frac{c \cdot D}{\nu} = \frac{0.994 \; \frac{m}{s} \cdot 0.16 \; m}{0.3 \cdot 10^{-6} \; \frac{m^2}{s}} = 530,133$$

The flow is naturally turbulent; it is shown in the Moody−diagram, that the corresponding D/k and Re values belong to the transition region. Applying the Colebrook equation, the starting value of the iteration $\lambda_o = 0.02$. Thus, the first step of the iteration is:

$$\lambda_1 = \frac{1}{\left[-2\lg\left(\frac{2.51}{530,133\sqrt{0.02}} + \frac{1}{3.715}\right)\right]^2} = 0.02020$$

The second step obtains:

$$\lambda_o = 0.02020$$

thus, further iteration is unnecessary. Knowing the friction factor the pressure loss can be calculated:

$$\Delta p' = \lambda \frac{H}{D} \rho \frac{c^2}{2} = 0.0202 \cdot \frac{2000}{0.16} \cdot 958 \cdot \frac{0.994^2}{2} = 119.500 \ \frac{N}{m^2} \cong 1.195 \ bar$$

The next example is the following; consider a vertical closed geothermal well filled with hot water. The depth of the well is 2000 m, the geothermal gradient is $0.05°C/m$, and the water density at the surface temperature is $999 \ kg/m^3$. The temperature decrease of the water between the reservoir and the wellhead can be neglected. Calculate the excess wellhead pressure.

This problem can be solved by substituting the data for Eq. (7.30):

$$p_{wh} - p_0 = \rho_0 g \left(\frac{1}{2} A\gamma H^2 + \frac{2}{3} B\gamma^2 H^3 \right)$$

$$= 999 \ \frac{kg}{m^2} 9.81 \ \frac{m}{s^2} \left(\frac{1}{2} \cdot 1.712 \cdot 10^{-4} \ \frac{1}{°C} 5 \cdot 10^{-2} \ \frac{°C}{m} \cdot 4 \cdot 10^6 \ m^2 \right.$$

$$\left. + \frac{2}{3} 3.232 \cdot 10^{-6} \ \frac{1}{°C^2} - 25 \cdot 10^{-4} \ \frac{°C^2}{m^2} \cdot 8 \cdot 10^9 \ m^2 \right) = 590,069 \ \frac{N}{m^2} \cong 5.90 \ ba$$

Another problem follows. Determine the wellhead pressure for a flowing hot water producing well. The necessary data are:

Depth of the well	$M = 2000 \ m$
Diameter of the tubing	$D = 0.16 \ m$
Friction factor	$\lambda = 0.0202$
Geothermal gradient	$\gamma = 0.05°C/m$
Water density at surface temperature	$\rho_0 = 999 \ kg/m^3$
Flow rate	$Q = 0.02 \ m^3/s$
Radius of the drainage area	$R_e = 400 \ m$
Radius of the inflow section of the well	$D = 0.08 \ m$
Reservoir thickness	$h = 26 \ m$
Permeability	$K = 1000 \ md = 10^{-12} \ m^2$
Skin factor is neglected	$s = 0$
Dynamic viscosity	$\mu = 3 \cdot 10^{-4} \ N/m^2$

The wellhead pressure can be calculated by Eq. (7.17):

$$p_e = p_0 + \rho_0 g \left(H - A\gamma \frac{H^2}{2} - B\gamma^2 \frac{H^3}{3} \right) = 101,325 + 999 \cdot 9.81 \cdot$$

$$\left(2000 - 1.72 \cdot 10^{-4} \cdot 5 \cdot 10^{-2} \cdot \frac{4 \cdot 10^6}{2} - 3.232 \cdot 10^{-6} \cdot 25 \cdot 10^{-4} \cdot \frac{8 \cdot 10^9}{3} \right)$$

$$= 19,322,790 \; \frac{N}{m^2} \cong 193.23 \; bar$$

$$\rho g H = \rho_0 g H \left(1 - A\gamma H - B\gamma^2 H^2 \right) = 999 \cdot 9.81 \cdot 2000 \cdot$$

$$\left(1 - 1.72 \cdot 10^{-4} \cdot 5 \cdot 10^{-2} \cdot 2 \cdot 10^3 - 3.232 \cdot 10^{-6} \cdot 25 \cdot 10^{-4} \cdot 4 \cdot 10^6 \right)$$

$$= 18,795,960 \; \frac{N}{m^2} \cong 187.6 \; bar$$

$$\Delta p' = 312.390 \; \frac{N}{m^2} \cong 3,12 \; bar$$

$$\Delta p'_w = \lambda \frac{H}{D^5} \frac{8}{\pi^2} \rho Q^2 = 0.0202 \cdot \frac{2000}{0.16^5} \cdot \frac{8}{3.14^2} \cdot 958 \cdot 0.02^2$$

$$= 119500 \; \frac{N}{m^2} \cong 1.12 \; bar$$

Thus the wellhead pressure is obtained as:

$$p_{wh} = 193.22 - 187.96 - 3.12 - 1.19 = 0.92 \; bar.$$

7.3 TWO-PHASE FLOW IN WELLS INDUCED BY DISSOLVED GAS

Two-phase flows frequently occur in the upper section of hot water production wells. There are two different types of these two-phase flows. Both of them arise from the pressure decrease of the upflowing fluid as it comes close to the wellhead.

One of them occurs as dissolved, non-condensable gases are released from the hot water. The other type is induced by flashing as the upflowing fluid pressure achieves the saturated steam pressure at the actual temperature.

As it is well known that geothermal waters comprise non-condensable gases in solution. Waters of the Upper Pannonian aquifer contain mainly methane, carbon dioxide, and nitrogen. Most Hungarian natural gas reservoirs are close to geothermal aquifers. The geothermal wells in Hajdúszoboszló, Debrecen, or Karcag produce $1000-2000$ m^3 hot water per day, together with 3000 m^3/day methane. The dissolved gas/water ratio is usually obtained by empirical correlation or diagrams in Nm^3/m^3 unit. The solubility depends on the pressure, the temperature, and the total dissolved solids. An approximate formula for the solved methane content in hot water is (Culbertson and McKetta, 1951):

$$R_{sw} = A \cdot (lgp)^B + C \tag{7.37}$$

It must substitute the pressure values in bars. The coefficients A, B, and C are temperature-dependent:

$$A = 34.8666 - 3.54688 \cdot T + 1.36463 \cdot 10^{-3} \cdot T^2 - 2.3441 \cdot 10^{-6} \cdot T^3$$
$$+ 1.51639 \cdot 10^{-9} \cdot T^4$$

$$B = -94.191 + 1.08167 \cdot T - 4.49302 \cdot 10^{-3} \cdot T^2 + 8.3335 \cdot 10^{-6} \cdot T^3$$
$$- 5.77818 \cdot 10^{-9} \cdot T^4$$

$$C = 3.79411 - 3.52815 \cdot T^{-2} + 1.08467 \cdot 10^{-4} \cdot T^2 - 1.09238 \cdot 10^{-7} \cdot T^3$$

Eichelberger (1988) has given a correction factor to consider the total dissolved solids in the thermal water:

$$lgK_E = \left(-1.66892 \cdot 10^{13} \cdot T^{-536321} - 1.5486 \right) \cdot x \tag{7.38}$$

where x is the total dissolved solids.

The total dissolved gas content of the geothermal fluid in solution of the reservoir fluid is homogeneous liquid phase. As the fluid upflows, its pressure decreases substantially, and its temperature is reduced slightly. Although the pressure of the bubble point the gas is released, at the beginning small bubbles can occur. As the pressure decrease is continued, gas bubbles expand and various flow patterns develop in the two-phase flow.

The determination of flow patterns is mostly carried out by direct visual observation, occasionally complemented with high-speed photography. These methods are rather subjective; nevertheless sufficiently consistent flow pattern ordination becomes possible. The most common interpretation of flow patterns is a diagram plotted in terms of the superficial velocities of each phase, i.e. the flow rate of that phase divided by the total cross-section of the pipe. The thusly obtained diagram is called a flow pattern map, in which certain regions correspond to characteristic flow patterns. A great number of flow pattern data are known from the literature, but unfortunately most relate to water–air flows, which is the most suitable fluid-pair for visual observation. The flow of a dark opaque crude oil—natural gas mixture cannot be investigated by visual methods. In this case measurements of pressure fluctuations or X-radiography may be used. The transition between two flow patterns is not as sharp as the laminar—turbulent transition of homogeneous fluid flow. A wide variety of flow patterns can be observed and defined.

In the following, various types of two-phase flow will be considered and the flow patterns which occur in them discussed. Consider first a liquid—gas mixture flowing upward in a vertical transparent section of pipe. When a homogeneous liquid flows by itself, direct visual observation cannot provide any information about the flow.

Introducing gas into the liquid at progressively increasing flow rates, a series of consecutive, changing flow patterns can be distinguished. At the lowest gas flow rates, the liquid phase is continuous and small, and spherical gas bubbles move upward near the pipe axis, faster than the liquid. A short time exposure photograph generally shows a bubble flow pattern, as depicted in Fig. 7.8A. As the gas flow rate is increased, the

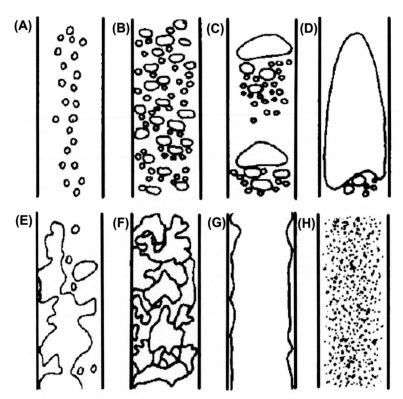

FIGURE 7.8   Flow patterns of vertical fluid—gas mixture flow.

number of bubbles increases, and owing to coalescence, the average bubble size increases. These larger, lens-shaped bubbles are pushed along in the liquid, with their largest cross-section normal to the flow, in periodically occurring groups. This is the bubble group flow pattern (see Fig. 7.8B.). A further increase in the gas flow rate causes an increase in the volume fraction of the bubbles, up to 30%, while bubble coalescence leads to the occurrence of large, mushroom-shaped bubbles which nearly span the entire cross-section of the pipe (see Fig. 7.8C.). These large, mushroom-shaped bubbles are followed by regions containing dispersions of smaller bubbles, and periodically bubble-free liquid plugs. This marks the beginning of the slug flow pattern. With a further increase in the gas flow rate, the large bubbles become longer having a bullet shape. These bullet-shaped bubbles are called Taylor bubbles. The slug flow pattern is characterized by periodically alternating Taylor bubbles and liquid regions containing a number of smaller spherical bubbles (see Fig. 7.8D.). The liquid phase flows down the outside of the Taylor bubbles as a falling film although the resultant flow of both liquid and gas is upward. In these flow patterns, the liquid phase is always continuous, and the gas phase is dispersed.

The slug flow pattern with long Taylor bubbles corresponds to an increase in the pressure loss. The increasing pressure gradient now tends to collapse the Taylor bubbles. Surface tension acts against this tendency, but large gas bubbles become unstable and finally collapse. At this point, the interfaces between the phases become highly distorted, both phases become dispersed, and the froth flow pattern develops (see Fig. 7.8E). Froth flow is highly unstable; an oscillatory upward—downward motion occurs in the liquid phase, particularly in pipes of large diameter. This is known as churn flow (see Fig. 7.8F). In small-diameter pipes, the breakdown of the Taylor bubbles is not so abrupt; the transition is more gradual without the occurrence of churn.

As the gas flow rate is increased still further, an upward moving wavy annular liquid layer develops at the pipe wall, and the gas flows with a substantially greater velocity in the center of the pipe. This is known as annular flow. The gas core flow may carry small fluid droplets ripped from the annular liquid layer, as is shown in Fig. 7.8G. With a further increase of the gas flow rate, the liquid film becomes progressively thinner while the number of the droplets in the core flow increases. Finally, the film will be removed from the wall and a pure mist flow occurs (see Fig. 7.8H.). The observed flow patterns are interpreted in Fig. 7.8. Repeating the previous experiment at a higher flow rate of the liquid, the number of flow patterns developed decreases. First the froth flow is omitted, later with further increasing liquid flow rates, bubble flow changes directly into annular or mist flow, without the intermediate stage of slug flow. All this is evident from Fig. 7.9.

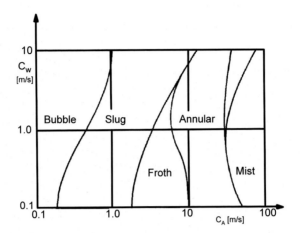

FIGURE 7.9   Flow pattern map for vertical fluid—gas mixture flow.

Note that such a flow-pattern map is valid only for a given fluid pair, pipe diameter, pressure level, and temperature. In spite of continuing efforts, generalized flow-pattern maps are not available. Although particular flow-pattern maps show the same sequences of flow regimes, there are no such flow variables by which congruent flow-pattern maps would be obtained.

A remarkable diagram is obtained by plotting the pressure-loss gradient against the gas flow rate, while liquid flow rate is taken as a parameter. This is shown in Fig. 7.10. It is seen that the pressure-loss gradient curve goes through a local minimum, a maximum, and a second minimum as the gas flow rate increases. The occurrence of these minimums and maximums corresponds to flow pattern transitions. As the liquid flow rate increases, these minimums and maximums become less distinct, though the slope of the curve still shows changes. The first minimum corresponds to the transition between the bubble-group- and slug-flow patterns. The maximum marks the collapse of the Taylor bubbles and the development of froth flow. The second minimum corresponds to the transition between the froth- and annular flow patterns. This phenomenon may be used to indicate changes in the flow pattern where other experimental facilities are not available.

It was previously mentioned that when the phases differ in density and viscosity one of them, usually the denser phase, flows with a lower in situ average velocity than the other. This velocity difference results in a change in the concentration of the phases along the length of the pipe. In the entrance section of the pipe, the less mobile phase concentrates; this concentration gradually decreases in the direction of flow. This

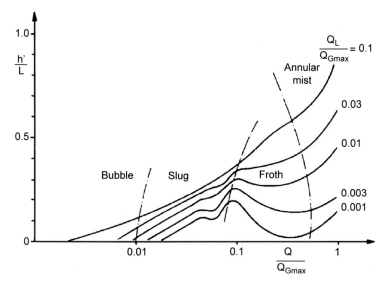

FIGURE 7.10 Pressure-loss gradient depending on gas flow rate.

phenomenon is called holdup. For a quantitative description of holdup a number of convenient parameters can be defined.

The local in situ volume fraction of phase i ε_i and its cross-sectional average value E_i are convenient measures of the holdup (see Fig. 7.11).

The average slip velocity is defined as the difference between the cross-sectional average velocities of the two phases:

$$c_s = c_i - c_j \tag{7.39}$$

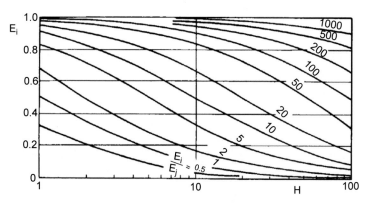

FIGURE 7.11 In situ volume fraction and holdup ratio.

Sometimes it is convenient to use the apparent phase velocity. This can be defined as the volumetric rate of flow of the phase divided by the total cross-sectional area of the pipe:

$$c_{oi} = \frac{Q_i}{A}, \quad c_{oj} = \frac{Q_j}{A} \tag{7.40}$$

It is clear that the cross-sectional average velocity of the mixture can be written as:

$$c = \frac{Q}{A} = \frac{Q_i + Q_j}{A} = c_{oi} + c_{oj} \tag{7.41}$$

The average slip velocity can be expressed in terms of the apparent phase velocities as:

$$c_s = \frac{c_{oi}}{E_i} - \frac{c_{oj}}{E_j} \tag{7.42}$$

A further convenient parameter is the flow-rate ratio:

$$\xi_i = \frac{Q_i}{Q}; \quad \xi_j = \frac{Q_j}{Q} \tag{7.43}$$

By combining Eqs. (7.41)–(7.43), the average in situ volume fractions can be expressed in terms of the average velocity of the mixture and the average slip velocity:

$$E_i = \frac{c + c_s}{2c_s} = \sqrt{\left(\frac{c + c_s}{2c_s} - \xi_i \frac{c}{c_s}\right)} \tag{7.44}$$

$$E_j = \frac{c_s - c}{2c_s} + \sqrt{\left(\frac{c_s - c}{2c_s}\right)^2 + \xi_j \frac{c}{c_s}} \tag{7.45}$$

Another measure of holdup is the ratio of the cross-sectional average in situ velocities at a given cross section; the so-called holdup ratio:

$$H = \frac{c_i}{c_j} \tag{7.46}$$

It is obvious that:

$$H = \frac{c_i}{c_j} = \frac{c_{oi} E_j}{c_{oj} E_i} = \frac{\xi_j E_j}{\xi_j E_i} \tag{7.47}$$

from which the average in situ volume fractions are obtained as:

$$E_i = \frac{c_{oi}}{c_{oi} + H c_{oj}} = \frac{\xi_j}{\xi_i + H\xi_j} \tag{7.48}$$

and

$$E_j = \frac{Hc_{oj}}{c_{oi} + Hc_{oj}} = \frac{H\xi_j}{\xi_i + H\xi_j} \tag{7.49}$$

The relationship of the in situ volume fraction of the less dense phase and the holdup ratio to the input volume ratio of the phases is shown in Fig. 7.12.

The relationship between the average slip velocity and the holdup ratio is obtained as:

$$c_s = \frac{H-1}{H}(c_{oi} + Hc_{oj}) \tag{7.50}$$

These relationships have general validity for any two-phase flow. The perfectly suspended two-phase flow pattern with axi-symmetrical velocity and concentration profiles, without holdup can be characterized by $H = 1$.

The holdup relations of various particular two-phase flows may be represented by a variety of diagrams.

In Fig.7.12, the holdup ratio for a vertical flow of a water—air mixture is plotted against the superficial air velocity, while the superficial water velocity is taken as a parameter. In the bubble flow region, the holdup ratio increases with the air velocity. An increase in the water velocity decreases the holdup ratio. The holdup ratio increases in the slug flow region, and reaches a maximum in the froth flow domain. As the air-flow

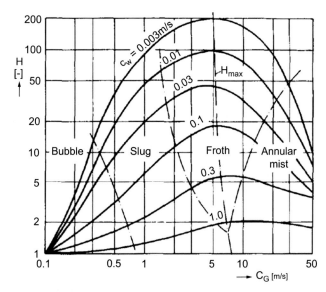

FIGURE 7.12   Holdup ratio of water—air mixture flow in vertical pipe.

rate is increased further, the holdup ratio decreases with the occurrence of annular-mist flow.

An important task is the determination of pressure losses of two-phase flow in the tubing. The mechanical energy equation for a two-phase flow in pipes obviously contains the dissipation terms for each of the phases. In the following form of the equation the dissipation terms are represented as pressure losses $\Delta p_i'$ and $\Delta p_j'$:

$$p_1 - p_2 = \frac{\rho_2 c_2}{c_1} U_2 - \rho_1 U_1 + \left(\frac{c_2}{c_1} - 1\right)p_2 + \Delta p_i' + \Delta p_j' \tag{7.51}$$

It can be recognized that holdup influences the changes of both the potential and the pressure energy. The pressure losses $\Delta p_i'$ and $\Delta p_j'$ may be combined into a two-phase pressure loss $\Delta p'$. Because of the holdup, this does not equal the pressure drop, even for horizontal pipes.

The pressure loss can be determined by purely analytical methods for homogeneous Newtonian fluid flow only. For homogeneous fluid flow, the friction factor is a function of a single similarity invariant, or the Reynolds number. In contrast, for a two-phase flow, the pressure loss, which can be calculated with a friction factor, depends on at least six variables: Reynolds number, Froude number, Weber number, density ratio, viscosity ratio, and the flow rate ratio. It is obvious that the variety of flow types and flow patterns cannot be encompassed within a single formula. It is necessary to develop semi-empirical methods of specific validity, applicable to certain types of flow. In order to obtain pressure-loss equations suitable for direct application in petroleum engineering practice, the various types of two-phase flows should be considered separately.

For vertical two-phase liquid—gas mixture flow, Ros (1961) developed a method to determine the pressure loss. There are two separate expressions for the pressure loss. If the liquid phase is continuous, i.e. for bubble, slug, and froth flow, the pressure loss is given by:

$$\Delta p' = \frac{2\lambda_R \rho_L c_{0L}^2}{gD}\left(1 + \frac{c_{0G}}{c_{0L}}\right)\Delta z \tag{7.52}$$

in which λ_R is the so-called Ros friction factor, c_{0L} and c_{0G} are the superficial velocities of the liquid and the gas, Δz is a finite, short length of the vertical pipe, and ρ_L is the density of the liquid. If the gas phase is continuous, i.e. for annular-mist and mist flows, the pressure loss can be calculated as:

$$\Delta p' = \frac{2\lambda_R \rho_G c_{0G}^2}{gD}\left(1 + \frac{c_{0L}}{c_{0G}}\right)\Delta z \tag{7.53}$$

where ρ_G now represents the density of the gas.

The Ros friction factor is defined as:

$$\lambda_R = \frac{\lambda_1 \lambda_2}{\lambda_3} \tag{7.54}$$

where λ_1, λ_2, and λ_3 are friction factors discussed in the following section.

For continuous liquid phase flows, λ_1 is a conventional Fanning friction factor, which is a function of the Reynolds number, defined as:

$$Re_{0L} = \frac{c_{0L} D}{\nu_L} \tag{7.55}$$

and the relative roughness of the pipe wall. The familiar friction factor relations are modified by Ros in the interval:

$$700 < Re_{0L} < 3000.$$

The modified friction factor interval is shown in Fig. 7.13.
Outside of this region, for laminar flow:

$$\lambda_1 = \frac{64}{Re_{0L}} \tag{7.56}$$

and for turbulent flow:

$$\frac{1}{\sqrt{\lambda_1}} = -2\lg\left(\frac{k}{3.715D} + \frac{2.51}{Re_{0L}\sqrt{\lambda_1}}\right) \tag{7.57}$$

In this modified region, the friction factor can be approximated as:

$$\lambda_1 = \frac{0.5999}{Re_{0L}^{0.506}} \tag{7.58}$$

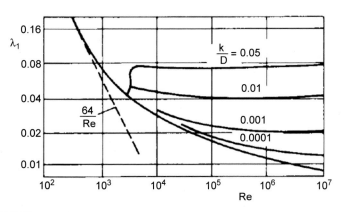

FIGURE 7.13 Ros friction factor λ_1.

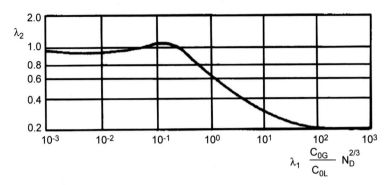

FIGURE 7.14   Ros friction factor λ_2.

The coefficient λ_2 can be obtained from diagram in Fig. 7.14, where it is plotted against the parameter:

$$X = \lambda \frac{c_{0G}}{c_{0L}} D^{\frac{2}{3}} \left(\frac{\rho_L g}{\sigma}\right)^{\frac{1}{3}} \tag{7.59}$$

This formula applies when:

$$c_{0G} \left(\frac{\rho_L}{g\sigma}\right)^{0.25} \leq 50 + c_{0L} \left(\frac{\rho_L}{\rho\sigma}\right)^{0.25} \tag{7.60}$$

In the annular and mist flow regions, the gas phase is continuous. In this case:

$$c_{0G} \left(\frac{\rho_t}{g\sigma}\right)^{0.25} \geq 75 + 84 c_{0L}^{0.75} \left(\frac{\rho_L}{g\sigma}\right)^{0.1875} \tag{7.61}$$

while both correction coefficients λ_2 and λ_3 are equal to unity, thus $\lambda_R = \lambda_1$.

The relative roughness in this case is not calculated as the wall roughness k, but it is obtained as the "film roughness" k^*. The film roughness k^* is obtained from a curve such as the one shown in Fig. 7.15.

The calculated $\lambda_1 = \lambda_R$ after substitution into Eq. (7.44) leads to the pressure loss for annular flow in a short pipe of length Δz. For a deep well, it should be used in the form of a series of step calculations.

7.4 TWO-PHASE FLOW IN WELLS INDUCED BY FLASHING

The two-phase flow in wells induced by the released gases is also a two-component flow. The water and released gas are chemically two

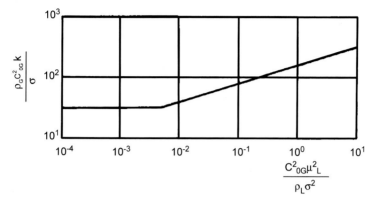

FIGURE 7.15   The film roughness of annular flow.

different materials. The reason for the occurrence of the gas bubbles is the pressure decrease until the bubble point, but in this situation the physical meaning of the pressure is the normal stress component only. In this case, the water is considered to incompressible, thus the pressure isn't a thermal state variable. Mechanical variables govern the phenomenon.

The pressure decrease also induces the process of flashing. Nevertheless, the pressure has another physical meaning. It is at once the normal stress component, and a thermal state variable as the pressure decrease attains the saturated steam pressure at the actual temperature. A sudden change of phase occurs as the liquid-phase hot water becomes wet steam; a water–steam mixture. This phenomenon is governed by thermodynamic variables.

An important problem in the determination of the depth in the well where the liquid flashes into vapor H_F, is to locate the depth of the flashing we need to consider the flow along a streamline from the reservoir through the well until the point where the hot water pressure decreases to the saturation pressure corresponding to the water temperature.

It is obvious that taking the mechanical energy equation in the form of (Eq. 7.17), substituting the elevation of the flashing horizon H_F and the saturated steam pressure p_s instead of the wellhead pressure p_{wh}, we obtain that:

$$p_s = p_e - \rho g H_F - \frac{\mu Q}{2\pi h K} \ln \frac{R_e}{R_w} - \lambda \frac{H_F}{D^5} \frac{8}{\pi^2} \rho Q^2 \tag{7.62}$$

The flashing horizon can be expressed from Eq (7.62):

$$H_F = \frac{p_e - p_s - \frac{\mu Q}{2\pi h K} \cdot \ln \frac{R_e}{R_w}}{\rho g + \frac{\lambda \rho}{D^5} \frac{8}{\pi^2} \cdot Q^2} \tag{7.63}$$

From Eq. (7.63), it can be seen that the increase of the flow rate decreases the elevation of the flash horizon. The location of flash point moves downward in the well; if flow rate is sufficiently large it can attain even the entrance section of the well, or the reservoir. Note that Eq. (7.63) was obtained by applying one-phase flow equations between the reservoir and the flash level. In this section, the liquid phase is homogeneous. The energy equation between the flash point and the wellhead can be written in modified form, because the water-stem mixture isn't barotropic.

The process undergone by the geothermal fluid can be represented in a thermodynamic state diagram. The fluid-specific entropy is plotted on the abscissa, while its temperature is plotted on the ordinate. The so-called temperature–entropy diagram is especially suitable to demonstrate the change of state of the water–steam mixture. It is shown in Fig. 7.16. The critical point of the water is at the peak of the typical bell-shaped curve. Its left-hand side branch separates the domains of the compressed liquid and the liquid–vapor mixtures. This is the so-called saturation curve; the corresponding saturation pressure and temperature values fall on this curve. The right-hand side branch separates the domains of the vapor–liquid mixture and the superheated steam. One accent mark designates the state variables belonging to the left-hand side branch, the so-called lower boundary curve, while the two accent mark designates the variables belonging to the right-hand side, the so-called upper boundary curve.

The flashing begins at the point $1'$ in the lower boundary curve. In this point the water content is 100%, while the steam is 0. The flashing process

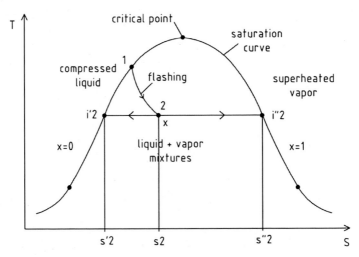

FIGURE 7.16   Flashing on temperature–entropy diagram.

is considered to isenthalpic, because it occurs steadily, without any heat transfer, and with no work involvement. The change of the kinetic and potential energy of the fluid is also neglected as it undergoes the flashing. Thus, we may write:

$$i_1 = i_2 \tag{7.64}$$

that is the enthalpy of the fluid is constant during flashing. This isenthalpic change of state is represented by a hyperbola-like curve as it is shown in Fig. 7.16. The mass fraction of the steam is denoted by x.

Under the bell-shaped saturation curve, in the two-phase domain the temperature and pressure of the mixture is the same as the corresponding saturation temperature and pressure. The isotherms and isobars are horizontal lines in this region. As a result of heat transfer, the temperature and pressure remain constant, the mass fraction of steam x changes only. This parameter plays a great part in description of the flashing process.

The change of state of the water–steam mixture cannot be described by analytical means. Well-known steam tables are used for this purpose. In the steam tables, the entropy, enthalpy, and density values are given at the saturation curve both on the lower boundary curve: s', i', and ρ', and the upper one (s'', i'', ρ'') as the function of the saturation temperature and pressure.

Consider now a flashing process between the saturation pressure p_1, and the end pressure p_2. The point 1, when the flashing begins falls to the lower boundary curve, where s'_2, i'_2, and ρ'_2, and s''_2, i''_2 and ρ''_2 belong to the end pressure p_2. The enthalpy is constant in the process, thus $i_1 = i_2$. As it is known, the following relations are exist:

$$s = s' + x(s'' - s') \tag{7.65}$$

$$i = i' + x(i'' - i') \tag{7.66}$$

$$\rho = \rho' + x(\rho'' - \rho') \tag{7.67}$$

In accordance:

$$i_2 = i'_1 = i'_2 + x\left(i''_2 - i'_2\right) \tag{7.68}$$

The mass fraction of the steam is obtained as:

$$x = \frac{i'_1 - i'_2}{i''_2 - i'_2} \tag{7.69}$$

Knowing the values of x, the entropy, the density can be obtained:

$$s_2 = s'_2 + x\left(s''_2 - s'_2\right) \tag{7.70}$$

$$\rho_2 = \rho'_2 + x\left(\rho''_2 - \rho'_2\right) \tag{7.71}$$

The velocity at the flash point is taken as equal to that at the well feed zone since the water is incompressible up to the flash point. The outflow velocity is the velocity that generates the same kinetic energy carried by the two-phase mixture:

$$\frac{c_2^2}{2} = x_2 \frac{c_{s2}^2}{2} + (1 - x_2) \frac{c_{w2}^2}{2} \tag{7.72}$$

where c_w, the water, and c_s is the steam velocity.

The steam mass fraction can be expressed as:

$$x_2 = \frac{\rho_{sz} A_{s2} \cdot c_{s2}}{\rho_{w2} A_{w2} c_{w2} + \rho_{s2} A_{s2} c_{s2}} = \frac{\dot{m}_s}{\dot{m}_s + \dot{m}_w} \tag{7.73}$$

The area terms account for the parts of the exit area occupied by the two phases. The holdup ratio H_2 relates to the velocities of the two phases:

$$H_2 = \frac{c_{s2}}{c_{w2}} \tag{7.74}$$

Its value may include the range of 4–5.

Thus, the exit velocity can be expressed as:

$$c_2 = c_{s2} \sqrt{x_2 + \frac{1 + x_2}{H_2^2}} \tag{7.75}$$

The average density for a lumped-parameter model is expressed as:

$$\rho_{av} = \frac{1}{2} \left[\rho_1' + x_2 \rho_{s2}'' + (1 - x_2) \rho_2' \right] \tag{7.76}$$

The mass flow rate based on the continuity equation is equal to the product of the density, the cross-sectional area, and the velocity. Thus, we can define an expression for the average two-phase velocity:

$$c_{av} = \frac{\dot{m}}{\rho_{av} \cdot A} = \frac{4\dot{m}}{\rho_{av} D^2 \pi} \tag{7.77}$$

The two-phase friction factor λ_{av} cannot be obtained as the λ for homogeneous fluids. Experimental data show that it is larger than the one-phase λ.

An acceptable empirical formula is:

$$\lambda_{av} = 1.05 (Re_w)^{-\frac{1}{5}} \tag{7.78}$$

where Re_w is the Reynolds number, and is the homogeneous flow in the well before the flashing point.

Now we can assemble the elements of the flow model the mechanical energy equation to obtain the wellhead pressure including the flashing within the well. The calculation of the pressure drop from the undisturbed reservoir to the flash horizon is unchanged except from the flash horizon to the wellhead. Finally we get:

$$p_{wh} = p_e - \rho g H_F - \frac{\mu Q}{2\pi hK} \ln \frac{R_e}{R_w} - \lambda \frac{H_F}{D^5} \frac{8}{\pi^2} \rho Q^2 - \rho_{av} g(H - H_F)$$

$$- \lambda_{av} \frac{H - H_F}{D^5} \frac{8}{\pi^2} \cdot \frac{\dot{m}^2}{\rho_{av}} - \rho_{av} \frac{c_2^2 - c_{av}^2}{2} \qquad (7.79)$$

It can be recognized that the wellhead pressure is developed by the following effects.

The undisturbed reservoir pressure p_e is reduced by the hydrostatic pressure drop of the water column until the flashing level, the pressure drop of the flow toward the well in the reservoir, the pressure drop of the upflowing water until the flash horizon. These obtain the pressure at the flashing level:

$$p_F = p_e - \rho g H_F - \frac{\mu Q}{2\pi hK} \ln \frac{R_e}{R_w} - \lambda \frac{H_F}{D^5} \frac{8}{\pi^2} \rho Q^2 \qquad (7.80)$$

Above the flashing level, the further pressure drop is developed by the hydrostatic pressure drop of the well in the two-phase section, the pressure loss of the two-phase flow, and the pressure drop due the acceleration of the flow in the two-phase section.

It can be recognized that Eq. (7.80) is an implicit expression. There are many terms referring to the two-phase flow depending on the wellhead pressure. The properties of the two-phase water-steam mixture depend on the mass fraction of steam x. The value of the mass fraction of steam can be determined only knowing the wellhead pressure. Thus, an iteration procedure must be carried out. It is necessary to choose a starting value of the wellhead pressure to obtain x. The properties of the two-phase system can be calculated, and we get the next value of p_{wh}. We must repeat the successive steps of iteration until the difference of $(p_{wh})_{i+1} - (p_{wh})_i$ decreases to a suitable small value.

Riley (1980) presented several numerical examples to obtain the relationship between mass flow rate and wellhead pressure. The mass flow rate plotted against the wellhead pressure obtains the productivity curve of the well that is its performance curve. Comparison of the calculated and measured values shows a satisfactory agreement. The general shape of the curves is not affected by the values of the friction factors, both λ and λ_{av}, but as the friction factor increases the mass flow rate decreases at a given wellhead pressure. It is remarkable that at a certain wellhead pressure, further lowering of the pressure does not result the increase the flow rate. In this case the flow is choked. The explanation of this phenomenon is that in the flashing fluid, the lowering of the pressure does not increase the velocity (as it is generally), but increases the size of the bubbles only. The large bubbles narrow down the cross-section, thus the mass flow rate remains unchanged.

References

Culberson, O.L., McKetta Jr., J.J., 1951. Phase equilibria in hydrocarbon-water systems, IV-vapor- liquid equilibrium constants in the methane-water and ethane-water systems, Trans AIME Pet. Div 192, p. 297–300.

Eichelberger, J.W., Behymer, T.D., Budde, W.L., 1988, Method 525. Determination of organic compounds in drinking water by liquid-solid extraction and capillary GC/MS, in Methods for the determination of organic compounds in drinking water: U.S. Environmental Monitoring and Support Laboratory, Cincinnati.

Ferrandon, J., 1948. Les Lois de I'ecoulemcnt de filtration. Genie Civil 125 (2), 24–28.

Riley, W. D., 1980. Well Engineering and Sampling Variables in the Evaluation of Geobrines, United States, Bureau of Mines.

Ros, N.C.J., 1961. Simultaneous flow of gas and liquid as encountered in well tubing. J. Petr. Tech 13, 1037.

8.1 MAIN TYPES OF DOWNHOLE PUMPS

The most expensive elements of a geothermal project are the drilling and completion of the wells. The rentability can be improved by minimizing the number of the wells, increasing their flow rates. The produced flow rate can be enhanced applying downhole pumps in the well.

Wells drilled for geothermal water operate in many cases by natural upflowing. With the progress of time, water production lowers the pressure in hydrostatic reservoirs. The elastic expansion is depleted as the excess pressure provided by the thermal lift is exhausted. In this case, downhole pumps are needed to induce a larger pressure drop at the

161

Copyright © 2017 Elsevier Inc. All rights reserved.

bottom-hole, enhancing the elastic expansion of the reservoir fluid. Downhole pumps are installed not only to lift the fluid to the surface, but also to prevent the release of non-condensable gases, which can lead to scale formation. An important benefit is the possibility to the control the flow rate and wellhead pressure.

Downhole pumps are multistage centrifugal pumps of restricted diameter in order to have place in the tubing of the production well. Centrifugal pumps are machines that do work on the fluid flowing through a rotating impeller. This work increases the mechanical energy content of the fluid maintaining the flow.

The downhole pumps can be classified into two categories. The so-called line-shaft pumps operate in the well under the water level. The multistage centrifugal pump is driven by a long shaft; the electric motor is at the surface. The shaft can be as long as 100−120 m. This restricts the depth of the installation. The motor operates at the surface temperature; its diameter is not restricted. Such a line shaft pump system is shown in Fig. 8.1.

Submersible pumps are the second category of downhole pumps. A typical submersible pump is sketched in Fig. 8.2, as it is installed to a well, together with its complete range of accessories. Down below is the electric motor which turns the pump. It is usually a two-pole, three-phase squirrel cage induction type unit. The next component is the so-called seal or protector section to separate the produced hot water from the motor oil. Above the protector there is the intake section. The electric motor is cooled by the upflowing fluid, thus the intake section must be above the protector. The pump itself is a multi-stage centrifugal pump. Each stage consists of a rotating impeller and a stationary diffuser with guide vanes. The impeller and the diffuser may be radial or mixed flow type as it is shown in Fig. 8.3. For smaller flow rates radial, for larger flow rates mixed flow impellers are used. The number of stages determines the manometric head of the pump. It may be a built in centrifugal gas separator between the protector and the pump. It separates the free gas from the produced fluid and leads it away from the pump intake. The complete system is hung from the wellhead with a tubing section. The electric cable, the switchboard, transformers, and valves are additional accessories of the submersible pump.

8.2 THEORETICAL HEAD OF THE CENTRIFUGAL IMPELLER

The purpose of applying centrifugal pumps is to increase the mechanical energy of the fluid to maintain flow. The mechanical energy increase happens as the fluid flows through the blade rows of the

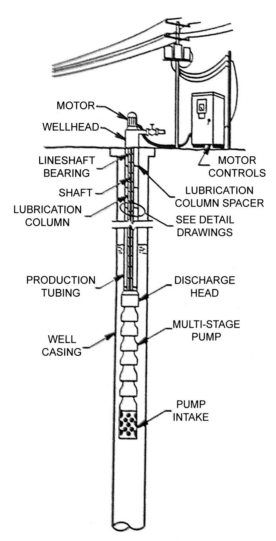

FIGURE 8.1   Lineshaft pump (Gudmundsson, 1988).

impeller. Since the only displacement of the blades is in a tangential direction, work is done by the displacement of tangential component of force acting on the blades. The velocity triangles are generally used to represent the flow through the rotating impeller. The absolute velocity $\vec{c}$ is obtained in a stationary coordinate system, while $\vec{w}$ is the velocity relative to the impeller. The peripheral velocity of the impeller is $\vec{u}$. Subscript 1 refers to the inflowing fluid, subscript 2 the outflowing fluid. It is shown in Fig. 8.4.

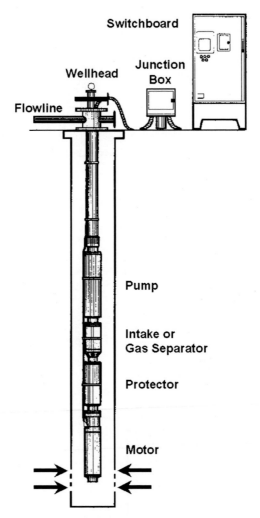

FIGURE 8.2 Submersible pump (Takács, 2009).

The absolute velocity $\vec{c}$ is unsteady, while the relative velocity $\vec{w}$ is steady. The fluid is considered to be inviscid; pressure loss is neglected. The relation:

$$\vec{c} = \vec{w} + \vec{u} \tag{8.1}$$

is obvious. The specific mechanical energy of the fluid can be expressed by the unsteady Bernoulli equation as:

$$\frac{\partial \phi}{\partial t} + \frac{c^2}{2} + \frac{p}{\rho} + zg = K \tag{8.2}$$

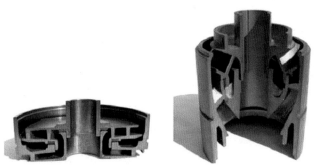

FIGURE 8.3 Radial and mixed flow impellers (Centrilift).

where ϕ is the velocity potential of the perfect fluid. The mechanical energy of the unit fluid mass is:

$$E = \frac{c^2}{2} + \frac{p}{\rho} + zg = K - \frac{\partial \phi}{\partial t} \tag{8.3}$$

It can be recognized, that the mechanical energy level of the flow is not steady. This equation was considered by Csanady (1965) to be the basic equation describing the performance of turbomachines. It is obvious that energy transfer in a perfect fluid can be possible for an unsteady flow only. Consider Fig. 8.4, in which a centrifugal impeller is depicted together with the velocity triangles at the inlet and outlet. The peripheral component of

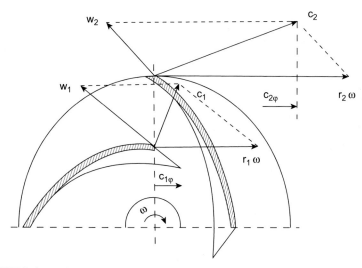

FIGURE 8.4 Centrifugal impeller with velocity triangles.

the absolute velocity is c_φ, which can be expressed in terms of the velocity potential:

$$c_\varphi = \frac{1}{r}\frac{\partial \phi}{\partial \varphi} \tag{8.4}$$

Applying the chain rule, we can this write as:

$$\frac{\partial \phi}{\partial t} = \frac{\partial \phi}{\partial(\varphi - \omega t)} \cdot \frac{\partial(\varphi - \cot)}{\partial t} = -\omega\frac{\partial \phi}{\partial(\varphi - \omega t)}, \tag{8.5}$$

where ωt is the periodicity of the transient flow. Since this unsteady flow is periodic:

$$\frac{\partial \phi}{\partial(\varphi - \omega t)} = \frac{\partial \phi}{\partial \varphi} \tag{8.6}$$

Combining these expressions, the time derivative of the velocity potential ϕ can be expressed in terms of the velocity component c_φ as:

$$\frac{\partial \phi}{\partial t} = -\omega r c_\varphi \tag{8.7}$$

in which ω is the angular velocity of the impeller. The energy increase between the outlet and inlet is obtained as:

$$\Delta E = K - \left(\frac{\partial \phi}{\partial t}\right)_2 - \left[K - \left(\frac{\partial \phi}{\partial t}\right)_1\right] = \omega(r_2 c_{2\varphi} - r_1 c_{1\varphi}) \tag{8.8}$$

This energy increase refers to the unit mass fluid; its dimension is (m^2/s^2). The engineering practice to use the energy increase of the unit weight fluid is more useful:

$$H = \frac{\Delta E}{g} = \frac{u_2 c_{2\varphi} - u_1 c_{1\varphi}}{g} \tag{8.9}$$

Its dimension is (m), and it is called to head. The energy increase of the perfect fluid is determined assuming a homogeneous velocity distribution along the outlet and the inlet cross-section. It is equivalent with the assumption that all streamlines are congruent, as infinitely thin, and infinitely many blades would form the shape of the streamlines. Thus, we get the so-called theoretical head belonging to an impeller of infinite number of blades. A theoretical head-flow rate curve can be obtained by using Eq. (8.9) and the velocity triangle of Fig. 8.4.

Since all streamlines are congruent with the blades, the angle the relative velocity w_2 makes with the peripheral velocity u_2 is the same as the blade angle β_2'. The so-called meridional component of the absolute velocity c_{2m} is normal to the periphery. Thus we can be written:

$$c_{2\varphi} = u_2 - c_{2m} \operatorname{ctg}\beta_2' \tag{8.10}$$

$$c_{2m} = \frac{Q}{D_2 \pi b_2} \tag{8.11}$$

where D_2 is the outlet diameter of the impeller, and b_2 is the width of the impeller at the outlet. The blade thickness is neglected, and a further assumption is that $c_{1\varphi} = 0$. Substituting into Eq. (8.9) we get:

$$H_\infty = \frac{u_2^2}{g} - \frac{u_2 ctg\beta_2'}{D_2\pi b_2 g} \cdot Q \qquad (8.12)$$

It can be recognized that a given impeller and speed the theoretical head H_∞ varies linearly with flow rate Q. In the usual design of centrifugal pumps, the blade angle at the outlet is 20–30 degrees, thus decreasing theoretical head belongs to an increasing flow rate. The subscript t_∞ marks the assumption that the congruent streamlines are obtained with perfect guidance, i.e., an infinite number of blades.

It can be recognized that the circulation of the absolute velocity field around the impeller:

$$\Gamma = 2\pi\left(r_2 c_{2\varphi} - r_1 c_{1\varphi}\right) \qquad (8.13)$$

The circulation around a single blade is:

$$\Gamma_b = \frac{\Gamma}{N} = \frac{2\pi}{N}\left(r_2 c_{2\varphi} - r_1 c_{1\varphi}\right) \qquad (8.14)$$

where N is the blade number of the impeller. The theoretical head can be expressed by the circulation:

$$H_\infty = \frac{N\Gamma\omega}{2\pi g} \qquad (8.15)$$

Pfleiderer (1959) assumed, that the circulation decrease caused by the finite blade number can be approximated by a single proportionality factor, the so-called circulation decrease coefficient λ:

$$\lambda = \frac{H_{th}}{H_\infty} \qquad (8.16)$$

As the result of a more sophisticated calculation of Czibere (1960) and Bobok (1970), it is obtained that the finite number of blades imparts the relative velocity with angle $\beta_2 < \beta_2'$ of the blade angle. This inability of the blades for perfect guidance reduces $c_{2\varphi}$ and hence decreases the theoretical head produced. This decrease of the theoretical head is not a real hydraulic loss. The difference between H_∞ and H_t does not occur in the power consumption of the pump.

8.3 HEAD LOSSES OF CENTRIFUGAL PUMPS

The actual flow rate versus head curves are different from the theoretical head lines. There are many reasons of this discrepancy. One of them is the fluid friction in the boundary layers along the rotating and the fixed passages. Since the flow is turbulent, this type of head loss is proportional to the square of the flow rate:

$$h_f' = C_1 \cdot Q^2 \tag{8.17}$$

Another major head loss is the form drag of the blades of the impeller and the guide vanes of the diffuser. The form drag depends on the direction of the inflow relative to the blade angle at the inlet. The ideal inflow direction is parallel to the blade angle, in this case the form drag can be neglected. The flow rate belonging to this performance state is Q_*. This is the point of the best efficiency of the pump at a given speed. For other flow rates, this loss varies approximately as the square of the difference of the actual flow rate Q and the flow rate at best efficiency point Q_*:

$$h_d' = C_2(Q - Q_*)^2 \tag{8.18}$$

There are further head losses proportional to the square of the flow rate. These are the secondary flow loss, the velocity equalization loss after the blades, and recirculation loss at the inlet. Finally it is obtained that the actual head curve depends on the flow rate. This is the so-called performance or manometric head curve, the lowest line as it is shown in Fig. 8.5.

The performance head curve can be determined experimentally, together with the brake horsepower and the efficiency. Typical performance curves for centrifugal pumps are shown in Fig. 8.6.

An important problem is to determine the necessary energy addition to the fluid in order to maintain the flow through the well and the pipelines; that is to determine the manometric head of the pump.

It is obvious that the mechanical energy decrease maintains the flow against viscous and turbulent forces. Natural flows through channels or riverbeds always take place from an inlet with a high mechanical energy level to an outlet with a low mechanical energy level. This mechanical energy difference is dissipated by the irreversible process of friction. In engineering practice, most flows are directed from a low energy inlet to a higher energy outlet while at the same time a considerable amount of mechanical energy is converted into heat. It is obvious that such a flow can exist only by adding mechanical energy to the flowing fluid. This energy addition is possible only by introducing unsteady flow resulting from the rotating blading of a pump or a compressor. The work done by a rigid body on the fluid can be determined if the velocity distribution on the blading is known.

8.4 FLOW IN PIPES WITH MECHANICAL ENERGY ADDITION

It is also possible to calculate the mechanical energy required to maintain the flow against the increasing energy level and fluid friction

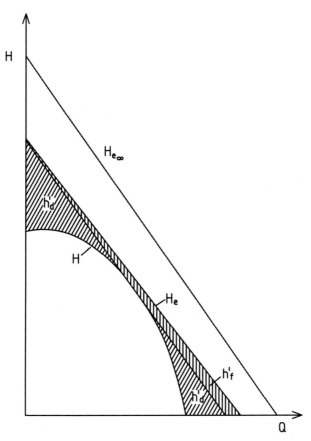

FIGURE 8.5 Theoretical and actual head curves.

from the variables of one-dimensional steady flow in the pipe. At first a general arrangement is studied.

Consider the flow system depicted in Figs. 8.7 and 8.8. Fluid is pumped from reservoir A to reservoir B. The flow is steady between reservoir A and the pump inlet 1, and also between the pump outlet 2 and reservoir B. For the two steady-flow sections, the mechanical energy equation can be written:

$$\frac{p_A}{\rho g} + z_A + \frac{c_A^2}{2g} = \frac{p_1}{\rho g} + z_1 + \beta_1 \frac{c_1^2}{2g} + h'_{A-1} \tag{8.19}$$

$$\frac{p_2}{\rho g} + z_3 + \beta_2 \frac{c_2^2}{2g} = \frac{p_B}{\rho g} + z_B + \frac{c_2^2}{2g} + h'_{2-B} \tag{8.20}$$

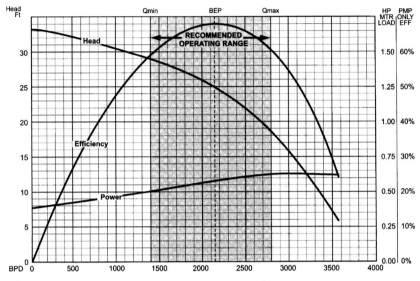

FIGURE 8.6 Performance curves of a submersible pump (Takács, 2009).

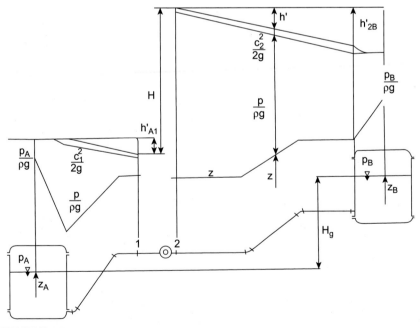

FIGURE 8.7 Flow system of a centrifugal pump.

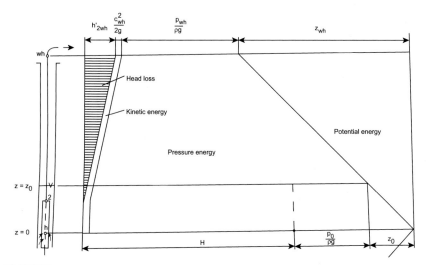

FIGURE 8.8   Energy diagram of a submersible pump system.

Since the diameters of the reservoirs are much greater than the diameters of the pipes, the kinetic energy terms $c_A^2/2g$ and $c_B^2/2g$ may be neglected. It can be seen from the energy diagram that the sum of the mechanical energy and the head loss is constant for both sections A-1 and 2-B of the pipe. The discontinuity in the energy curve represents the amount of mechanical energy added to the fluid by the pump. Using the energy equation this can be expressed as:

$$H = \frac{p_2 - p_1}{\rho g} + z_2 - z_1 + \frac{\beta_2 c_2^2 - \beta_1 c_1^2}{2g} \tag{8.21}$$

Using Eqs. (8.20) and (8.21), this energy difference can be rewritten in the form:

$$H = \frac{p_B - p_A}{\rho g} + z_B - z_A + h'_{A1} + h'_{2B} \tag{8.22}$$

Since at the free surfaces of the reservoirs the pressure is atmospheric, it is obvious that:

$$p_B = p_A = p_0 \tag{8.22a}$$

The major part of the energy increase takes the form of a pressure increase. The potential energy difference between the pump outlet and inlet, as well as the kinetic energy difference there, is negligibly small compared to the pressure difference. Thus the energy increase expressed by Eq. (8.22) is called the manometric head, H, of the pump. The potential energy difference between the levels of the fluids in the reservoirs is called the geodetic head H_g. Thus:

$$H_g = Z_B - Z_A \tag{8.23}$$

Using this notation, the manometric head can be written:

$$H = H_g + \sum h'_{AB} \tag{8.24}$$

In this expression, the geodetic head is independent of the flow rate, but the sum of the head losses does depend on the flow rate. For a laminar flow, this dependence is linear; for a fully developed turbulent flow it is parabolic. For a smooth pipe, or in the transition region between laminar and turbulent flow, head losses vary approximately to the 1.8th power of the flow rate.

Consider now the flow system of a submersible pump built in the well, as it is shown in Fig. 8.8. The fluid flows up through the casing, then in the annulus between the casing and the electric motor of the submersible pump. The fluid inflows to the pump across the intake screen 1. Over the intake in the annulus, the fluid is in hydrostatic state. The sum of the atmospheric pressure p and the hydrostatic pressure ρgz actuates the fluid into the pump, where in front of the first impeller stage, the pressure is lower than at the intake. Energy transfer can occur only in unsteady flow on the impeller bladings according to Euler's turbine equation. At the outflow cross-section of the pump, the flow is steady again, the energy level is constant between the pump outlet and the wellhead. The zero level of the potential energy is chosen at the intake of the pump. Thus the following Bernoulli equations can be written:

$$\frac{p_{wf}}{\rho g} + z_{BH} + \frac{c_{wf}^2}{2g} = \frac{p_1}{\rho g} + z_1 + \frac{c_1^2}{2g} + h'_{B1} \tag{8.25}$$

Between the pump outlet and the wellhead it is:

$$\frac{p_2}{\rho g} + z_2 + \frac{c_2^2}{2g} = \frac{p_{wh}}{\rho g} + z_{wh} + \frac{c_{wh}^2}{2g} + h'_{2wh} \tag{8.26}$$

The energy increase is:

$$H = e_2 - e_1 = \frac{p_{wh} - p_{wf}}{\rho g} + z_{wh} - z_{BH} + \frac{c_{WH}^2 - c_{wf}^2}{2g} + h'_{2wh} + h'_B \tag{8.27}$$

Using the Eqs. (8.25) and (8.26), we obtain:

$$H = h + \frac{p_{wh} - p_{wf}}{\rho g} + \sum h'_w \tag{8.28}$$

where $h = h - h_1$ is the so-called dynamic fluid level, depending on the flowing bottom-hole pressure and the head loss in the well. It is obtained as:

$$h_D = h - \frac{P_{wf} - P_0}{\rho g} + h'_{B1} \qquad (8.29)$$

The manometric head H is the energy increase of the unit-weight fluid. In the energy diagram, it shows the change of each kind of mechanical energy. The hydraulic power of the pump can be determined as:

$$P = \rho g Q H \qquad (8.30)$$

This value is lower than the brake horsepower of the pump P. The efficiency of the pump is obtained as the ratio of the hydraulic and the brake horsepower.

8.5 DIMENSIONLESS PERFORMANCE COEFFICIENTS

The experimentally determined performance curves, manometric head, brake horsepower and the efficiency depending on the flow rate, all at constant speed, are shown in Fig. 8.6.

There are several relationships that permit the data of the performance curves to be adapted to other speeds or sizes of impellers. If the speed of the pump changes, it is necessary to correct the performance data. The theoretical base of the correction is the approximation that the flowing fluid is considered inviscid. The dynamic similarity of two inviscid flows is attained if their velocity distributions are similar. This condition is fulfilled approximately if the velocity triangles are similar in the corresponding performance states. It can be defined with dimensionless parameters to express this. The capacity coefficient:

$$\varphi = \frac{c_{2m}}{u_2} \qquad (8.31)$$

and the head coefficient:

$$\psi = \frac{2gH}{u_2} \qquad (8.32)$$

must be equal in the corresponding performance states. According to these changing speeds of the pump, the ratio of the flow rates can be expressed as:

$$\frac{Q_1}{Q_2} = \frac{D_{21}}{D_{22}} \frac{b_{21}}{b_{11}} \frac{c_{m1}}{c_{m2}} = \left(\frac{D_{21}}{D_{22}}\right)^3 \frac{n_1}{n_2} \qquad (8.33)$$

Thus since the diameters are the same for two different speeds:

$$\frac{Q_1}{Q_2} = \frac{n_1}{n_2}$$ (8.34)

It follows from the equality of the head coefficients, that:

$$\frac{H_1}{H_2} = \left(\frac{n_1}{n_2}\right)^2$$ (8.35)

The correction of the brake horsepower is possible using the equation:

$$\frac{P_1}{P_2} = \left(\frac{n_1}{n_2}\right)^3$$ (8.36)

There is no such a simple expression for the efficiency correction. Finally, an important parameter, the so-called specific speed, is obtained based on the dimensionless performance coefficients:

$$n_q = n\frac{Q^{\frac{1}{2}}}{H^{\frac{3}{4}}}$$ (8.37)

The physical meaning of specific speed is revolution per minute to produce the flow rate of one cubic meter per second at 1 m of head with a similar impeller reduced in size. However, the physical meaning of specific speed has no application in practice, and specific speed is used only as a type number for best efficiency point of all similar impellers irrespective of their size or rotative speed. For multistage pumps, as all submersible units, specific speed is referred to the head per stage. The specific speed of all geometrically similar impellers is the same.

8.6 CAVITATION IN SUBMERSIBLE PUMPS

Cavitation is the process in which the vapor phase of a liquid is generated due the pressure reduction by hydrodynamic reasons at a constant ambient temperature. According to this definition, it is necessary to distinguish between the phase transition caused by pressure decrease, and the process of boiling caused by addition of heat. Cavitation has paramount importance in hot water flows near their boiling point. Submersible pumps built in thermal water production wells operate in high temperature water in which the connected pressure and temperature values are close to the saturation curve. In the first stage of submersible pumps, the pressure reduction can be attained at the saturation pressure of the water.

The presence of the steam bubbles changes the designed flow pattern drastically. This phenomenon induces a series of unwanted adverse effects in the operation of the pump. The bubble flow changes the uniform mass distribution of the water flowing through the impeller. The dynamically unbalanced rotating impeller causes vibration and noise. This causes damage to the lifetime of the bearings. The presence of bubbles produces a random pulsation of the torque, the produced head, and the velocities. This pulsation is ceased by intensive turbulent momentum transfer, producing high pressure and efficiency losses. The hydraulically active surfaces of the impeller exposed to cavitation are subject to severe damage especially in zones where the bubbles are collapsed. The collapsing bubbles preserve their spherical shape as long as it is possible, while the effect of the surface tension and the decreasing radius of the bubbles produce local pressure maximums as the radius tends to zero. These local pressure maximums are certainly high enough to account for brittle or fatigue failure of most materials. The deterioration of the impeller blades can occur after a few hours of cavitating operation. Thus, the operation of a pump in a cavitating state is strictly forbidden.

There are two different ways to investigate the cavitation in centrifugal pumps. One of them represents the point of view of the designers, analyzing the flow in the blading by a sophisticated theoretical approach using comprehensive numerical calculations. Another way to investigate cavitation represents the standpoint of operators, studying the performance variables outside of the pump as the inlet and outlet pressures, the flow rate, the speed, and the depth of submergence.

Consider the energy diagram in Fig. 8.9 as it is obtained in a coordinate system rotating together with the impeller. This diagram is the graphical representation of the Bernoulli equation in the rotating coordinate system. It can be recognized that the pressure distribution is different at the front (pressure) side and the back (suction) side of the impeller blades. The pressure minimum is obtained obviously at the suction side of the blades. At the beginning of the cavitation, this pressure minimum is equal to the saturation pressure belonging to the actual temperature. The Bernoulli equation can be written using the usual notation as it is shown in Fig. 8.9:

$$\frac{p}{\rho g} = \frac{p_i}{\rho g} + \frac{u^2 - u_i^2}{2g} + \frac{w_i^2 - w^2}{2g} \tag{8.38}$$

This pressure distribution can be determined correctly by the method of hydrodynamic singularities (Bobok, 1970). Knowing the pressure minimum, the so-called blade depression is obtained as the difference of the pressure heads in front of the leading edge and the pressure minimum:

$$\Delta h = \frac{p_i - p_{min}}{\rho g} = \frac{w_{min}^2 - w_i^2}{2g} + \frac{u_i^2 - u_{min}^2}{2g} \tag{8.39}$$

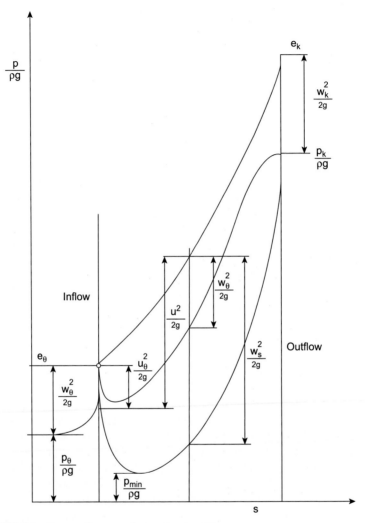

FIGURE 8.9 Energy diagram of a rotating impeller.

where w and u are the relative and peripheral velocities at the point of the pressure minimum. The expression is usual:

$$\Delta h = \left\{ \left(\frac{w_{min}}{w_i} \right)^2 - 1 + \left[1 - \left(\frac{R_{min}}{R_i} \right)^2 \right] \left(\frac{u_i}{u_{min}} \right)^2 \right\} \frac{w_i^2}{2g} \qquad (8.40)$$

in which the sum in the curly bracket depends on the shape of the impeller blading and the relative velocity at the inlet. The parameter characterizes the impeller in point of view the cavitation:

$$\lambda w = \frac{2g\Delta h}{w_i^2} \qquad (8.41)$$

The parameters of cavitation can also be determined experimentally by measuring the flow variables outside the pump in the pipe fitted to it and in the well. The Bernoulli equation is written between the free water surface in the well, the so-called dynamic water level, and the inlet cross-section of the impeller:

$$\frac{p_0}{\rho g} + z_0 = \frac{c_i^2}{2g} + \frac{p_i}{\rho g} + h'_{1i} \tag{8.42}$$

in which h'_{1i} is the head loss between the suction connection and the eye of the impeller of the first stage.

On the other hand, between the dynamic level and the suction connection, the following equation can be written:

$$\frac{p_0}{\rho g} + z_0 = \frac{c_1^2}{2g} + \frac{p_1}{\rho g} \tag{8.43}$$

It can be recognized considering Fig. 8.9 that:

$$\frac{p_i}{\rho g} = \frac{p_{min}}{\rho g} + \Delta h' \tag{8.44}$$

Substituting into Eq. (8.42) we get:

$$\frac{p_0}{\rho g} + z_0 = \frac{c_i^2}{2g} + \frac{p_{min\,i}}{\rho g} + \Delta h' + h'_{1i} \tag{8.45}$$

The so-called net positive suction head, NPSH is defined as:

$$\text{NPSH} = \Delta H = \frac{p_1}{\rho g} + \frac{c_i^2}{2g} - \frac{p_{min}}{\rho g} \tag{8.46}$$

Considering Eq. (8.43)

$$\text{NPSH} = \frac{p_0 - p_{min}}{\rho g} + z_0 \tag{8.47}$$

It is obvious that in the cavitating operation point, the pressure minimum is equal to the saturation pressure of the water:

$$p_{min} = p_{sat} \tag{8.48}$$

The critical NPSH value belonging to this state of operation is obtained as:

$$\text{NPSH}_{crit} = \Delta H_{crit} = \frac{p_0 - p_{sat}}{\rho g} + z_0 \tag{8.49}$$

Based on this equation the necessary depth of submergence to avoid cavitation can be determined as:

$$(z_0)_{crit} = \Delta H_{crit} - \frac{p_0 - p_{sat}}{\rho g} \tag{8.50}$$

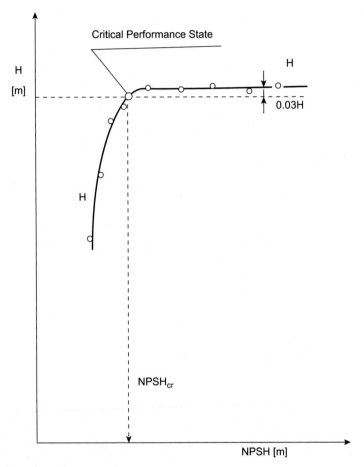

FIGURE 8.10   Initiation of cavitation.

It is obvious that NPSH depends on the flow rate and the speed of rotation. Thus the dependence of NPSH on the flow rate belongs to the performance curves of the pump as the head H (Q), the brake horsepower P (Q), and the efficiency η (Q) curves characterizing its operation. The depth of the built-in pump can be determined in the knowledge its NPSH (Q) curve at a given rpm value.

The critical value of NPSH can be determined by experiments. It is known that the manometric head decreases as cavitation occurs. It is generally accepted to consider the 3% manometric head reduction at the beginning of the cavitation in a certain operating point. The experiment is carried out in the following way; in some steady operating point holding the flow rate at a constant value, the submergence of the pump is decreased step by step. Thus the NPSH obviously decreases. As it is shown in Fig. 8.10, the decrease of NPSH in the beginning is not influenced the manometric head. When the least pressure in the impeller attains the saturation pressure, vapor bubbles occur producing the rapid decrease of the manometric head. By repeating this procedure at different flow rates, the critical NPSH dependence is obtained as it is shown in Fig. 8.11.

As the hot water upflows in the well, its pressure decreases. The pressure attaining the so-called bubble point, non-condensable gases deliberate from the geothermal water. Above the depth belonging to the bubble point growing gas bubbles occur in the flow. The gas/water volume ratio continuously increases in the upflow. This two-phase flow entering the pump disturbs its normal operation similarly to cavitation in many aspects. The presence of the gas phase is augmented by the flow rate, thus there are hydraulic losses too. Because of this, the manometric head and the

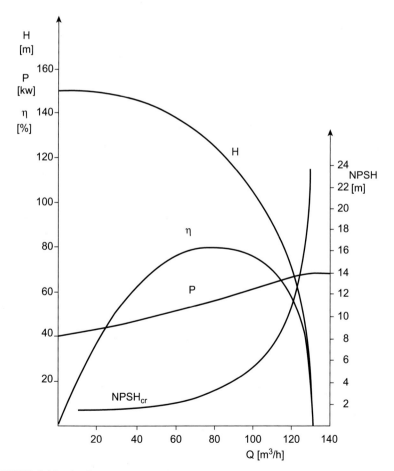

FIGURE 8.11 Performance curves.

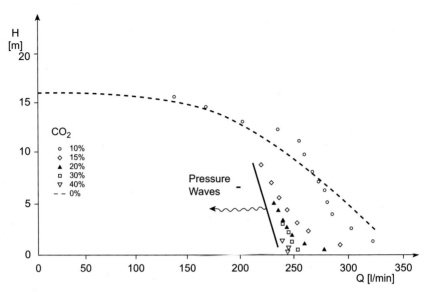

FIGURE 8.12   Head reduction at different CO_2 content.

efficiency of the pump decrease drastically. The effect of the bubble flow on the performance curve of the pump is shown in Fig. 8.12. Another effect is the unbalanced mass distribution of the impeller, inducing unwanted vibration. The growing gas bubbles tend in opposite directions of the centrifugal force, flowing through the impeller, even blocking the water flow. As the water flow ceases, the submersible motor can be overheated.

The main difference between the bubble flow and the cavitation is that the gas bubbles are not collapsed suddenly as the pressure increases again. Thus, the erosion of the hydraulically active surfaces fails to come about. Another effect is that the bubble point occurs at a greater depth in the well than the saturation pressure. It is an effective protection against this adverse phenomenon to build in a centrifugal separator in front of the first stage of the pump.

References

Bobok, E., 1970. An approximate solution of the second principal problem of the hydrodynamical cascade theory. Proc. Univ. Heavy Ind. 31, 397–417.
Csanady, T., 1965. Theory of Turbomachines. McGraw Hill, New York.
Czibere, T., 1960. Berechnungsverfahren zum Entwurfe ge3rader Flügelgitter mit stark gevölbten Profilschaufeln I-II. Acta Tech. Hung. 28.
Gudmundsson, J.S., 1988. The elements of direct use. Geothermics 17 (1), 119–136.
Pfleiderer, C., 1959. Die Kreiselpumpen. Springer, Berlin.
Takács, G., 2009. Electrical Submersible Pumps. Elsevier, Amsterdam.

Heat Transfer in Wells

9.1 TEMPERATURE DISTRIBUTION OF PRODUCTION WELLS

The temperature of the produced geothermal fluid can substantially decrease from the reservoir to the wellhead. The reason of this phenomenon is that the temperature of the upflowing fluid is higher than the adjacent rock around the well. This temperature inhomogeneity induces a radial heat flow from the well toward the adjacent rock. The temperature of the upflowing fluid decreases as it heats the rock around the well. The temperature of the adjacent rock increases slowly as the inhomogeneity of the temperature field decreases together with the outward radial heat flow. Thus, the temperature of the produced fluid gradually increases with the time until the steady state will be achieved. This heat transfer process will be investigated in the following (D'mour, 1995).

Consider a vertical geothermal well, producing hot water. The sketch of the well completion is shown in Fig. 9.1. A cylindrical coordinate system is chosen in accordance the geometry of the well. The z-axis of the coordinate system is directed downward; its origin z = 0 is at the surface. A cylindrical control surface is chosen coaxially with the axis of the well, at an arbitrary depth of z. Its upper and lower boundaries are

Copyright © 2017 Elsevier Inc. All rights reserved.

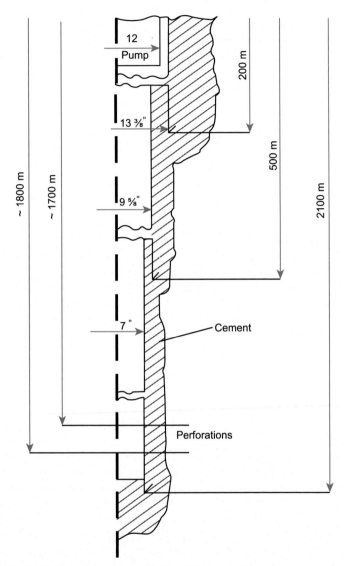

FIGURE 9.1   Typical hot water production well completion.

two parallel planes; their distance is dz from one another. The outer boundary of the cylinder is R_e, which is the time-dependent radius of the domain of the undisturbed natural geothermal temperature distribution. The control surface is shown in Fig. 9.2.

It is suitable to divide the system into two sub-systems. One of them is the upflowing fluid in the well, the other is the adjacent rock around the well.

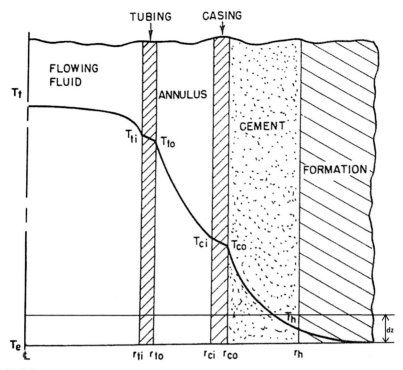

FIGURE 9.2 Control surface.

The outstanding feature of the flowing fluid is the convective heat transfer; in the adjacent rock the transient heat conduction is the dominating phenomenon. Since the governing differential equations are different for the two sub-systems, it is suitable to deal them separately. The joint condition of them is the same temperature at their boundary surface. Thus, we can write the balance equations of the internal energy for the water as flows up through the tubing:

$$\dot{m}cdT = 2R_{ti}\pi U_{ti}(T - T_h)dz, \qquad (9.1)$$

The second equation expresses the equality of the radial heat fluxes at the boundary surface between the well completion and the surrounding rock:

$$2R_{ti}\pi U_{ti}(T - T_h) = \frac{2\pi k_R}{f}(T_h - T_e), \qquad (9.2)$$

where $\dot{m}$ is the mass flow rate of the upflowing water, R_{ti} is the inner radius of the tubing, U_{ti} is the overall heat transfer coefficient of the well completion referring R_{ti}, T_h is the temperature at the wall of the borehole, k_R is the heat conductivity of the rock, T_e is the undisturbed temperature of the rock, and f is the so-called transient heat conduction function. This dimensionless parameter depends on the time, the thermal properties of the rock and the overall heat transfer coefficient.

Values of f are calculated by Jesse (1990) and experimentally determined and tabulated by Willhite (1967) as it is shown in Table 9.1.

The overall heat transfer coefficient can be determined considering the radially outward heat flow as it is passed through the serially connected elements of the well completion. The mechanism of the heat transfer is different in the different elements of the system. The internal surface of the tubing there is a thermal boundary layer with forced convection across it. The heat flux across the cylindrical surface of unit thickness is:

$$Q = 2\pi R_{ti} h_{ti}(T - T_{ti}) \tag{9.3}$$

Through the tubing wall the heat propagates by conduction. The heat flux Q remains naturally the same:

$$Q = 2\pi k_s \frac{T_{ti} - T_{to}}{\ln \frac{R_{to}}{R_{ti}}} \tag{9.4}$$

If the annulus is filled with fluid, the heat is transferred by free convection. In this case, the heat flux is obtained as:

$$Q = 2\pi R_{to} h_a (T_{to} - T_{oi}) \tag{9.5}$$

Through the casing wall, the conductive heat flux is:

$$Q = 2\pi k_s \frac{T_{ci} - T_{co}}{\ln \frac{R_{co}}{R_{ci}}} \tag{9.6}$$

Finally, the conductive propagating heat through the cement sheet can be calculated as follows:

$$Q = 2\pi k_c \frac{T_{oo} - T_h}{\ln \frac{R_h}{R_{co}}} \tag{9.7}$$

In this equation, h_{ti} is the heat transfer coefficient of the thermal boundary layer, k_s is the heat conductivity of the steel, h_a is the heat transfer coefficient of the annulus, and k_c is the heat conductivity of the cement sheet. Expressing the temperature differences from these five equations and summing them, we get:

$$T - T_h = \frac{Q}{2\pi R_{ti}} \left(\frac{1}{h_{ti}} + \frac{R_{ti}}{k_a} \ln \frac{R_{to}}{R_{ti}} + \frac{R_{ti}}{R_{to}} \frac{1}{h_a} + \frac{R_{ti}}{k_a} \ln \frac{R_{co}}{R_{ci}} + \frac{R_{ti}}{k_c} \ln \frac{R_h}{R_{co}} \right) \tag{9.8}$$

Let's compare this to the following equation:

$$Q = 2\pi R_{ti} U_{ti}(T - T_h) \tag{9.9}$$

TABLE 9.1 Transient Heat Conduction Function

$\dfrac{r_{to}U_{to}}{k_e} = \dfrac{\alpha t}{r_h^2}$	Time Function f(t) for the Radiation Boundary Condition Model													
	0.01	0.02	0.05	0.1	0.2	0.5	1.0	2.0	5.0	10	20	50	100	∞
0.1	0.313	0.313	0.314	0.316	0.318	0.323	0.330	0.345	0.373	0.396	0.417	0.433	0.438	0.445
0.2	0.423	0.423	0.424	0.427	0.430	0.439	0.452	0.473	0.511	0.538	0.568	0.572	0.578	0.588
0.5	0.616	0.617	0.619	0.623	0.629	0.644	0.666	0.698	0.745	0.772	0.790	0.802	0.806	0.811
1.0	0.802	0.803	0.806	0.811	0.820	0.842	0.872	0.910	0.958	0.984	1.00	1.01	1.01	1.02
2.0	1.02	1.02	1.03	1.04	1.05	1.08	1.11	1.15	1.20	1.22	1.24	1.24	1.25	1.25
5.0	1.36	1.37	1.37	1.38	1.40	1.44	1.48	1.52	1.56	1.57	1.58	1.59	1.59	1.59
10.0	1.65	1.66	1.66	1.67	1.69	1.73	1.77	1.81	1.84	1.86	1.86	1.87	1.87	1.88
20.0	1.96	1.97	1.97	1.99	2.00	2.05	2.09	2.12	2.15	2.16	2.16	2.17	2.17	2.17
50.0	2.39	2.39	2.40	2.42	2.44	2.48	2.51	2.54	2.56	2.57	2.57	2.57	2.58	2.58
100.0	2.73	2.73	2.74	2.75	2.77	2.81	2.84	2.86	2.88	2.89	2.89	2.89	2.89	2.90

The so-defined overall heat transfer coefficient can be obtained as:

$$\frac{1}{U_{ti}} = \frac{1}{h_{ti}} + \frac{R_{ti}}{k_s} \ln \frac{R_{to}}{R_{ti}} + \frac{R_{ti}}{R_{to}} \cdot \frac{1}{h_a} + \frac{R_{ti}}{k_s} \ln \frac{R_{co}}{R_{ci}} + \frac{R_{ti}}{k_c} \ln \frac{R_h}{R_{co}} \tag{9.10}$$

It still remains to determine the two heat transfer coefficients h_{ti} and h_a. There is an empirical relationship for the similarity invariants of the heat transfer:

$$Nu = 0,015 \cdot R_e^{0,83} \cdot Pr^{0,42}, \tag{9.11}$$

where the Nusselt number is:

$$Nu = \frac{h_{ti}2R_{ti}}{k} \tag{9.12}$$

the Reynolds number is:

$$Re = \frac{v \cdot 2R_{ti}}{v} \tag{9.13}$$

and the Prandtl number is:

$$Pr = \frac{\rho c v}{k} \tag{9.14}$$

Now v is the cross-sectional average velocity in the tubing, c is the specific heat capacity, v is the kinematic viscosity, and k is the heat conductivity of the fluid.

For the annulus, the Nusselt number depends on the Grashof number and the Prandtl number. The experimental relationship can be expressed by the following formula:

$$Nu = 0,52 \cdot (Gr \cdot Pr)^{0,25} \tag{9.15}$$

where the Grashof number is:

$$Gr = \frac{\alpha g(T_{to} - T_{ci})(R_{ci} - R_{to})^3}{v^2} \tag{9.16}$$

where α is the thermal expansion coefficient of the fluid. The heat transfer coefficient of the annulus is:

$$h_a = \frac{k \cdot Nu}{2R_{to}} \tag{9.17}$$

Since T_{to} and T_{ci} are temporarily unknown, we must be carry out an iteration, but a good estimation of $(T_{to} - T_{ci})$ may be suitable. This brief derivation was presented to indicate how various heat transfer mechanisms are included in an overall heat transfer coefficient. Combining Eqs. (9.1) and (9.2) we get:

$$\frac{dT}{dz} = \frac{2\pi k_R R_{ti} U_{ti}(T - T_e)}{\dot{m}c(k_R + f R_{ti} U_{ti})} \tag{9.18}$$

Considering that the undisturbed geothermal temperature is:

$$T_e = T_o + \gamma z, \qquad (9.19)$$

and that k_R is constant along the depth. It is shown in Fig. 9.1 that the well completion is different through different intervals of depth. Especially the overall heat transfer coefficient and the heat conduction function changes by sections. It is suitable to introduce the so-called performance parameter, including that variables which are independent or weakly dependent on the depth:

$$A = \frac{\dot{m}c(k_R + f \cdot R_{ti}U_{ti})}{2\pi R_{ti}U_{ti}k_R} \qquad (9.20)$$

The performance parameter has a dimension of length (m). Variation of A with depth is usually small because of the overall heat transfer coefficient U_{ti} appears both in the numerator and the denominator. Thus an approximation taking the performance parameter to constant causes negligible inaccuracy only. Frequently used the inverse of A, the so-called relaxation length parameter L_R having a dimension of $1/m$. Assuming the constant value of A, we obtain a first-order, linear, inhomogeneous differential equation as:

$$A\frac{dT}{dz} = T - T_o - \gamma z \qquad (9.21)$$

Its solution can be attained in closed form. One of the possible methods is the superposition of particular solutions of Eq. (9.21). In this case, the general solution can be obtained without integration as the sum of three particular solutions. At first it is taken the solution of the homogeneous equation:

$$A\frac{dT}{dz} = T \qquad (9.22)$$

It is obtained after the separation of variables in the form:

$$T_1 = Ce^{\frac{z}{A}} \qquad (9.23)$$

The next step to look for the particular solution of the inhomogeneous equation:

$$A\frac{dT}{dz} - T = -\gamma z \qquad (9.24)$$

The solution is obtained as:

$$T_2 = \gamma(A + z) \qquad (9.25)$$

Finally the particular solution of the equation:

$$A\frac{dT}{dz} - T = -T_o \qquad (9.26)$$

is the expression of:

$$T_3 = T_o \qquad (9.27)$$

The general solution is the sum of these three particular solution:

$$T = Ce^{\frac{z}{A}} + \gamma(z + A) + T_o \qquad (9.28)$$

The constant of integration can be determined by the boundary condition $z = H$; $T = T_o + \gamma H$. It means that the temperature of the inflowing water is the same as the formation temperature at the perforation. Thus we get:

$$C = -\gamma A e^{-\frac{H}{A}} \qquad (9.29)$$

Finally, the temperature distribution of the upflowing water along the depth is:

$$T = T_o + \gamma(z + A) - \gamma A e^{\frac{z-H}{A}} \qquad (9.30)$$

This solution refers to an instantaneous moment of the process, a snapshot of the time-dependent temperature distribution. Eq. (9.30) shows that the fluid temperature decreases exponentially from the bottom hole to the wellhead. The difference in the temperature between the produced hot water and the formation is generally increases as the water ascends the well.

The dominant performance variable is the mass flow rate that influences the velocity distribution. As the mass flow rate increases, the upflowing fluid temperature also increases. Decreasing flow rate leads to a substantial temperature drop.

Another strong influence is made by the time of operation. The upflowing fluid heats the adjacent rock mass around the well. As the temperature difference between the fluid and the rock decreases the outward heat flux is also diminished. Thus, the outflowing fluid temperature gradually increases with time until the steady state will be attained, mainly after 30 days. It is obvious that the warming up of the surrounding rock around the well is a very slow transient process. The transient heat conduction function $f(Fo, U)$ has a very weak change with time, especially the late times.

For very deep wells, the temperature difference between the produced water and the undisturbed geothermal temperature might asymptotically approach a constant value. The magnitude of this temperature difference depends on the value of A. Thus, if the asymptotic approach holds, temperature logs can be used to determine the value of A. The wellhead temperature is obtained from Eq. (9.30) as:

$$T_{WH} = T_o + \gamma A - \gamma A e^{-\frac{H}{A}} \qquad (9.31)$$

Expressing A from this equation, we get an implicit formula:

$$A = \frac{H}{-\ln\left(1 - \frac{T_{WH} - T_o}{\gamma A}\right)} \qquad (9.32)$$

The calculation of A can be possible by iteration.

Temperature logs are also suitable to estimate flow rates from various producing zones if thermal properties needed to calculate the performance parameter A are available. This calculation is possible because A is directly proportional to the mass flow rate.

The boundary condition $z = H$; $T = T_o + \gamma H$ is valid only if the inflowing section of the well is a relatively narrow interval close to $z = H$. Most geothermal wells in the Upper Pannonian aquifer are having a broad discharge interval, even a few hundred meters with interbedded impermeable layers. If the difference of the depth between the lowest and the uppermost discharge formations is 300 m, and the geothermal gradient is 0,05°C/m, the difference of the inflowing water temperatures is 15°C. The inflowing waters of different temperatures are mixed rapidly in the upflow developing a common temperature accordingly the law of calorimetry. This common temperature after the mixing and the depth belonging to it form the real boundary condition solving the differential Eq. (9.21).

Consider many discharge formations at the depths $z_1, z_2, \ldots z_n$, having thicknesses $h_1, h_2 \ldots h_n$, and permeabilities $K_1, K_2, \ldots K_n$. The discharges belonging to them are $Q_1, Q_2, \ldots Q_n$ with the averaged temperatures $T_1, T_2 \ldots T_n$. The total flow rate of the well is:

$$Q = Q_1 + Q_2 + \ldots + Q_n = \sum_{i=1}^{n} Q_i \qquad (9.33)$$

The thermal power of the well is:

$$P = \rho c(Q_1 T_1 + Q_2 T_2 + \ldots + Q_n T_n) = \sum_{i=1}^{n} \rho a Q_i T_i \qquad (9.34)$$

After mixing it is the same:

$$P = \rho c Q T \qquad (9.35)$$

Thus the common temperature T is:

$$T = \frac{\sum_{i=1}^{n} Q_i T_i}{Q} \qquad (9.36)$$

The discharge of every single permeable formation may be approximated based on the $K_i h_i$ values, thus the connected temperature is:

$$T = T_o + \gamma \frac{\sum_{i=1}^{n} K_i h_i z_i}{\sum_{i=1}^{n} K_i h_i} \qquad (9.37)$$

The C constant of integration in Eq. (9.28) can be obtained as:

$$C = \gamma \left[\frac{\Sigma K_i h_i z_i}{\Sigma K_i h_i} - H - A \right] e^{-\frac{H}{A}} \qquad (9.38)$$

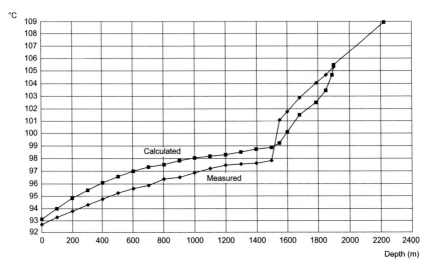

FIGURE 9.3 Calculated and measured temperature distribution in the hot water well of SZR 21.

Thus, the temperature of the upflowing water is obtained as:

$$T = T_o + \gamma(A + z) + \gamma\left(\frac{\Sigma K_i h_i z_i}{\Sigma K_i h_i} - H - A\right) \cdot e^{\frac{z-H}{A}} \qquad (9.39)$$

As an illustration, the comparison of the calculated and measured temperature distribution of the geothermal well SZR21 is Szarvas is demonstrated. There is no greater difference of 1,5°C along the entire depth of the well. It can be seen in Fig. 9.3.

9.2 TEMPERATURE DISTRIBUTION OF INJECTION WELLS

Reinjection of the utilized geothermal fluids is necessary for several reasons. The dissolved solid content of the geofluids represents a serious environmental problem. It is hazardous to let in fresh waters at the surface. Another problem is the pressure level decrease of the reservoirs resulting from the thermal water production. Reinjection helps to maintain the pressure level in the reservoir. A further important benefit of reinjection is that with the continuous flushing of the rock matrix by the cooled water, the recoverable geothermal energy from the reservoir substantially increases. The recovery factor can be increased even 10 times, relating to the technology based on the elastic expansion.

The design of the injection process makes it necessary to calculate the temperature distribution of the injected fluid as it flows down in the wellbore.

The differential equations describing the heat transfer between the reinjected water and the surrounding rock can be written considering the control volume outlined in Fig. 9.4. The system is divided into two subsystems. One of them is the flowing fluid in the well, the other is the adjacent rock around the well. For the downflowing water it is:

$$\dot{m}cdT = 2R_{ti}\pi U_{ti}(T_h - T)dz \tag{9.40}$$

The heat fluxes through the surrounding rock and the well completion are obviously equal:

$$2R_{ti}\pi U_{ti}(T_h - T) = \frac{2\pi k_R}{f}(T_e - T_h) \tag{9.41}$$

The only difference obtained in the heat transfer between the production and the injection well is in the direction of the heat fluxes. The heat flux in the production well is directed radially outward toward the rock, while in the injection well it is directed radially inward toward the well. The consequence of this is the opposite sign of the temperature differences in the equations. Combining the two equations, the following differential equation is obtained:

$$\frac{dT}{dz} = \frac{T_o + \gamma z - T}{\frac{\dot{m}c}{2\pi}\left(\frac{k_R + fR_{ti}U_{ti}}{k_R R_{ti}U_{ti}}\right)} \tag{9.42}$$

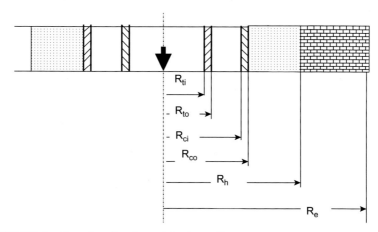

FIGURE 9.4   Control surface for an injection well.

It can be recognized that the performance coefficient A is the same for both production and injection wells. Thus, we obtain a first-order, linear, inhomogeneous differential equation:

$$A\frac{dT}{dz} = T_0 + \gamma z - T \tag{9.43}$$

Introducing the auxiliary variable:

$$\Theta = T - \gamma z \tag{9.44}$$

the modified differential equation can be written:

$$A\frac{d\Theta}{dz} = -\Theta + T_0 - \gamma A \tag{9.45}$$

It can be solved by variation of constants. At first we attain the solution of the homogeneous equation:

$$\Theta = Ce^{\frac{-c}{A}} \tag{9.46}$$

After this, the constant C is taken as it would be the function of z. Derivation the homogeneous solution by z we get:

$$\frac{d\Theta}{dz} = \frac{dC}{dz}e^{-\frac{z}{A}} + Ce^{-\frac{z}{A}}\left(-\frac{1}{A}\right) \tag{9.47}$$

Substituting it into the Eq. (9.45), a differential equation is obtained for C:

$$\frac{dC}{dz} = \left(\frac{T_0}{A} - \gamma\right)e^{\frac{z}{A}} \tag{9.48}$$

Its solution is:

$$C = (T_0 - \gamma A)e^{\frac{z}{A}} + K \tag{9.49}$$

The obtained expression of C is substituted into the solution of the homogeneous differential Eq. (9.46). Thus, we get the expression of:

$$T - \gamma z = T_0 - \gamma A + Ke^{-\frac{z}{A}} \tag{9.50}$$

In order to determine the constant of integration K, the boundary condition is if $z = 0$, then $T = T_{inj}$. This leads to the expression:

$$K = T_{inj} - T + \gamma A \tag{9.51}$$

Finally, the general solution for the temperature distribution along the depth is:

$$T = T_0 + \gamma(z - A) + (T_{inj} - T_0 + \gamma A)e^{-\frac{z}{A}} \tag{9.52}$$

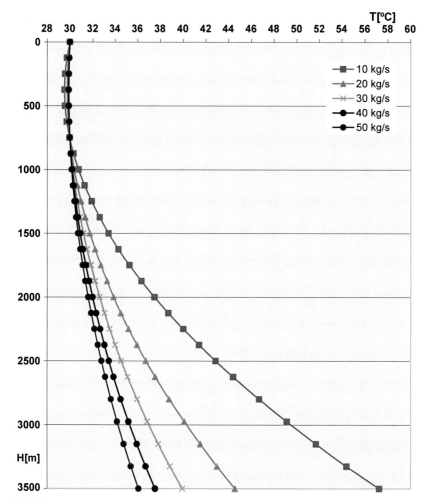

FIGURE 9.5   Temperature distribution along the depth at different mass flow rates.

The calculated temperature distributions are shown in Fig. 9.5, taking the constant values of the mass flow rate as a parameter. It can be seen the definitive influence of the mass flow rate on the injected water temperature at the bottomhole. The effect of the elapsed time on the temperature distribution is shown in Fig. 9.6. The bottomhole temperature becomes lower as time goes by. The importance of this effect is in that the temperature decrease results in a shrinking of the reservoir rock, the aperture of the fractures increases, and thus the permeability of the reservoir increases too.

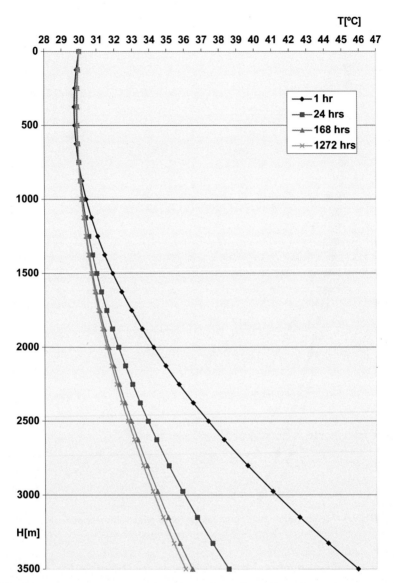

FIGURE 9.6   Temperature distribution along the depth at different times.

References

D'mour, H.N., 1995. Simulation of Heat Transfer in Boreholes (Ph.D. thesis). Miskolc.

Jesse, A.M., 1990. Thermal Geophysics. Elsevier, Amsterdam.

Willhite, G.P., 1967. Overall heat transfer coefficients in steam and hot water injection and production wells. J. Petr. Techn. 1509–1522.

10

Gathering System of Geothermal Fluids

10.1 ONE-DIMENSIONAL APPROXIMATION FOR FLOW IN PIPES

Recovering a geothermal resource there will be a number of wells tapping the reservoir. The capacity of geothermal fields may be as large as the geysers with more than 500 wells, or a small resource with a single well only. The site of the production and the utilization may be located a

Copyright © 2017 Elsevier Inc. All rights reserved.

considerable distance away. The transmitted fluid may be steam, a mixture of steam and hot water, or hot water only. In Hungarian geothermal practice the typical case is a small, hot water reservoir with at least a dozen wells. Thus we consider primarily the gathering system of hot water.

Problems of one-dimensional flow in pipes occur widely scattered throughout petroleum engineering practice. Design methods for drilling, cementing, hydraulic fracturing, production, and pipeline transportation of oil and gas require the handling of such problems. The equations obtained earlier in this book are totally adequate for calculating the flow variables for the one-dimensional case. The particular importance of this type of flow requires a detailed review of the fundamental principles, limitations, and recommended design methods.

The flow of a fluid along a streamline in the form of an infinitesimal stream tube of varying cross-section is the simplest example of one-dimensional flow. Any flow through a pipe forms an analogy with a stream-tube flow. The bounding surface of a stream tube consists of streamlines, thus the velocity has no component normal to it. Since the pipe wall is obviously impermeable, it may be considered to represent a stream tube of finite cross-section. An essential difference is that the flow variables are uniform over an infinitesimal cross-sectional area of the stream tube, while a pipe has a finite cross-section across which the flow variables may have predetermined nonuniform distributions, though it is always possible to take at least the integral mean values as uniform variables. In engineering practice the pipe flow is assumed to be one-dimensional. This assumption is only an approximation, and certain corrections have to be applied in order to get practical results.

Difficulties arise due to the entrance section of the pipe, the curvature of the flow, and changes in the cross-section. The longer the section of pipe, the better does the approximation of considering the flow to be one-dimensional apply. For large curvatures or moderate changes in cross-section, the deviation from the one-dimensional character of the flow may be neglected (Bobok, 1993).

It is known that any pressure change normal to the flow direction is hydrostatic. Consider now a flow in a cylindrical pipe in which all streamlines are parallel to the pipe axis. In any cross-section the hydrostatic equation for two arbitrary points 1 and 2 can be written:

$$p_1 + z_1 \rho g = p_2 + z_2 \rho g$$

From this it can be seen that although the pressure and the potential energy may vary considerably over the cross-section, their sum remains constant. Thus for any cross-section of a pipe with finite size, the single value represented by this sum applies to the whole flow or to any of the individual streamlines which compose it. Thus no change has to be made as an infinitesimal stream tube is expanded to encompass a pipe of finite size; the sum of the pressure and the potential energy are uniform over any cross-section, assuming, of course, that the streamlines are straight, parallel lines. It is obvious, that $p + z\rho g$ cannot remain constant if the streamlines are sharply convergent, divergent, or curved, so that the flow cannot be considered to be one-dimensional.

10.2 BASIC EQUATIONS FOR ONE-DIMENSIONAL FLOW IN PIPES

An arbitrary flow problem can be solved, at least in principle, by the simultaneous solution of a set of balance equations. These are the equations for the conservation of mass, the balance of momentum, balance of kinetic and internal energy, and the equation of state. Tacitly, the balance of angular momentum is taken into account by the symmetry of the stress tensor. Similarly, the constitutive relation of a Newtonian fluid is incorporated into the Navier–Stokes, Reynolds, or energy equations. Boundary conditions are added to complete the mathematical model. The integral forms of these basic equations are especially suitable to obtain the simpler expressions for pipe flow. Consider first the equation for the conservation of mass. As it is previously shown in Chapter 2:

$$\int_V \frac{\partial \rho}{\partial t} \, dV + \int_{(A)} \rho \vec{v} \, d\vec{A} = 0 \tag{10.1}$$

Let us apply the equation to the control volume shown in Fig. 10.1. The closed control surface (A) can be divided into three parts: the inlet surface A_1, the outlet surface A_2, and the pipe wall, treated as an impermeable stream surface A_3. As a result of this impermeability:

$$\vec{v} \, d\vec{A} = 0 \quad \text{at} \quad A_3 \tag{10.1a}$$

The inlet and outlet cross-sections are not necessarily perpendicular to the streamlines, thus the velocity vector $\vec{v}$ and the surface element vector $d\vec{A}$ are not necessarily parallel. At the inlet cross-section their scalar product must be negative; at the outlet cross-section it is obviously positive. Let the mass of fluid contained in V be designated by M

$$M = \int_V \rho \, dV \tag{10.2}$$

The integral mean of the velocity at any cross-section is

$$c = \frac{1}{A} \int_A \vec{v} \, d\vec{A} \tag{10.3}$$

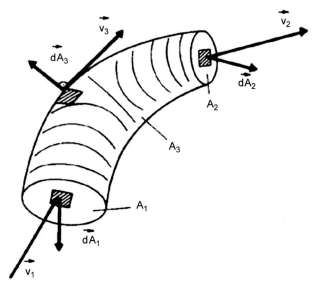

FIGURE 10.1   Control volume for one-dimensional flow in a pipe.

The area-averaged density is

$$\rho = \frac{1}{A} \int_A \rho dA \tag{10.4}$$

Thus the equation for the conservation of mass for the pipe section shown in Fig. 10.1 can be written

$$-\frac{\partial M}{\partial t} = \rho_2 A_2 c_2 - \rho_1 A_1 c_1 \tag{10.5}$$

For a steady flow

$$\rho_2 A_2 c_2 = \rho_1 A_1 c_1 = \dot{m} \tag{10.6}$$

thus the mass flow rate is the same across any cross-section. For incompressible fluids

$$A_2 c_2 = A_1 c_1 = Q \tag{10.7}$$

thus the flow rate Q at any cross-section is constant and given by the usual expression for the cross-sectional average velocity

$$c = \frac{Q}{A} \tag{10.8}$$

The momentum equation can be applied to a pipe section in the same way as the continuity equation. In its general form, we have

$$\int_V \frac{\partial(\rho\vec{v})}{\partial t} dV + \int_A \rho\vec{v}\left(\vec{v}d\vec{A}\right) = \int_V \rho\vec{g}dV + \int_{(A)} Td\vec{A} \tag{10.9}$$

For steady flow this simplifies to

$$\int\limits_{(A)} \rho \vec{v}(\vec{v}d\vec{A}) = \int\limits_{V} \rho \vec{g}dV + \int\limits_{(A)} T d\vec{A} \qquad (10.10)$$

The convective momentum flux through the pipe wall is obviously zero. Calculating the integral mean of the convective momentum flux it should be noted that the integral mean of the product of velocities is not equal to the product of the integral means of the velocity. This can be corrected by using a multiplier α defined by the equation

$$\alpha = \frac{\int\limits_{A} \rho|\vec{v}|(\vec{v}d\vec{A})}{Ac^2} \qquad (10.11)$$

The resultant of the body forces in a gravity field is obviously

$$\int\limits_{V} \rho \vec{g}dV = M\vec{g} \qquad (10.12)$$

The stress tensor can be split into three components,

$$T = -pI + T_V + T' \qquad (10.13)$$

where T_V is the viscous stress tensor and T' is the tensor of the Reynolds stresses.

For incompressible flow the shear stress distribution, as well as the Reynolds-stress distribution, is the same at any cross-section. Since the surface normal at the inlet and at the outlet cross-sections are opposite, the integrals of $(T_V + T')$ over A_1 and A_2 will cancel. For compressible flow this condition is only an approximation. At A_1 and A_2 the pressure will give nonzero integrals only. At the pipe wall, the normal stresses will cancel due to the symmetry, thus only the $(T_V + T')$ term produces nonzero results.

Thus, we finally obtain

$$\rho Q(\alpha_2 \vec{c}_2 - \alpha_1 \vec{c}_1) = M\vec{g} + \vec{f}_1 + \vec{f}_2 + \vec{f}_3 \qquad (10.14)$$

where

$$\vec{f}_1 = -\int\limits_{A_1} p_1 d\vec{A} \qquad (10.14a)$$

$$\vec{f}_2 = -\int\limits_{A_2} p_2 d\vec{A} \qquad (10.14b)$$

$$\overrightarrow{f}_3 = \int_{A_3} \left(T_V + T'\right) d\overrightarrow{A} \tag{10.14c}$$

In the expression $\overrightarrow{f}_1$ is the resultant pressure force at the inlet, $\overrightarrow{f}_2$ is the same at the outlet, and $\overrightarrow{f}_3$ is the resultant shear force acting on the system. The vector polygon which represents Eq. (10.14) geometrically is shown in Fig. 10.2.

In the case of a one-dimensional flow in a pipe inclined at an angle φ to the horizontal, the equation simplifies to

$$p_1 A_1 - p_2 A_2 - f_3 = Mg \sin\varphi \tag{10.15}$$

The mechanical energy equation for a barotropic fluid in a conservative body-force field can be written as

$$\int_V \frac{\partial}{\partial t}\left(\frac{v^2}{2} + U + \wp\right) \rho dV + \int_{(A)} \rho\left(\frac{v^2}{2} + U + \wp\right) \overrightarrow{v} d\overrightarrow{A}$$

$$== \int_{(A)} \overrightarrow{v}\left(T_V + T'\right) d\overrightarrow{A} - \int_V \left(T_V + T'\right) : S dV \tag{10.16}$$

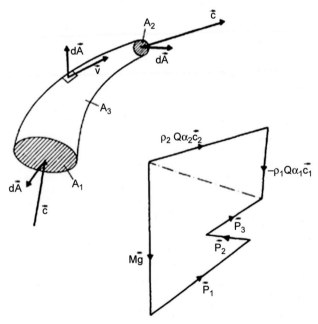

FIGURE 10.2   Forces acting on a control volume of pipe flow.

In discussing the continuity or the momentum equation, it is of no consequence whether the flow is laminar or turbulent, but for the kinetic energy equation this is of importance. For a turbulent flow it is even possible to write different formulations for the balance of kinetic energy equation, based either on time-averaged velocities, or on the actual velocity fluctuations. For a one-dimensional flow in a pipe the first type of equation is most suitable. We shall confine ourselves to steady flows only. At the pipe wall $\vec{v}\,\mathrm{d}\vec{A} \equiv 0$. Note, that on the right-hand side of the equation the product $(T_v + T')\,\mathrm{d}\vec{A}$ must first be obtained, thus in this surface integral the condition $\vec{v}\,\mathrm{d}\vec{A} \equiv 0$ is not fulfilled. In spite of this the velocity at the pipe wall is zero, thus the integral over A_3 must vanish. The result is the same, but the manner in which it is obtained is different. Thus we have

$$\int_{A_2} \rho \left(\frac{v^2}{2} + U + \wp \right) \vec{v}\,\mathrm{d}\vec{A} + \int_{A_1} \rho \left(\frac{v^2}{2} + U + \wp \right) \vec{v}\,\mathrm{d}\vec{A}$$

$$== \int_{A_2} \vec{v} \left(T_v + T' \right) \mathrm{d}\vec{A} + \int_{A_1} \vec{v} \left(T_v + T' \right) \mathrm{d}\vec{A} - \int_V \left(T_v + T' \right) : S\mathrm{d}V$$

$$(10.17)$$

Remember, that the turbulent shear stress is much greater than the viscous shear stress except within the laminar sublayer. Since the thickness of the latter is negligibly small relative to the diameter of the pipe, the viscous shear stress may be neglected. For a pipe of constant cross-section the two surface integrals will cancel, since the velocity distributions are equal and the surface normals opposite. If the cross-section changes, this condition is satisfied only approximately. For a long pipeline such an approximation is acceptable. For the integral mean of the convective kinetic energy flux a correction factor β is defined as

$$\beta = \frac{\int_A \left(\frac{v^2}{2} \right) \vec{v}\,\mathrm{d}\vec{A}}{\frac{c^3}{2} A} \qquad (10.18)$$

Thus we have

$$\rho_2 A_2 c_2 \left(\beta_2 \frac{c_2^2}{2} + U_2 + \wp_2 \right) + P_T = \rho_1 A_1 c_1 \left(\beta_1 \frac{c_1^2}{2} + U_1 + \wp_1 \right) \qquad (10.19)$$

where P_T is the mechanical power loss due to the turbulent shear stress:

$$P_T = \int_V T : S\mathrm{d}V \qquad (10.20)$$

Since

$$\rho_2 A_2 c_2 = \rho_1 A_1 c_1 = \dot{m} \qquad (10.20a)$$

subdividing by the mass flow rate $\dot{m}$ we obtain

$$\beta_2 \frac{c_2^2}{2} + U_2 + \wp_2 + \frac{P_T}{\dot{m}} = \beta_1 \frac{c_1^2}{2} + U_1 + \wp_1 \qquad (10.21)$$

For the case of a gravity field and an incompressible fluid we obtain, after subdividing by g, the so-called viscous Bernoulli equation

$$\beta_2 \frac{c_2^2}{2g} + z_2 + \frac{P_2}{\rho g} + h'_{1-2} = \beta_1 \frac{c_1^2}{2g} + z_1 + \frac{P_1}{\rho g} \qquad (10.22)$$

in which

$$h'_{1-2} = \frac{P_T}{\dot{m} g} \qquad (10.23)$$

is the so-called head loss, it is the mechanical energy decrease per unit mass of fluid flowing between cross Sections 10.1 and 10.2.

All terms of the mechanical energy equation in this Bernoulli-like form have the dimension of length. Because of this they may also be taken to represent vertical linear distances to visualize the equation as shown in Fig. 10.3. This visualization can be realized experimentally by using vertical piezometer tubes. In this case the slope of the energy line is parallel to the hydraulic grade line. This type of graphical representation, which is called an energy diagram, is widely used in the analysis of engineering problems. The energy line and the hydraulic grade line are also known as the total head line and the piezometric head line, respectively.

For a laminar flow the head loss can be determined in a purely analytical way, since the Hagen–Poiseuille equation is valid for such a flow.

For a turbulent flow an approximate, almost analytical computation, the so-called Weisbach equation, yields the head loss. This was originally determined as an empirical formula, and later confirmed by dimensional analysis. In this chapter it will be derived by solving the momentum equation for turbulent flow.

10.3 DETERMINATION OF THE APPARENT TURBULENT SHEAR STRESS ACCORDING TO THE MIXING LENGTH THEORY

On the preceding pages it has already been mentioned that when the Reynolds equation of motion is used to describe turbulent flow, the number of unknown variables is increased: even in the simplest case

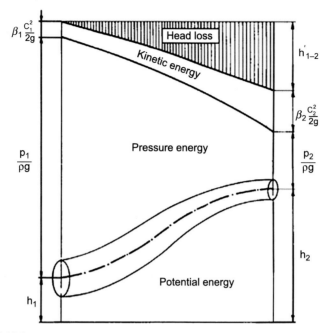

FIGURE 10.3 Energy diagram for pipe flow.

where the tensor of the apparent turbulent shear stress is symmetrical, it is necessary to define six new unknown variables. At the present time no system of correlation analogous to Stokes's law is known, according to which the Reynolds' stresses and the kinematic parameters characterizing the average velocity field can be related to each other.

For practical purposes it has thus been tried to find such correlations between approximations which are suitable, at least in the case of one-dimensional flows, to permit the determination of the distribution of the mean velocity and the flow resistance. The simplest of these approximation procedures was worked out by Prandtl (1956).

Consider the mean velocity profile outlined in Fig. 10.4. At the wall the mean velocity is zero and it increases gradually with increasing distance away from the wall. At a distance y from the wall the mean velocity is $v_x(y)_p$ and the fluctuation components v'_x and v'_y are superimposed on it. The turbulent momentum transfer due to the fluctuations may be approximated in the following way. Consider that a fluid particle in the layer with a mean velocity v_x moves upward as a result of the transverse fluctuation v'_y to a region at a distance $y + h$ from the wall. Here the mean velocity is greater; it is obtained by the linearization of the velocity profile as $v_x + \frac{dv_x}{dy} h$. In the new surrounding the mean

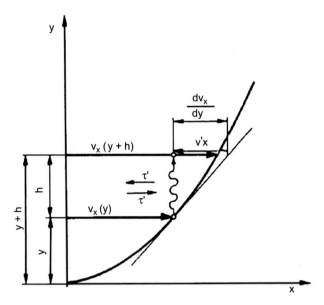

FIGURE 10.4   Turbulent velocity profile showing the mixing length.

velocity of the particle is smaller than the local average there. The difference

$$v_x(y) - v_x(y + h)$$

appears as the (negative) velocity fluctuation of the particle

$$v'_x = v_x - \left(v_x + \frac{dv_x}{dy}h\right) = -h\frac{dv_x}{dy} \tag{10.24}$$

Conversely, the particle which arrives from above the layer with a negative v'_y gives rise to a positive v'_x in it. On the average therefore, a positive v'_y is associated with a negative v'_x, and a negative v'_y is associated with a positive v'_x.

Assume that v'_x and v'_y are of the same order of magnitude, thus we obtain

$$v'_y = \alpha h\frac{dv_x}{dy} \tag{10.25}$$

where α is a dimensionless multiplier.

Since the apparent turbulent shear stress for the one-dimensional mean flow is

$$\tau'_{xy} = -\rho\overline{v'_x v'_y} \tag{10.26}$$

it can be written, that

$$\tau'_{xy} = \rho \alpha h^2 \overline{\left|\frac{dv_x}{dy}\right| \frac{dv_x}{dy}} \qquad (10.26a)$$

Since α, h, and the mean velocity gradient are nonfluctuating quantities, they remain unchanged during averaging, so that we can simply write

$$\tau'_{xy} = \rho \alpha h^2 \left|\frac{dv_x}{dy}\right| \frac{dv_x}{dy} \qquad (10.27)$$

It is obvious that there is a certain distance h between the mixing layers, at which the absolute values of the velocity fluctuations v'_x and v'_y are equal, so that $\alpha = 1$. This length is obtained to a characteristic quantity of the turbulent flow. It may be regarded as a correlation factor, and it is called the mixing length l. The mixing length can be considered as the average distance perpendicular to the mean flow covered by the mixing particles. If the turbulent momentum transfer is intense, the value of l increases. If, however, the degree of fluctuations decreases, l tends to zero. The mixing length changes with the distance from the wall. Prandtl assumed that

$$l_p = \kappa y \qquad (10.28)$$

where κ is a dimensionless constant, which is obtained indirectly from observations of the mean velocity distribution.

Kármán introduced another concept by means of the similarity theory. He assumed that mixing length had a real physical sense only, if the correlation is the same in all points of the flow region considered. This assumption seems to be valid in all cases in which mixing length is small in comparison with the dimensions of the region. Kármán's mixing length expression is

$$l_\kappa = \kappa \frac{\left|\frac{dv_x}{dy}\right|}{\left|\frac{d^2 v_x}{dy^2}\right|} \qquad (10.29)$$

where κ is the same constant as in Eq. (10.28). Thus the apparent turbulent shear stress is obtained as

$$\tau'_{xy} = \rho \kappa^2 \left|\frac{dv_x}{dy}\right|^3 \left|\frac{d^2 v_x}{dy^2}\right|^{-2} \frac{dv_x}{dy} \qquad (10.30)$$

This relation yields velocity distributions and pipe resistance factors which are in excellent agreement with experimental results for flow in pipes or between parallel plates.

10.4 TURBULENT FLOW THROUGH PIPES

Consider a steady flow through a straight cylindrical pipe of infinite length. The fluid is incompressible and the mean velocity has only one component; in the direction of the pipe axis. It is assumed that the distribution of the velocity fluctuations v_r' and v_z' are also axisymmetric. This assumption seems at first rather arbitrary, but the velocity distribution obtained from this solution is in good agreement with experimental data. The gravitational force has a potential

$$\vec{g} = -\text{grad}(gh) \qquad (10.31)$$

Let us use cylindrical coordinates as shown in Fig. 10.5. Then the control volume of the integration is a cylinder of radius r and length L, coaxial with the pipe axis. The mean velocity field is steady, thus the local derivative of the momentum is zero. Hence:

$$\int_{(A)} \rho \vec{v}(\vec{v}d\vec{A}) = \int_V \rho \vec{g} dV - \int_{(A)} pd\vec{A} + \int_{(A)} Vd\vec{A} - \int_{(A)} \rho(\overline{\vec{v}' \circ \vec{v}'})dA$$

$$(10.32)$$

The surface integral on the left-hand side represents the convective change of momentum due to the mean velocity field. Along the cylindrical stream surface $\vec{v}d\vec{A} = 0$. The velocity distributions are identical at the inlet and the outlet cross-section but the unit normal vectors are in opposite directions. Thus this integral vanishes on account of the continuity condition. We now transform the surface integral of the pressure force into a volume integral, thus the first and second terms of the right-hand side can be added up to form a volume integral of a gradient vector.

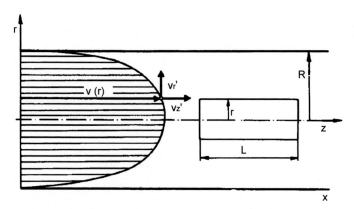

FIGURE 10.5   Turbulent flow in a cylindrical pipe.

Eq. (10.32) thus becomes:

$$0 = -\int_V \text{grad}(\rho gh + p)dV + \int_{(A)} T_v d\vec{A} - \int_{(A)} \rho(\overline{\vec{v'} \circ \vec{v'}})d\vec{A} \quad (10.33)$$

The viscous shear stress generated on the cylindrical surface is constant since the flow pattern is axisymmetric and does not change in the z-direction.

At the inlet and the outlet cross-section the flow patterns are the same but the unit normal vectors are in opposite directions, thus these two terms vanish. This is obviously also valid for the apparent turbulent shear stress. Consequently the second and third integrals should only be taken over the cylindrical surface of radius r.

The evaluation of the integrals is done in the following way. The force maintaining the motion is:

$$-\int_V \text{grad}(\rho gh + p)dV = -\vec{k}\int_V \frac{d}{dz}(\rho gh + p)dV = Jr^2\pi L\vec{k} \quad (10.34)$$

The sum $\rho gh + p$ is a linear function of z only, thus its gradient is constant $(-J)$. Instead of integration we can multiply it simply by the volume. This does not require any special explanation. The viscous force acting on the cylindrical surface is:

$$\int_{Ap} T_v d\vec{A} = \int_{Ap} \rho\frac{dv}{dr}\vec{k}\,dA = \mu\frac{dv}{dr}2\pi rL\vec{k} \quad (10.35)$$

Since the velocity decreases in radial direction, the value of the derivative dv/dr is negative and the sign of the viscous friction force is opposite to that of the force maintaining the motion.

To determine the apparent force due to the turbulent exchange of momentum the following integral needs to be evaluated:

$$-\int_{Ap} \rho\left(\overline{\vec{v'} \circ \vec{v'}}\right)d\vec{A} \quad (10.35a)$$

Let us assume an axisymmetric distribution for v_r' and v_z' though, of course, it also changes both with the radius and with z.

Based on Taylor's correlation theorem it is possible to replace the time average of the product of the velocity fluctuations with the mean integrated value of the same over the length L. This is acceptable since we can regard the mean integrated value of $\overline{v_r'v_z'}$ over the length L as a time average formed during the time $t_0 = L/v$ over which the averaging is carried out. Accordingly:

$$-\int_{Ap} \rho\left(\overrightarrow{v'} \circ \overrightarrow{v'}\right) d\overrightarrow{A} = -\int_{Ap} \overline{\rho v'_r v'_z} \, \overrightarrow{k} \, dA = -\overline{\rho v'_r v'_z} 2\pi r L \, \overrightarrow{k} \qquad (10.36)$$

If the evaluated integrals are now added and their sum divided by $2\pi r L$ we obtain the differential equation for turbulent flow in a pipe

$$\frac{\rho g J r}{2} + \mu \frac{dv}{dr} - \overline{\rho v'_r v'_z} = 0 \qquad (10.37)$$

Solving this equation we can distinguish two domains in which the flow behavior is totally different.

Experimental data show that all turbulent fluctuations vanish at the pipe wall, while they are very small in its immediate neighborhood. Thus the Reynolds stresses are zero in a very thin layer. Consequently, in every turbulent flow there exists this so-called laminar sublayer, in which the motion is laminar; thus for this laminar sublayer the third term of Eq. (10.37) may be omitted so that the differential equation for this sublayer is:

$$\frac{\rho g J r}{2} + \mu \frac{dv}{dr} = 0 \qquad (10.38)$$

Integrating this equation the velocity profile may be obtained in this domain.

The thickness of the laminar sublayer δ is very small in comparison to the radius of the pipe, thus Prandtl introduced an additional assumption, namely that the shear stress τ is constant throughout the laminar sublayer and equal to the shear stress at the wall τ_R:

$$\tau = \tau_R = \frac{\rho g J R}{2} \qquad (10.39)$$

Dividing the shear stress at the wall by the density, and then extracting the square root results in a parameter called the friction velocity

$$v_* = \sqrt{\frac{g J R}{2}} \qquad (10.40)$$

which is not a real velocity, having only the dimension of velocity. Thus the differential equation

$$\frac{dv}{dr} = -\frac{v_*^2}{v} \qquad (10.41)$$

is obtained from which, after integration, we obtain the velocity distribution in the laminar sublayer:

$$v = \frac{v_*^2}{\nu}(R - r) \tag{10.42}$$

Outside the laminar sublayer the effect of the viscous shear stress diminishes and the role of the turbulent momentum flux, the so-called Reynolds stress, increases. Under these conditions we can omit the second term, the viscous shear stress, from Eq. (10.37). In order to determine the velocity distribution of the turbulent core flow it is necessary to find an expression for the apparent turbulent shear stress which would relate it to the mean velocity. Prandtl's mixing-length theory is one of the simplest methods of estimating the Reynolds stress. As an extension of the mixing-length theory, Kármán proposed the following expression

$$l_K = \frac{\left|\frac{dv}{dr}\right|}{\left|\frac{d^2v}{dr^2}\right|} \tag{10.43}$$

which leads to the expression

$$\tau' = -\rho\overline{v_r' v_z'} = -\rho\kappa^2 \frac{\left(\frac{dv}{dr}\right)^4}{\left(\frac{d^2v}{dr^2}\right)^2} \tag{10.44}$$

for the turbulent momentum flux. After substitution we obtain:

$$\frac{gJr}{2} = \kappa^2 \frac{\left(\frac{dv}{dr}\right)^4}{\left(\frac{d^2v}{dr^2}\right)^2} \tag{10.45}$$

This equation requires some manipulation, while care should also be taken with the sign of the square roots.

The derivative d^2v/dr^2 must be negative, since the velocity distribution reaches a maximum value along the centerline of the pipe. Thus, again using the friction velocity, we obtain:

$$-v_*\sqrt{\frac{r}{R}} = \kappa \frac{\left(\frac{dv}{dr}\right)^2}{\left(\frac{d^2v}{dr^2}\right)} \tag{10.46}$$

Taking the reciprocal of both sides, we have

$$-\frac{\frac{d^2v}{dr^2}}{\left(\frac{dv}{dr}\right)^2} = \frac{\kappa}{v_*}\sqrt{\frac{R}{r}} \tag{10.47}$$

which can be readily integrated. After integration we obtain:

$$\frac{1}{\frac{dv}{dr}} = \frac{2\kappa}{v_*}\sqrt{Rr} + K_1 \tag{10.48}$$

In order to determine the constant of integration K_1 we may prescribe as a boundary condition that the velocity gradient at the wall becomes infinite:

$$\left(\frac{dv}{dr}\right)_{r=R} = \infty \tag{10.48a}$$

This assumption is permitted since it is outside the validity interval of the solution.

Thus:

$$K_1 = -\frac{2\kappa R}{v_*} \tag{10.48b}$$

Substituting the constant K_1, the following expression is obtained:

$$\frac{1}{\left(\frac{dv}{dr}\right)} = -\frac{2\kappa R}{v_*}\left(1 - \sqrt{\frac{r}{R}}\right) \tag{10.49}$$

Taking the reciprocal of both sides and integrating yields

$$v = \frac{v_*}{\kappa}\left[\sqrt{\frac{r}{R}} + \ln\left(1 - \sqrt{\frac{r}{R}}\right)\right] + K_2 \tag{10.50}$$

Now we need one more boundary condition to determine the constant of integration K_2. It is a known experimental fact that the maximum of the velocity distribution is at the pipe axis. Therefore, at $r = 0$, $v = v_{max}$, thus $K_2 = v_{max}$.

Thus we obtain the following dimensionless velocity profile:

$$\frac{v_{max} - v}{v_*} = -\frac{1}{\kappa}\left[\sqrt{\frac{1}{R}} + \ln\left(1 - \sqrt{\frac{r}{R}}\right)\right] \tag{10.51}$$

This relation offers a good description of the velocity distribution, but does contain v_{max} as an unknown arbitrary additional term. It is obvious that the velocities are equal on both sides of the surface which forms the boundary between the turbulent core flow and the laminar sublayer adjacent to the wall. Thus the turbulent velocity profile must match that of the laminar sublayer at the position $r = R - \delta$, i.e.,

$$\frac{v_* \delta}{v} = \frac{v_{max}}{v_*} + \frac{1}{\kappa}\left[\sqrt{\frac{R - \delta}{R}} + \ln\left(1 - \sqrt{\frac{R - \delta}{R}}\right)\right] \tag{10.52}$$

A very important observation of Prandtl is that

$$\frac{v_* \delta}{v} = \alpha = \text{const.} \tag{10.52a}$$

Since $\delta/R \ll I$, the following approximation based on the binominal theorem is convenient:

$$\sqrt{1 - \frac{\delta}{R}} = 1 \tag{10.52b}$$

and

$$\ln\left(1 - \sqrt{\frac{\delta}{R}}\right) = \ln\left[1 - \left(1 - \frac{\delta}{2R}\right)\right] = \ln\frac{\delta}{2R} \tag{10.52c}$$

Substituting these expressions into the velocity profile which matches that of the laminar sublayer as given by Eq. (10.42), the following result is obtained:

$$\frac{v_{max}}{v_*} = \frac{1}{\kappa}\ln\frac{v_* 2R}{v} - \frac{1}{\kappa}(1 + \ln\alpha) + \alpha \tag{10.53}$$

The constant α and κ can be determined from experimental data. One possible method is to evaluate these constants from velocity profile measurements, another one, which has less uncertainty, is by measuring the pressure drop for different flow rates. These methods yield the following values:

$$\alpha = 12.087$$
$$\kappa = 0.407 \tag{10.53a}$$

The velocity distribution along a radius is:

$$\frac{v}{v_*} = \frac{1}{\kappa}\left[\sqrt{\frac{r}{R}} + \ln\left(1 - \sqrt{\frac{r}{R}}\right)\right] + \frac{1}{\kappa}\ln\left(\frac{Rev_*}{c}\right) - \frac{1}{\kappa}(1 + \ln\alpha) + \alpha \tag{10.54}$$

This dimensionless velocity profile satisfies the boundary conditions. From this expression, the influence of the Reynolds number (Re) on the radial velocity distribution is obvious. The cross-sectional average velocity may be obtained by equation

$$\frac{c}{v_*} = \frac{1}{R^2\pi}\int_0^R \frac{v}{v_*} 2\pi r \, dr \tag{10.55}$$

Substituting v/v_* from Eq. (10.54) into Eq. (10.55), using the values for α and κ, and integrating we have:

$$\frac{c}{v_*} = \frac{v_{max}}{v_*} - \frac{1}{\kappa}\frac{77}{60} = \frac{v_{max}}{v_*} - 3.153 \tag{10.56}$$

Finally, the averaged velocity is obtained as

$$\frac{c}{v_*} = 2.49 \cdot \ln\left(\frac{Rev_*}{c}\right) - 2.811 \tag{10.57}$$

Another dimensionless velocity profile is:

$$\frac{v}{v_{max}} = 1 + \frac{\sqrt{\frac{r}{R}} + \ln\left(1 + \sqrt{\frac{r}{R}}\right)}{1.42 + \ln Re\frac{v_*}{c}} \tag{10.58}$$

In Fig. 10.6 velocity distributions are demonstrated along the dimensionless pipe radius (r/R), taking the Reynolds number as a parameter. This shows that the velocity distribution becomes more uniform over the cross-section as the Reynolds number increases, while the averaged velocity tends to the hypothetical velocity distribution of a perfect fluid. As a consequence, during kinematic investigations, the inviscid, perfect fluid model can be used satisfactorily in the range of high Reynolds numbers. Naturally the perfect fluid model cannot be used for pressure-drop calculations because of the intensive turbulent dissipation.

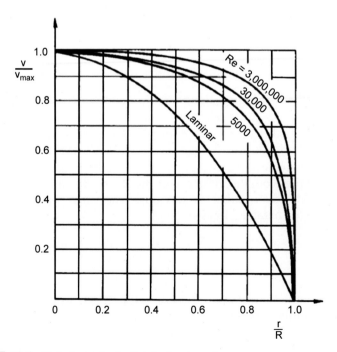

FIGURE 10.6 Turbulent velocity distribution in a pipe.

10.5 HEAD LOSS IN STRAIGHT CYLINDRICAL PIPES

The determination of head loss as a function of the flow rate is perhaps the most common problem of fluid mechanics. In this section we shall only consider the case of an incompressible Newtonian fluid. It is obvious that the flow through a pipe can be either laminar or turbulent. For a laminar flow the Hagen–Poiseuille equation gives the head loss as

$$h'_{1-2} = \frac{8vLc}{gR^2} \tag{10.59}$$

For practical applications this equation has to be somewhat modified. In engineering practice the diameter of a pipe is given rather than the radius. We also wish to obtain an equation for the head loss which depends explicitly on the Reynolds number. Expressing Eq. (10.59) in terms of the diameter D and rearranging the terms, we have

$$h'_{1-2} = \frac{64}{\frac{cD}{v}} \frac{L}{D} \frac{c^2}{2g} \tag{10.60}$$

which may be written as

$$h'_{1-2} = \lambda \frac{L}{D} \frac{c^2}{2g} \tag{10.61}$$

where λ is called the friction factor. It can be calculated from the equation

$$\lambda = \frac{64}{Re} \tag{10.62}$$

This expression is valid for the range of $Re < 2300$, i.e., laminar flow only. If the Reynolds number of the flow exceeds the critical value of 2300, the flow becomes turbulent, and the friction factor increases abruptly. Thus the smallest value of the friction factor is obtained for laminar flow immediately before the transition. For $Re = 2300$ the friction factor is

$$\lambda = \frac{64}{2300} = 0.0278 \tag{10.62a}$$

If the laminar–turbulent transition could be retarded the friction factor, and thus the head loss, would be much smaller. This is a frequently applied technique in the petroleum industry. Long-chain polymer additives can reduce the friction factor of a "solvent," thus retarding the laminar–turbulent transition. This phenomenon will be discussed in detail in Chapter 11.

Turbulent flow is a more complex phenomenon compared to laminar flow. Thus the determination of the head loss for such flows is a rather difficult problem. For turbulent flow in a pipe two cases can be distinguished.

If the laminar sublayer covers the surface roughness of the pipe wall the turbulent flow is not affected by the roughness, as the pipe can be considered to be absolutely smooth. This case is referred to as that of a turbulent flow in a hydraulically smooth pipe. On the other hand there are flows, where the laminar sublayer cannot cover the surface roughness of the pipe wall, or there is no laminar sublayer at the pipe wall. Such a flow is independent of the Reynolds number; all flow variables are functions of the surface roughness only. When a pipe wall exhibits this behavior, it is called hydraulically fully rough.

Consider first the turbulent flow in a hydraulically smooth pipe. The mechanical energy equation can be written as

$$h'_{1-2} = \frac{\beta_1 c_1^2 - \beta_2 c_2^2}{2g} + h_1 - h_2 + \frac{p_1 - p_2}{\rho g} \tag{10.63}$$

Since the cross-section of the pipe is constant and the fluid is incompressible

$$c_1 \equiv c_2 \tag{10.63a}$$

$$\beta_1 \equiv \beta_2 \tag{10.63b}$$

The head loss is obtained as

$$h'_{1-2} = h_1 - h_2 + \frac{p_1 - p_2}{\rho g} = JL \tag{10.64}$$

therefore it is the product of the hydraulic gradient J and the pipe length L. The hydraulic gradient can be expressed in terms of the friction velocity v_*, since by definition

$$J = \frac{2}{gR} v_*^2 \tag{10.65}$$

Substituting this into the equation for h' we get

$$h'_{1-2} = \frac{2L}{gR} v_*^2 \tag{10.66}$$

The friction velocity can be expressed in terms of the cross-sectional average velocity c:

$$\frac{c}{v_*} = \frac{1}{\kappa} \ln\left(Re \frac{v_*}{c} \right) - \frac{1}{\kappa} (2.283 + \ln\alpha) + \alpha \tag{10.67}$$

It is clear that v_*/c depends on the Reynolds number only, since α and κ are temporarily unknown constants. Therefore,

$$\frac{v_*}{c} = f(Re) \tag{10.68}$$

This relationship together with Eq. (10.66) leads to the equation for the head loss

$$h'_{1-2} = 8f^2 \frac{L}{D} \frac{c^2}{2g} \tag{10.69}$$

in which the group of coefficients $8f^2$ is designated by

$$\lambda = 8f^2(\text{Re}) \tag{10.70}$$

which is the friction factor for the turbulent flow.

We can write

$$\frac{v_*}{c} = \sqrt{\frac{\lambda}{8}} \tag{10.71}$$

This can be substituted into Eq. (10.67), thus resulting in an implicit equation for the friction factor:

$$\frac{1}{\sqrt{\frac{\lambda}{8}}} = \frac{1}{\kappa} \ln\left(\text{Re}\sqrt{\frac{\lambda}{8}}\right) - \frac{1}{\kappa}(2.283 + \ln\alpha) + \alpha \tag{10.72}$$

Assuming that α and κ are constants, we obtain the following values from friction factor measurements

$$\alpha = 12.087 \tag{10.72a}$$

$$\kappa = 0.407 \tag{10.72b}$$

Thus, using common logarithms, we can write

$$\frac{1}{\sqrt{\lambda}} = 2\lg(\text{Re}\sqrt{\lambda}) - 0.8 \tag{10.73}$$

Experimental results confirm this equation to be valid with a fair degree of accuracy. Nikuradse conducted experiments on the turbulent flow of water in smooth pipes for Reynolds numbers ranging from 4000 to 3,240,000. The data from his investigation are shown in Fig. 10.7, where the friction factor is plotted versus the Reynolds number. The curve representing Eq. (10.73) is plotted for comparison.

Since the equation for the friction factor is obtained in implicit form, λ can be calculated by iteration. The convergence of the iteration is rather fast, particularly when the starting value of λ is chosen using Fig. 10.7. In such a case two, or at the most three, iteration steps are sufficient.

The surface of the pipe walls is usually quite rough. Glass and PVC pipes or drawn steel pipes may be considered to be smooth, but standard pipes are generally rather rough, even when new. This boundary roughness is shown schematically in Fig. 10.8. The average height of the

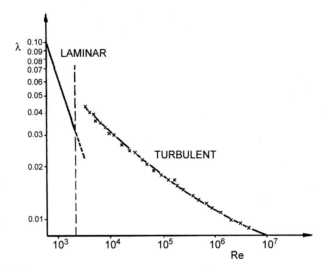

FIGURE 10.7   Friction factor data of Nikuradse for smooth pipes.

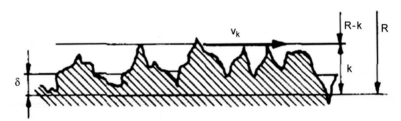

FIGURE 10.8   Pipe wall roughness and laminar sublayer.

roughness projections is expressed by the height k. The radius of the pipes is a fictitious length extending from the pipe axis to the average distance of the wall.

In hydraulically rough pipes there is no continuous boundary surface between the laminar sublayer and the turbulent core flow. For very large Reynolds numbers or very rough pipe walls there is no continuous laminar sublayer at all. The derived velocity distribution for the turbulent core flow is connected to the velocity of the laminar sublayer. It is obvious that this velocity profile equation cannot be valid for turbulent flow in rough pipes. Thus for rough pipes a new equation for the velocity distribution is required in order to derive the friction factor equation.

Examining the microgeometry of a commercial pipe wall (Fig. 10.8), it is clear that this random surface profile is too complex to be characterized by a single parameter. A full characterization of the roughness would require a complete description of its geometry, including the height,

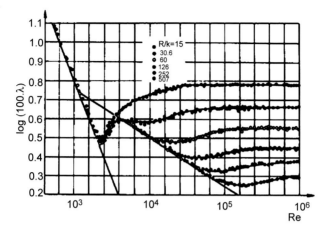

FIGURE 10.9   Friction factor data of Nikuradse for artificially roughened pipes.

length, width, and shape of all protrusions or indentations, together with their distribution. This would be a hopeless task, thus for experimental purposes an artificially created roughness is used. Nikuradse (1933) applied uniform sand grains to cover the pipe wall; in his experiments the parameter k represents the grain size of the sand. The roughness of the wall in this case may be characterized by a single roughness parameter, the relative roughness k_s/R. Nikuradse's results are shown in Fig. 10.9, where the friction factor is plotted against the Reynolds number. It is seen that for each value of R/k_s an individual friction factor curve is obtained. In the turbulent region of the flow each friction factor curve eventually tends to be horizontal as the Reynolds number increases.

This shows that in this region the friction factor is independent of the Reynolds number and is solely a function of the relative roughness. This region, where the friction factor curves are horizontal, is called the region of fully developed turbulent flow. Nikuradse obtained experimentally a friction factor equation for fully developed turbulent flow:

$$\frac{1}{\sqrt{\lambda}} = 2\lg\frac{R}{k_s} + 1.74 \tag{10.74}$$

In Nikuradse's experiments the sand roughness k, could be measured directly. The natural roughness of a commercial pipe can be determined indirectly as follows. If, for a given section of pipe, the flow rate Q and the pressure difference Δp are measured, the friction factor can be calculated as

$$\lambda = \frac{2D\Delta p}{L\rho c^2} \tag{10.75}$$

The value of Reynolds number can be obtained from its definition

$$Re = \frac{cD}{\nu} \qquad (10.75a)$$

Thus the point corresponding to the measured values of λ and Re can be determined in the friction factor chart. If it is in the fully developed turbulent region, we obtain

$$k = R \, 10^{-\frac{1}{2}\left(\frac{1}{\sqrt{\lambda}}-1.74\right)} \qquad (10.76)$$

Note, that k is the equivalent sand grain roughness and that the natural roughness of the pipe is expressed in terms of the sand grain roughness which would result from the same friction factor. The only way this can be done is to compare the behavior of a naturally rough pipe with an artificially roughened pipe. Moody has made these comparisons; his results are plotted in Fig. 10.10.

The friction factor and the relative roughness can also be related theoretically. Kármán derived such an equation for fully developed turbulent flow. Consider the rough pipe wall in Fig. 10.8. The thickness δ of the laminar sublayer is smaller than the average height of the roughness. Since the height k is much smaller than the radius of the pipe R, it seems to be an acceptable approximation to consider the turbulent shear stress near the wall to be constant within a layer of thickness k.

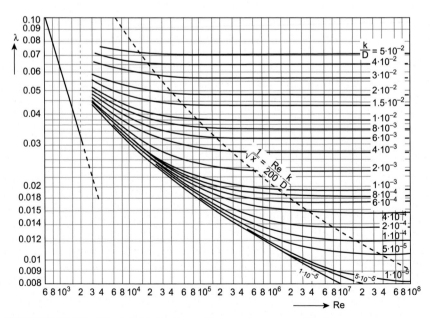

FIGURE 10.10   Moody's friction factor diagram for commercial pipes.

Thus it is assumed, that

$$\tau'_R = \tau'_k = f\left(\frac{v_k k}{\nu}\right)\rho v_k^2 \tag{10.77}$$

where v_k is the velocity at the edge of the constant shear stress layer. Introducing the friction velocity v_*, we can write

$$\frac{v_*^2}{v_k^2} = f\left(\frac{v_k}{v_*}\frac{v_* k}{\nu}\right) \tag{10.78}$$

It is clear that the v_k/v_* ratio depends only on the so-called roughness Reynolds Re_k number, i.e.

$$\frac{v_k}{v_*} = \phi\left(\frac{v_* k}{\nu}\right) \tag{10.79}$$

It is well known that the integration constant of the velocity profile of a turbulent flow in a pipe is an additional unknown. It can be determined using v_k/v_*. At the radius $r = R - k$ we have

$$\frac{v_k}{v_*} = \frac{v_{max}}{v_*} + \frac{1}{\kappa}\left[\sqrt{\frac{R-k}{R}} + \ln\left(1 - \sqrt{\frac{R-k}{R}}\right)\right] \tag{10.80}$$

Since $k \ll R$, expanding the expression in binomial form we get

$$\sqrt{\frac{R-k}{R}} \cong 1 \tag{10.80a}$$

and

$$\ln\left[1 - \sqrt{\frac{R-k}{R}}\right] \cong \ln\frac{k}{2R} \tag{10.80b}$$

Applying these formulas, we obtain

$$\phi = \frac{v_k}{v_*} = \frac{v_{max}}{v_*} + \frac{1}{\kappa}\left(1 + \ln\frac{k}{D}\right) \tag{10.81}$$

Alternatively, the cross-sectional average velocity c, given by

$$\frac{c}{v_*} = \frac{v_{max}}{v_*} - \frac{1}{\kappa}\left(\frac{25}{12} - \frac{4}{5}\right) \tag{10.82}$$

It can be used to eliminate v_{max}/v_*. Thus we get

$$\frac{c}{v_*} = \phi - \frac{1}{\kappa}\left(2.283 + \ln\frac{k}{D}\right) \tag{10.83}$$

From Eq. (10.71) we have

$$\frac{c}{v_*} = \sqrt{\frac{8}{\lambda}}$$

(10.83a)

so that the friction factor is obtained as

$$\sqrt{\frac{8}{\lambda}} = \frac{1}{\kappa}\ln\frac{D}{k} - \frac{2.283}{\kappa} + \phi$$

(10.84)

The coefficient κ has the same value for both smooth and rough pipes. The relationship between ϕ and the roughness Reynolds Re_k number can be determined experimentally. The result is shown in Fig. 10.11. In the interval $Re_k < 3$ a linear relation is evident. In the region $Re_k > 70$ the horizontal line shows that ϕ is independent of Re_k. In this region ϕ is constant, thus λ is a function of the relative roughness D/k only. In the interval $3 < Re_k < 70$ the curve of ϕ attains a maximum. This is the transition zone between the smooth and the fully developed turbulent behavior.

If $Re_k < 3$, as long as the roughness is covered by the laminar sublayer, it has no effect on the friction factor.

It is clear from Fig. 10.11 that the roughness makes itself felt only at values of $Re_k > 3$. The measured points start to plot away from the straight line for a smooth pipe. Up to a value of $Re_k = 70$, the friction factor depends on both the relative roughness and the Reynolds number.

For values of $Re_k > 70$ the laminar sublayer vanishes and the friction factor depends solely on the relative roughness. For this fully developed

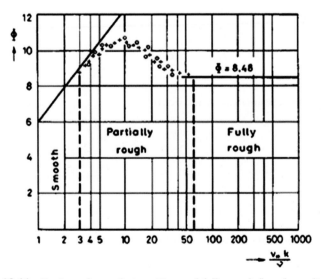

FIGURE 10.11   Regions of smooth, transition, and fully rough flow depending on Re_k.

turbulent flow, after substitution of κ and ϕ, the friction factor equation is obtained as

$$\frac{1}{\sqrt{\lambda}} = 2\lg\frac{3.715D}{k} \tag{10.85}$$

For the transition region, where the friction factor is a function of both the Reynolds number and the relative roughness, Colebrook's equation is applicable:

$$\frac{1}{\sqrt{\lambda}} = -2\lg\left(\frac{k}{3.715D} + \frac{2.51}{Re\sqrt{\lambda}}\right) \tag{10.86}$$

Colebrook's equation is an implicit expression, thus iteration is required to calculate the friction factor in the transition zone.

Finally a series of curves is presented (Fig. 10.12) to estimate the relative roughness of a pipe as a function of the diameter and the material of the pipe. These protrusions are representative of the materials indicated, but cannot be expected to be sufficiently accurate for actual calculations. It is recommended that whenever possible and economic, the results of an actual flow test through a certain length of the pipe in question be used to calculate the friction factor and the actual relative roughness. These values are sufficient for further engineering calculations.

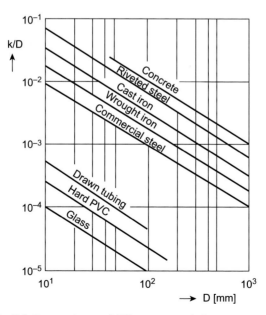

FIGURE 10.12 Relative roughness of different types of pipes.

10.6 FLOW PATTERNS IN HORIZONTAL STEAM–WATER MIXTURE FLOW

Transporting hot water at or near its boiling point through a pipeline it is suitable that the pressure will be higher along the pipe length than the saturated pressure at the actual temperature of the fluid. If the pressure were lower the flowing fluid would flash and characteristic two-phase flow patterns occur. Gas–liquid mixtures flowing in horizontal pipes tend to be somewhat more complex than vertical flows. If the density difference between the phases is pronounced, the flow is asymmetric: the more dense phase tends to accumulate at the bottom of the pipe (Govier and Aziz, 1972).

Consider a horizontal transparent pipe with a constant liquid flow rate, into which gas is introduced. If the superficial liquid velocity is sufficiently high, say 2–3 m/s, the introduced gas is present as small spherical bubbles, as shown in Fig. 10.13A. The finely dispersed bubbles have no symmetrical concentration profile; the maximum concentration occurs in the upper part of the pipe cross-section. Increasing the gas flow rate leads to the formation of larger bubbles, and the occurrence of bubble groups. The larger bubbles occupy the uppermost part of the pipe, while the smaller bubbles are dispersed asymmetrically, as shown in Fig. 10.13B. As the gas flow rate is increased further, still larger bubbles occur at the upper pipe wall. Here elongated bubbles are followed by smaller

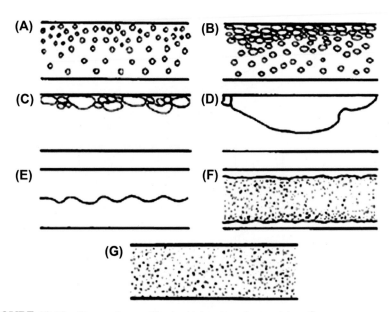

FIGURE 10.13   Flow patterns of horizontal water–steam mixture flow.

spherical bubbles, and they frequently coalesce (see Fig. 10.13C). With the further increase in the gas flow rate, very large elongated bubbles slide along the upper pipe wall and suffer distortion resulting in unstable shapes (see Fig. 10.13D). Each deformed elongated bubble is followed by a liquid plug which may contain trains of gas bubbles. A further increase in the gas flow rate leads to a separated stratified flow pattern. The horizontal phase interface may be smooth, wavy or become ripply as the gas flow rate increases (see Fig. 10.13E). As the gas flow rate is increased further, the waves on this interface become so large that the thinned liquid layer can no longer support them, the liquid spreads along the walls of the pipe, forming an annular film, with some liquid droplets dispersed in the gas core flow (see Fig. 10.13F). Finally, a mist flow pattern develops with the further increase in the gas flow rate, as shown in Fig. 10.13G. A flow-pattern map for the flow of a water−air mixture is shown in Fig. 10.14.

If the track of the pipeline is located at an uneven surface the pressure can be increased at certain points. The pressure rise produces the collapse of the steam phase. This could cause serious waterhammer, or even rupture of the pipe with disastrous consequences. To avoid such an event it must be maintained at an adequate pressure level to suppress flashing by pumping.

Shock waves occur when the velocity changes, as may result from operating a control valve at the outflowing end of a pipeline. Every movement of a valve, opening or closing, causes pressure waves, the amplitude of which depends on the rate of velocity change.

$$\Delta p = \rho a \Delta c, \tag{10.87}$$

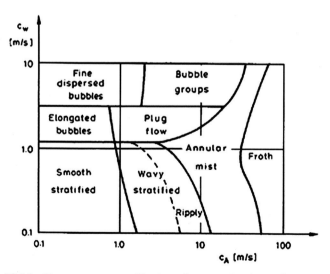

FIGURE 10.14   Flow pattern map of horizontal water−air mixture flow.

where a is the wave velocity in the fluid confined within the pipe, and Δc is the sudden change of the velocity. As it is known

$$a = \sqrt{\frac{B/\rho}{1 + \frac{D}{\delta} \cdot \frac{B}{E}}}$$ (10.88)

where B is the bulk elasticity modulus of the fluid, ρ is its density, D is the diameter of the pipe, δ is the wall thickness, and E is the elasticity modulus of the pipe material. A typical example: water flows in steel pipe $B = 2.10^5 \cdot 10^9 \, N/m^2$, $\rho = 998 \, kg/m^3$, $E = 2 \cdot 10^{11} \, N/m^2$, $D/\delta = 100$. In this case

$$a = \sqrt{\frac{\frac{2,105 \cdot 10^9}{998}}{1 + 100 \cdot \frac{2,105 \cdot 10^9}{2 \cdot 10^{11}}}} = 1012 \, m/s$$

If the flowing fluid has a velocity of 1.5 m/s and the flow is suddenly stopped the pressure rise

$$\Delta p = 998 \cdot 1012 \cdot 1.5 = 1514964 \, N/m^2 = 15.15 \, bar$$

It is a substantial pressure rise, and it can be recognized that in a long pipeline a sudden stopping of the moving water mass would cause a large rise dynamic effect. It is more dangerous when the sudden pressure decrease is caused by rapid acceleration with resultant formation of steam which subsequently collapses.

Valve movements must be carefully controlled, such that the time of the valve operation should be longer than the time of the refraction there and back of the waves.

Determination of the pressure loss for a homogeneous liquid flow is a conventional problem. The two-phase pressure loss in a horizontal pipe may be determined using the semiempirical method of Lockhart and Martinelli (1949). The method is based upon the assumption that the pressure loss of the liquid phase is equal to the pressure loss of the gas phase regardless of the actual flow pattern:

$$\Delta p'_L = \Delta p'_G$$ (10.89)

The gas phase is assumed to be incompressible. Thus the pressure losses are obtained as

$$\Delta p'_L = \lambda_L \frac{L}{4R_{HL}} \rho_L \frac{c_L^2}{2}$$ (10.90)

and

$$\Delta p'_G = \lambda_G \frac{L}{4R_{HG}} \rho_G \frac{c_G^2}{2}$$ (10.91)

where R_{HL} and R_{HG} are the hydraulic radii of the region of the pipe in which the liquid or the gas phase flows. The cross-sectional average velocities of the liquid and the gas are

$$c_L = \frac{Q_L}{\xi_L R_{HL}^2 \pi} \tag{10.92}$$

and

$$c_G = \frac{Q_G}{\xi_G R_{HG}^2 \pi} \tag{10.93}$$

where ξ_L and ξ_G are hydraulic radius correction factors.

The friction factors of the liquid and gas λ_L and λ_G are expressed as functions of the Reynolds number in the form of an approximate equation:

$$\lambda_L = \frac{B_L}{\left(\dfrac{Q_L}{\xi_L R_{HL} \pi v_L}\right)^m} \tag{10.94}$$

and

$$\lambda_G = \frac{B_G}{\left(\dfrac{Q_G}{\xi_G R_{HG} \pi v_G}\right)} \tag{10.95}$$

B_L and B_G can be evaluated experimentally.

Let us define two fictitious pressure losses for both the liquid and the gas, using the apparent velocities

$$c_{0L} = \frac{Q_L}{R^2 \pi}; \quad c_{0G} = \frac{Q_G}{R^2 \pi} \tag{10.96}$$

We can then write

$$\Delta p'_{0L} = \lambda_{0L} \frac{L}{2R} \rho_L \frac{c_{0L}^2}{2} \tag{10.97}$$

and

$$\Delta p'_{0G} = \lambda_{0G} \frac{L}{2R} \rho_G \frac{c_{0G}^2}{2} \tag{10.98}$$

It is obvious, that

$$\lambda_{0L} = \frac{B_L}{\left(\dfrac{Q_L}{R \pi v_L}\right)^m} \tag{10.99}$$

and

$$\lambda_{0G} = \frac{B_G}{\left(\dfrac{Q_G}{R \pi v_G}\right)^n} \tag{10.100}$$

Using these expressions the relationships between the actual and the fictitious pressure losses are obtained as

$$\Delta p'_L = \Delta p'_{0L} \xi_L^{m-2} \left(\frac{R}{R_{HL}} \right)^{5-m} \tag{10.101}$$

and

$$\Delta p'_G = \Delta p'_{0G} \xi_G^{n-2} \left(\frac{R}{R_{HG}} \right)^{5-n} \tag{10.102}$$

The ratio of the actual to the fictitious pressure loss is designated by

$$\phi_L^2 = \xi_L^{m-2} \left(\frac{R}{R_{HL}} \right)^{5-m} \tag{10.103}$$

and

$$\phi_G^2 = \xi_G^{n-2} \left(\frac{R}{R_{HG}} \right)^{5-n} \tag{10.104}$$

Lockhart and Martinelli experimentally determined the functions ϕ_L and ϕ_G. The results are plotted as a function of the dimensionless parameter

$$X = \sqrt{\frac{\Delta p'_{0L}}{\Delta p'_{0G}}} \tag{10.105}$$

Lockhart and Martinelli grouped their data into four separate groups as follows.

1. Both components flow laminar. In this case the Reynolds numbers obtained from the superficial velocities must be smaller than 1000.

$$Re_{0L} = \frac{c_{0L}D}{L} < 1000; \ Re_{0G} = \frac{c_{0G}D}{G} < 1000$$

2. The liquid flow is laminar, the gas flow is turbulent. In this case

$$Re_{0L} = \frac{c_{0L}D}{L} < 1000; \ Re_{0G} = \frac{c_{0G}D}{G} < 2000$$

3. The liquid flow is turbulent, the gas flow is laminar

$$Re_{0L} = \frac{c_{0L}D}{L} < 2000; \ Re_{0G} = \frac{c_{0G}D}{G} < 1000$$

4. Both phases flow turbulently, thus

$$Re_{0L} = \frac{c_{0L}D}{L} < 2000; \; Re_{0G} = \frac{c_{0G}D}{G} < 2000$$

These four flow regions are indicated by the subscripts 1, 2, 3, and 4 in Fig. 10.15.

The critical value of $Re_0 = 1000$ is chosen since the actual velocity is always greater than the apparent velocity, thus the actual Reynolds number is greater than Re_0.

For large, commercial-size pipes the error of the Lockhart–Martinelli method may be as much as 50%, particularly for stratified flow. This is because the basic assumption is not valid and the pressure loss depends on the actual flow pattern. Thus modified empirical relationships can be obtained for large pipes with diameters of up to 300 mm. The source of experimental data is a flow of gasoline–natural gas mixture in a 300-mm diameter pipeline.

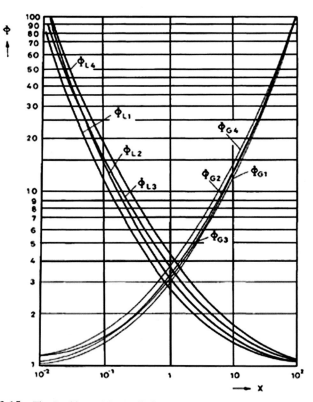

FIGURE 10.15 The Lockhart–Martinelli functions.

A comparison of the pressure losses calculated by the Lockhart–Martinelli method with experimental data indicates that the actual pressure drop is generally smaller than the calculated values. The difference varies with the flow pattern: it is least for dispersed bubble flow, and greatest for stratified flow. If both phases of a stratified flow are in turbulent motion, separate equations can be derived to modify the original Lockhart–Martinelli relationships for each flow pattern. These approximate equations, which are recommended for design purposes, are listed below.

For a dispersed bubble flow pattern

$$\phi_G = 4 - 12\lg X + 28(\lg X)^2 \qquad (10.106)$$

For elongated bubble flow

$$\phi_G = 4 - 15\lg X + 26(\lg X)^2 \qquad (10.107)$$

For a smooth surface stratified flow

$$\phi_G = 2 + 4.5\lg X + 3.6(\lg X)^2 \qquad (10.108)$$

For a stratified-wavy flow

$$\phi_G = 3 + 1.65\lg X + 0.45(\lg X)^2 \qquad (10.109)$$

For slug flow

$$\phi_G = 2.2 + 6.5\lg X \qquad (10.110)$$

Finally, for annular-mist flow the following equation is recommended:

$$\phi_G = 4 + 2.5\lg X + 0.5(\lg X)^2 \qquad (10.111)$$

The phenomenon of the holdup occurs in horizontal two-phase flows too. Fig. 10.16 shows the holdup ratio for a liquid–gas mixture flow in a horizontal pipe. It is noticeable that each curve has a local maximum and a minimum. The maximum corresponds to the ripply stratified flow, the minimum coincides with the occurrence of capillary waves.

10.7 PRESSURE LOSS OF A LOW-VELOCITY SUPERHEATED STEAM FLOW

The produced superheated steam is transported from the wellheads to the power plant through the gathering pipeline system. This system is generally rather complex because many wells produce the necessary amount of steam for the power plant. At first a single pipe section is studied as the fundamental element of the steam-gathering pipeline. The

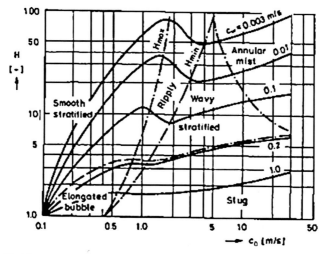

FIGURE 10.16   Holdup in horizontal two-phase flow.

temperature of the steam changes very slightly as it flows through an insulated pipe section. Thus it is an acceptable approximation to treat a low-velocity compressible flow as being isothermal.

Consider first a horizontal pipeline for which the change in potential energy is zero. Thus for an infinitesimal length of this pipe the mechanical energy equation can be written as

$$c\,dc + \frac{dp}{\rho} + gdh' = 0 \tag{10.112}$$

The infinitesimal change in kinetic energy caused by the expansion of the gas can be treated as being the effect of an additional friction factor $\bar{\lambda}$, i.e., we can write

$$c\,dc = \bar{\lambda}\frac{dL}{D}\frac{c^2}{2} \tag{10.113}$$

The infinitesimal friction loss is

$$g\,dh' = \lambda\frac{dL}{D}\frac{c^2}{2} \tag{10.114}$$

Using the above expressions the mechanical energy equation can be written as

$$\frac{dp}{\rho} + (\lambda + \bar{\lambda})\frac{dL}{D}\frac{c^2}{2} = 0 \tag{10.115}$$

For isothermal flow

$$\frac{p_0}{\rho_0} = \frac{p}{\rho} \tag{10.116}$$

Since the cross-section of the pipe is uniform the continuity equation becomes

$$\rho_0 c_0 = \rho c \tag{10.117}$$

Thus the kinetic energy at any cross-section can be expressed as

$$\frac{c^2}{2} = \frac{\rho_0 \rho_0}{\rho \rho} \frac{c_0^2}{2} \tag{10.118}$$

Substituting this into the mechanical energy equation we have

$$\frac{p_0}{\rho_0} \frac{dp}{p} + \frac{\lambda + \bar{\lambda}}{2D} \frac{\rho_0 \rho_0}{\rho \rho} c_0^2 dL = 0 \tag{10.119}$$

Multiplying this equation by $p\rho$, and using the relation

$$\frac{\rho}{\rho_0} = \frac{p}{p_0} \tag{10.120}$$

we get the differential equation

$$p \, dp = -(\lambda + \bar{\lambda}) \frac{p_0 \rho_0 c_0^2}{2D} dL \tag{10.121}$$

Before integrating this expression, consider the change in the friction factors λ and $\bar{\lambda}$ along the length of the pipe due to velocity and density changes. For a "fully developed" turbulent flow the friction factor does not depend on the Reynolds number, nor of course on the velocity. For smooth pipes λ depends on the Reynolds number only. Since

$$Re = \frac{cD}{\nu} = \frac{\rho c D}{\mu} \tag{10.121a}$$

and

$$\rho c = const, \tag{10.122}$$

the Reynolds number is constant along the pipe axis, and therefore, the friction factor is also constant:

$$\frac{dRe}{dL} = 0; \quad \frac{d\lambda}{dL} = 0 \tag{10.123}$$

The additional friction factor $\bar{\lambda}$ is not constant along the length of the pipe. From Eq. (10.114) it can be expressed as

$$\bar{\lambda} = \frac{2D}{c} \frac{dc}{dL} \tag{10.124}$$

For an isothermal flow

$$\frac{p}{\rho} = \text{const}; \quad \rho c = \text{const}$$

thus

$$\rho c = \text{const.} \tag{10.125}$$

This leads to the relation

$$\frac{dc}{c} = \frac{dp}{p} \tag{10.126}$$

thus the additional friction factor can be written as

$$\bar{\lambda} = -\frac{2D}{p}\frac{dp}{dL} \tag{10.127}$$

Substituting this into Eq. (10.121) and replacing p_0, ρ_0, c_0 by the inlet flow variables p_1, ρ_1, c_1 the following integral is obtained:

$$\int_{p1}^{p2} p \, dp = -\frac{p_1 c_1^2}{p_1}\ln\left(\frac{p_2}{p_1}\right)^2 = 1 - \frac{2\rho_1}{p_1}\lambda\frac{L}{D}\frac{c_1^2}{2} \tag{10.128}$$

This, after a little manipulation, leads to the following result

$$\left(\frac{p_2}{p_1}\right)^2 - \frac{p_1 c_1^2}{p_1}\ln\left(\frac{p_2}{p_1}\right)^2 = 1 - \frac{2\rho_1}{p_1}\lambda\frac{L}{D}\frac{c_1^2}{2} \tag{10.129}$$

This expression is implicit for p_2 or $p_2 - p_1$, but p_2 can be more readily determined using the Newton–Raphson method as described below.

Consider the error caused by neglecting the kinetic energy change due to expansion. The ratio of the two friction factors is

$$\frac{\bar{\lambda}}{\lambda} = \frac{\rho c^2}{p - \rho c^2} = \frac{c^2}{RT - c^2} \tag{10.130}$$

For superheated steam at 250°C and 50 m/s, it is 0.01045. Therefore, for a low-velocity compressible flow, e.g., superheated steam pipelines, the effect of expansion may be neglected. Thus Eq. (10.129) is obtained in a simpler form:

$$p_1^2 - p_2^2 = \lambda p_1 \rho_1 \frac{L}{D} c_1^2 \tag{10.131}$$

The pressure loss can be expressed as the function of the mass flow rate. It must be considered that the change of the thermal state of the superheated steam near to the upper boundary curve is different than the perfect gas. An acceptable approximation is obtained using a modified equation of state as:

$$\frac{P}{\rho} = ZRT \tag{10.132}$$

where Z is the compressibility factor, and R is the technical gas constant. In this case:

$$\dot{m} = c\rho\frac{D^2\pi}{4} \tag{10.133}$$

Substituting into Eq. (10.131) it is obtained that:

$$p_1^2 - p_2^2 = \lambda\left(\frac{4}{\pi}\right)^2 ZRT_1\frac{L}{D^5}\,\dot{m}^2 \tag{10.134}$$

This expression is more suitable than Eq. (10.131) for simulation of a complex gathering system.

For an inclined high-pressure pipeline or in a vertical well, gravity effects must be taken into account. In the mechanical energy equation, we need not use the acceleration term $c \cdot dc$. Instead, we can use the term c^2. So we can write:

$$\frac{p_1}{\rho_1}\frac{dp}{p} + gdz + \frac{dL}{D}\frac{c^2}{2} = 0 \tag{10.135}$$

Let α be the angle between the pipe axis and the horizontal direction, thus

$$dz = dL\sin\alpha \tag{10.136}$$

Substituting into the mechanical energy equation, and multiplying by $p_2/\rho_1/p_1$, and since $c^2p^2 = c_1^2p_1^2$, we obtain

$$pdp + g\frac{\rho_1}{p}p^2dL\sin\alpha + \lambda p_1\rho_1\frac{dL}{D}\frac{c_1^2}{2} = 0 \tag{10.137}$$

For an infinitesimal length of pipe we thus obtain the expression

$$dL = \frac{pdp}{g\frac{\rho_1}{p_1}p^2\sin\alpha + \lambda\frac{p_1\rho_1}{2D}c_1^2} \tag{10.138}$$

Integrating this expression we obtain

$$L = \frac{p_1}{2\rho_1 g\sin\alpha}\ln\left(\frac{g\sin\alpha + \frac{\lambda c_1^2}{2D}}{\left(\frac{p_2}{p_1}\right)^2 g\sin\alpha + \frac{\lambda c_1^2}{2D}}\right) \tag{10.139}$$

Rearranging, the pressure ratio (p_2/p_1) can be expressed as

$$\left(\frac{p_2}{p_1}\right)^2 = \left(1 - \frac{\lambda c_1^2}{2Dg\sin\alpha}\right)\exp\left(-\frac{2L\rho_1 g\sin\alpha}{p_1}\right) - \frac{\lambda c_1^2}{2Dg\sin\alpha} \qquad (10.140)$$

For modest temperature variations along the length of the pipe, it is convenient to perform the calculations assuming an average temperature. If the temperature variation is considerable, the pipeline should be divided along its length into a finite number of isothermal sections, each of which is then treated individually.

A complex gathering system consists of one or more large-diameter main pipelines to which smaller-diameter branch pipes connect coming from the wellheads. The mass flow rate is increasing in the direction of the flow as the main pipeline approaches the powerhouse. The main consideration in the design of the gathering system is to determine the pipe diameters. The reservoir, the wells, and the gathering pipeline form a synergetic flow system. The change of the performance of any elements influences the performance of the whole system. The wellhead pressure, the mass flow rate, the length of the branch line, and its elevation are generally different. The reservoir, the well, and the branch pipe are in serial connection; their pressure losses must be added. On the other hand all branch pipes are in parallel connection, thus their mass flow rates must be added. Because of this the diameter of the main gathering line is increased in the direction of the flow. There is an allowable velocity maximum based on the experience of everyday industrial practice. This velocity limit is necessary since the water droplets and fine solid particles carried by the high-velocity steam flow cause erosion of elbows, valve seats, or other exposed parts. In large-diameter pipes this velocity limit can even be 50 m/s, in the branch pipes it is at most 20 m/s.

10.8 HEAT TRANSFER OF HOT WATER TRANSPORTING PIPELINES

The energy content of the produced geofluid decreases as it flows from the well to the site of utilization. It is necessary for lagging of the pipes with an insulating material in order to minimize heat loss. The steel pipe is lagged with the isolating material coaxially, which is surrounded by a thin, mainly alumina, protecting cover. The isolated hot water transporting pipe is surrounded by the ambient air.

Consider now an infinitesimal length dl of the pipe. Neglecting the head loss, the interval energy balance can be written to the control volume sketched in Fig. 10.17. It is assumed that the flowing fluid is incompressible, its density ρ, specific heat capacity c, viscosity μ, and heat conductivity k is constant in the actual temperature interval. The flow is steady, turbulent, and one-dimensional. The temperature field is also

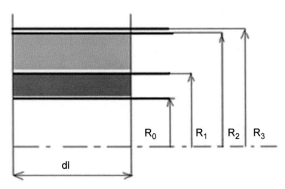

FIGURE 10.17 Control surface of an isolated pipe.

steady p. The conductive heat flux in the direction of the flow can be neglected. In this case

$$-\dot{m}cdT = 2R_o\pi U_o(T - T_a)dl \qquad (10.141)$$

where $\dot{m}$ is the mass flow rate, R_o is the internal radius of the pipe, U_o is the overall heat transfer coefficient referring to R_o, and T_a is the temperature of the ambient air. It can be recognized that the internal energy decrease of the flowing fluid equals the radially outward heat flux across the pipe and the insulation. The overall heat transfer coefficient can be calculated as

$$\frac{1}{U_o} = \frac{1}{h_o} + \frac{R_o}{k_s}\ln\frac{R_1}{R_o} + \frac{R_o}{k_i}\ln\frac{R_2}{R_1} + \frac{R_o}{k_a}\ln\frac{R_3}{R_2} + \frac{R_o}{R_3 h_3}, \qquad (10.142)$$

where h_o is the heat transfer coefficient of the forced convection between the fluid and the pipe wall, k_s, k_i, and k_a are the heat conductivity of the steel, the insulation and the alumina, h_3 is the heat transfer coefficient of the free convection between the outer surface of the cover and the ambient air.

The heat transfer coefficient h_o can be determined as

$$h_o = \frac{Nu_1 k}{2R_o} \qquad (10.143)$$

where the Nusselt number is obtained as

$$Nu_1 = 0,015 \cdot Re^{0,83}Pr^{0,42}. \qquad (10.144)$$

The Reynolds number is

$$Re = \frac{v \cdot \rho 2R_o}{\mu}, \qquad (10.145)$$

in which v is the cross-sectional average velocity, and μ is the dynamical viscosity. The Prandtl number is

$$\Pr = \frac{c\mu}{k}. \tag{10.146}$$

The protecting cover thermal resistance is very small may be neglected. The heat transfer coefficient of the free convection is obtained as

$$h_3 = \frac{Nu_3 \cdot k_{air}}{2R_3} \tag{10.147}$$

where the Nu_3 Nusselt number can be calculated as

$$Nu_3 = 0,52(Gr \cdot Pr)^{0,25} \tag{10.148}$$

Now the Grashof number is obtained as

$$Gr = \frac{\alpha \cdot g \cdot (2R_3)^3 (T_3 - T_{air})}{\upsilon^2}, \tag{10.149}$$

in which α is the thermal expansion of the air $\left(\frac{1}{273}\right)$ v is the kinematical viscosity of the air. The $T_3 - T_{air}$ temperature difference can be estimated by a trial-and-error procedure. Since the temperature of the ambient air is constant along the pipe, its derivative

$$\frac{dT_{air}}{dl} = 0, \tag{10.150}$$

which may be subtracted from the left-hand side of Eq. (10.1)

$$-\dot{m}c\frac{d}{dl}(T - T_{air}) = 2\pi R_o U_o (T - T_{air}) \tag{10.151}$$

The inflowing hot water temperature at the entrance of the pipe is T_1. The temperature difference $T_1 - T_{air}$ is obviously constant, thus both sides of the equation can be divided by $T_1 - T_{air}$. Introducing the dimensionless temperature

$$\Theta = \frac{T - T_{air}}{T_1 - T_{air}} \tag{10.152}$$

the differential equation becomes clearly arranged for the integration

$$\frac{d\Theta}{\Theta} = -\frac{2R_o\pi U_o}{\dot{m}c}dl \tag{10.153}$$

After integration it is obtained that:

$$\ln \Theta = -\frac{2\pi R_o U_o l}{\dot{m}c} + C \qquad (10.154)$$

The boundary condition to determine the constant of integration is
$L = 0$; $T = T_1$, that is $\Theta = 1$.

Substituting into Eq. (10.154) we get that $C = 0$. Turning back to the
actual temperatures it is obtained that

$$\ln \frac{T - T_{air}}{T_1 - T_{air}} = -\frac{2\pi R_o U_o l}{\dot{m}c} \qquad (10.155)$$

After a little manipulation we get the formula obtaining the temperature of the hot water at distance l from the pipe inlet:

$$T = T_{air} + (T_1 - T_{air}) \cdot e^{-\frac{2R_o \pi U_o l}{\dot{m}c}} \qquad (10.156)$$

The arriving hot water temperature at the end of an insulated pipe of
length L and diameter D is

$$T_2 = T_{air} + (T_1 - T_{air})e^{-\frac{D\pi L U_o}{\dot{m}c}} \qquad (10.157)$$

The thermal power loss during the transportation is

$$\dot{P} = \dot{m}c(T_1 - T_{air})\left(1 - e^{-\frac{D\pi L U_o}{\dot{m}c}}\right) \qquad (10.158)$$

It is obvious, that the thermal properties of the water; ρ, μ, k slightly
vary as the temperature decreases. Since the value of U_o depends primarily on the thermal resistance of the insulation: there is no substantial
error neglecting the temperature dependence of ρ, μ, and k.

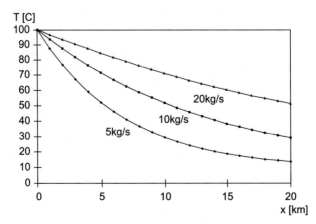

FIGURE 10.18   Effect of mass flow rate on temperature distribution.

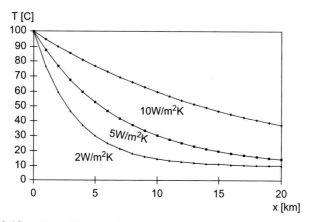

FIGURE 10.19   Effect of the overall heat transfer coefficient on temperature distribution.

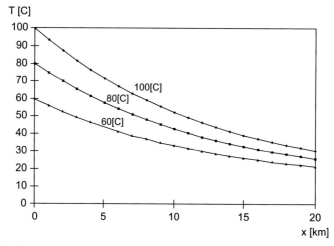

FIGURE 10.20   Effect of the initial temperature on temperature distribution.

The obtained results give a clear answer for the distance of transportation of geothermal energy. In Figs. 10.18–10.20 the effect of the parameters $\dot{m}$, U_o, and T_1 on the temperature distribution along the pipe length are demonstrated.

Another usual configuration is that the pipeline is buried by soil. In this case the overall heat transfer coefficient can be calculated regarding different conditions. Considering Fig. 10.21, it can be recognized that the soil cover is asymmetric around the pipe. The heat conduction through this asymmetric soil mass can be determined analytically by the method of thermal singularities. The heat flow patterns are two-dimensional in parallel plans which are perpendicular to the symmetry axis of the pipe.

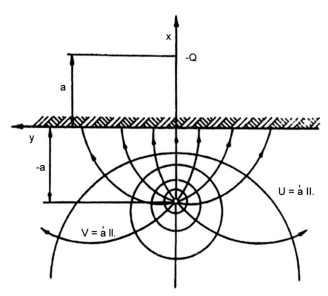

FIGURE 10.21 Heat conduction model of a buried pipe by method of thermal singularities.

The mathematical model is analogous to that used to describe two-dimensional potential flows.

The equation of heat conduction for a steady two-dimensional case can be written in the simple form:

$$\frac{\partial q_x}{\partial x} + \frac{\partial q_y}{\partial y} = 0 \tag{10.159}$$

where q_x and q_y are the orthogonal components of the heat flux vector. On the other hand, Fourier's law of heat conduction is obtained as

$$q_x = -k\frac{\partial T}{\partial x}; \; q_y = -k\frac{\partial T}{\partial y} \tag{10.160}$$

Assuming that the heat conductivity is constant, an obvious consequence of Eq. (10.159) and Eq. (10.160) is that

$$\frac{\partial q_y}{\partial x} - \frac{\partial q_x}{\partial y} = 0 \tag{10.161}$$

If these equations are satisfied in every points of the xy plane, it must require the existence of the following expressions

$$q_x = \frac{\partial U}{\partial x} \text{ and } q_y = \frac{\partial U}{\partial y} \tag{10.162}$$

$$q_x = \frac{\partial V}{\partial y} \text{ and } q_y - \frac{\partial V}{\partial x} \qquad (10.162a)$$

in which U and V are harmonic functions. These relationships conduce to the Cauchy–Riemann equations

$$\frac{\partial U}{\partial x} = \frac{\partial V}{\partial y} \qquad (10.163)$$

$$\frac{\partial U}{\partial y} = -\frac{\partial V}{\partial x} \qquad (10.163a)$$

Any analytical complex variable function is solution of the Cauchy–Riemann equations, in which U is the real V which is the imaginary part of this function:

$$W(z) = U(x, y) + iV(x, y) \qquad (10.164)$$

The lines along which

$$U = -kT + \text{const.} \qquad (10.164a)$$

are called thermal potential lines.

Actually, these are the isotherms of the temperature field. The lines along which V is constant are called the heat streamlines. The isotherms and the heat streamlines form an orthogonal net. Since both U and V are harmonic, namely potential functions, W is called the thermal complex potential function of the heat flux vector field. The singular points of the analytic function W are the thermal singularities. In the singular points the basic Eqs. (10.159) and (10.161) are not satisfied. These singularities modify the homogeneous heat flux field, inducing additional heat flux vector components. The mathematical model of the thermal singularities is perfectly analogous to the formulas referring to a two-dimensional potential flow. Knowing the complex potential W, all important variables of the heat flux field can be calculated.

The conjugate complex heat flux vector is obtained by a simple derivation as:

$$\overline{q} = \frac{dW}{dz} = q_x - iq_y \qquad (10.165)$$

The equation of the isotherms is the equipotential lines of the real part of W, while the equation of the heat streamlines is the equipotential lines of the imaginary part of W.

Based on the principle of superposition complicated heat flow patterns can be constructed by combining elementary complex potentials, for example sources, sinks, or doublets.

When the heat flux vector field has straight or circular boundaries, the method of images can be applied to obtain the complex potential of the system. This method consists of the finding of the images of singularities beyond the boundary of the system and the superposition of the heat flux generated these images on the original heat flow pattern.

The complex potential of the two-dimensional asymmetric heat flux vector field around the buried pipe can be constructed from a heat source which represents the pipe, and an image, a heat sink beyond the soil surface "suspending" over the straight line as a reflection of mirror. This apparent singularity represents the soil surface, which is an isothermal line. In order to ensure that this condition be satisfied, the heat flux field due to the mirror image is superimposed on that due to the heat source itself. Since the Cauchy–Riemann equations are linear, the combined complex potential satisfies again the Cauchy–Riemann equations. The boundary condition referring the soil surface is now automatically satisfied.

The complex potential of this combined system is the sum of complex potentials of the source and the sink:

$$W = \frac{Q}{2\pi} \ln(z+a) - \frac{Q}{2\pi} \ln(z-a) \tag{10.166}$$

This can be written in a brief form as:

$$W = \ln \frac{z+a}{z-a} \tag{10.167}$$

Let denote $z + a = r_2 e^{i\varphi_2}$ and $z - a = rr_1 e^{i\varphi_1}$. Substituting into Eq. (10.167) we get:

$$W = \frac{Q}{2\pi} \ln \frac{r_2 e^{i\varphi_2}}{r_1 e^{i\varphi_1}} \tag{10.168}$$

Its real and imaginary parts can be separated easily

$$U = \frac{Q}{2\pi} \ln \frac{r_2}{r_1} \tag{10.169}$$

$$V = \frac{Q}{2\pi} (\varphi_2 - \varphi_1) \tag{10.170}$$

The equation of the isotherms is obtained as:

$$\frac{r_2}{r_1} = e^{\frac{2\pi n}{Q}} \tag{10.171}$$

After some manipulation we get:

$$\left(\frac{r_2}{r_1}\right)^2 = \frac{(x+a)^2 + y^2}{(x-a)^2 + y^2} = C \tag{10.172}$$

The $C = 1$ value belongs to the coordinate axis, which is the soil surface. One of the circle-shaped isotherms coincides with the contour of the pipe.

The conjugate of the heat flux vector is obtained as:

$$\overline{q} = \frac{Q}{2\pi}\left(\frac{1}{z+a} - \frac{1}{z-a}\right) \tag{10.173}$$

The equation of the heat streamlines is based on the expression of

$$\varphi_2 - \varphi_1 = \frac{2\pi n}{Q} \tag{10.174}$$

Regarding that

$$\text{tg}\varphi_2 = \frac{y}{y-a} \quad \text{and} \quad \text{tg}\varphi_1 = \frac{y}{x+a} \tag{10.174a}$$

finally the equation of the heat streamlines is obtained as:

$$x^2 + \left(y - \frac{a}{C}\right)^2 = a^2\left(1 + \frac{1}{C^2}\right) \tag{10.175}$$

Regarding that

$$h = \frac{1+C}{1-C}a \tag{10.176}$$

and

$$R^2 = \frac{4Ca^2}{(1-C)^2} \tag{10.177}$$

the parameter a can be eliminated. Thus we get:

$$\frac{h^2}{(1+C)^2} = \frac{R^2}{4C} \tag{10.178}$$

The canonical form of Eq. (10.178) is

$$C^2 + \left(2 - \frac{4h^2}{R^2}\right)C + 1 = 0 \tag{10.179}$$

Its solution is

$$C_{12} = 2\left(\frac{h}{R}\right)^2 - 1 - 2\frac{h}{R}\sqrt{\left(\frac{h}{R}\right)^2 - 1} \qquad (10.180)$$

Obviously, the relevant isotherm equation at the soil surface is

$$-kT + \text{const} = \frac{Q}{2\pi}\ln C_0 = 0$$

The isotherm coinciding to the pipe contour is obtained as

$$-kT_{R3} + \text{const} = \frac{Q}{4\pi}\ln\left[2\left(\frac{h}{R}\right)^2 - 1 - 2\frac{h}{R}\ln\sqrt{\left(\frac{h}{R}\right)^2 - 1}\right] \qquad (10.181)$$

Thus the overall heat flux between the pipe and the soil surface is

$$Q = -2\pi ks\frac{T_3 - T_s}{\ln\left[2\left(\frac{h}{R}\right)^2 - 1 - 2\frac{h}{R}\ln\sqrt{\left(\frac{h}{R}\right)^2 - 1}\right]^{0.5}} \qquad (10.182)$$

Based on this equation a form-parameter can be introduced, denoted by

$$A = \left[2\left(\frac{h}{R}\right)^2 - 1 - 2\frac{h}{R}\ln\sqrt{\left(\frac{h}{R}\right)^2 - 1}\right]^{0.5} \qquad (10.183)$$

The overall heat transfer coefficient for the buried gathering pipe is obtained as

$$\frac{1}{U_0} = \frac{1}{h_0} + \frac{R_0}{k_s}\ln\frac{R_1}{R_0} + \frac{R_0}{k_i}\ln\frac{R_2}{R_1} + \frac{R_0}{k_p}\ln\frac{R_3}{R_2} - \frac{R_0}{k_s}\ln A \qquad (10.184)$$

The following steps of the calculation are the same as the preceding case.

References

Bobok, E., 1993. Fluid Mechanics for Petroleum Engineers. Elsevier, Amsterdam.
Govier, G.W., Aziz, K., 1972. The Flow of Complex Mixtures in Pipes. Van Nostrand Reinhold, New York.
Lockhart, R.W., Martinelli, R.C., 1949. Proposed correlation of data for isothermal, two-phase, two-component flow in pipes. Chem. Eng. Prog. 45, 39.
Nikuradse, J., 1933. Strömungsgesedze in rauhen Rohren. VDI Forschungsh. 361.
Prandtl, L., 1956. Führer durch dieStrömungslehre, Braunschweig.

11

Geothermal Power Generation

11.1 CHANGE OF STATE OF WET STEAM

The law of conservation and conversion of energy is the fundamental general law of nature. This law states that energy does not vanish nor appear anew, it only passes from one form of energy into another through various physical and chemical processes. Geothermal energy is produced as the enthalpy content of the recovered reservoir fluids. The conversion of heat into work is realized in steam power cycles. The distinguishing feature of steam power cycles is that the state of aggregation of the working medium used in the cycle changes from liquid to a two-phase mixture (wet steam), then to superheated steam. The state of superheated steam is usually so close to the saturation region that the laws for an ideal gas are not applicable. In general, it is not possible to express the

Copyright © 2017 Elsevier Inc. All rights reserved.

equation of state in a simple analytic form. It can be tabulated or plotted against the variables of state. It is obvious that an equation such as:

$$p = p(V, T)$$

can be represented by a surface in the coordinate system p,V,T. This surface of state consists of piecewise continuous surface parts as shown in Fig. 11.1. It is customary to plot this relation as projected onto any ones of the three planes, p-V, p-T, and V-T. In such a projection, the third variable is treated as a parameter. The shape of the state surface is characteristic of a particular material.

Other thermal state variables, such as enthalpy (i) or entropy (s), are also single-valued functions of the saturation temperature or pressure along the boundary curves. The experimentally determined values of i′ and s′ along the lower boundary curve are tabulated depending on the saturation pressure. Similarly, i″ and s″ values obtained along the upper boundary curve are embedded this so-called steam table. It can be found in the Appendix. Steam tables are used to calculate thermodynamic processes of steam–water mixtures, while a temperature–entropy (T–s) diagram helps with understanding by visualization of them. Consider certain properties of the T–s diagram in the following.

The so-called T–s diagram is shown in Fig. 11.2. Only the regions of steam and liquid states of water that are of great interest in engineering are plotted in the diagram. Consider the main properties of this diagram. The entropy is plotted in the abscissa, and the temperature is plotted in the ordinate. The characteristic bell-shaped piece of the thermal state surface occurs in the center of this diagram. The contour of this domain is the boundary curve. At the top of the curve, the critical point can be

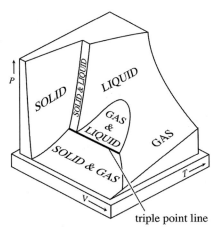

FIGURE 11.1   Thermal state surface. *http://www.met.reading.ac.uk/pplato2/h-flap/phys7_3.html.*

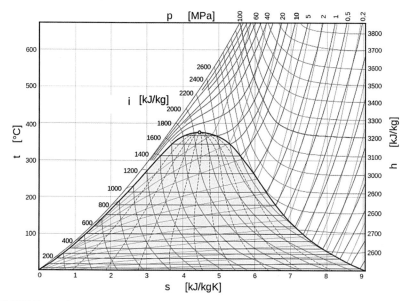

FIGURE 11.2 Temperature–entropy (T–s) diagram of water. *https://commons.wikimedia. org/wiki/File:T-s_diagram.svg.*

found. The critical parameters of water are $p_{cr} = 217.7$ bar, $T_{cr} = 374°C$, $V_{cr} = 0.00318 \text{ m}^3/\text{kg}$, $s_{cr} = 4.4296 \text{ kJ/kg °C}$, and $i_{cr} = 2099.7 \text{ kJ/kg}$.

The left-hand branch of the contour line is the so-called lower boundary curve, while the right-hand branch is the upper boundary curve. The lower boundary curve separates the regions of the liquid water and the two-phase water-steam mixture. The latter is called wet steam. Along the lower boundary curve, the saturated liquid state can be found. The upper boundary curve separates the two-phase region from the dry, superheated steam region.

The values of the thermodynamic variables along the boundary curves are functions of only one variable. For example, the value of the specific volume of a boiling liquid v' is uniquely determined by the value of saturation pressure or temperature. Similarly, the specific volume of the saturated steam v'' is determined by the saturation temperature or pressure.

It can be recognized that the difference between the specific volumes of the coexisting phases $v'' - v'$ decreases as the temperature increases. It is clear that this difference is zero in the critical point. In order to determine the state of the wet steam unambiguously, the ratio of the saturated steam mass to the mass of the mixture must be known. The so-called dryness fraction is defined as:

$$x = \frac{m_s}{m_e + m_s} \tag{11.1}$$

Then, the content of liquid in the mixture is:

$$1 - x = \frac{m_e}{m_e + m_s} \tag{11.2}$$

The quantity $1 - x$ is called the degree of wetness of the two-phase mixture.

The value $x = 1$ corresponds to the dry saturated steam at the upper boundary curve, and the value $x = 0$ to the saturated liquid water at the lower boundary curve.

The shape of the isobars in the two-phase region is horizontal lines coincident with the isotherms. The supercritical isobars are increasing as entropy increases, having an inflexion. The isobar passes through the critical point has a horizontal tangent at this point. The area below the subcritical isobars equal to $T(s'' - s')$ represents the latent heat of vaporization at a certain pressure.

The lower and upper boundary curves are almost symmetrical.

Thermodynamic variables of wet steam can be calculated easily, because of volume, density, enthalpy, entropy, and internal energy are extensive thus additive quantities. Therefore the following equations are valid in the two-phase domain:

$$v = (1 - x)v' + xv'' \tag{11.3}$$

$$\rho = (1 - x)\rho' + x\rho'' \tag{11.4}$$

$$i = (1 - x)i' + xi'' \tag{11.5}$$

$$s = (1 - x)s' + xs'' \tag{11.6}$$

$$u = (1 - x)u' + xu'' \tag{11.7}$$

The entropy is assumed to be zero at the triple point of water, namely at $p = 611.7 \, N/m^2$, and $T = 0.01°C$. This is an arbitrarily chosen reference point.

Isobars and isotherms are *horizontal lines* across the wet steam region. *Vertical straight lines* represent the isentropic processes, while isenthalpic change of state occurs as a hyperbolic curve on the diagram. It is significant that the enthalpy maximum of the wet steam domain doesn't coincide with the highest temperature point. It is more apparent in the so-called i–s (enthalpy–entropy) diagram, which is shown in Fig. 11.3. Consider the main properties of this diagram.

Interestingly, the critical point is located far to the left of the boundary curve's maximum point of enthalpy. The slope of the isobars is always positive, as the diagram shows. The isobars do not have an inflexion in the supercritical region. The higher the saturation pressure and the temperature, the steeper the isobar in the two-phase regions. The isotherms

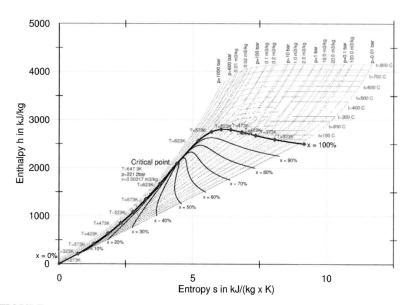

FIGURE 11.3 Enthalpy–entropy diagram. *https://en.wikipedia.org/wiki/Enthalpy–entropy_chart#/media/File:HS-Wasserdampf_engl.png.*

coincide with isobars in the wet steam domain, but their slope has a discontinuity crossing the boundary curve. Moving away from the upper boundary curve, the isotherms approach the horizontal line asymptotically, as the superheated steam tends to behave like a perfect gas.

11.2 THE CLAUSIUS–RANKINE CYCLE

A power cycle consists of a series of repeating thermodynamic processes along a closed process path, while heat is converted into mechanical work. The most widespread working medium is water. The power cycle involves the water's change of phase from a liquid state into superheated steam. The expanding steam performs work on its surroundings, then it returns to its initial state, changing its phase into liquid. Such a cycle was invented by Clausius and Rankine, and is usually called the Clausius–Rankine cycle (in the English-speaking world, Rankine cycle only). A schematic diagram of the power plant in which the Clausius–Rankine cycle is realized is shown in Fig. 11.4.

The boiler feed pump (1) does work on the water increasing its pressure. The high-pressure water flows into the steam boiler (2). The liquid water is warmed up to its saturation temperature, then vaporization

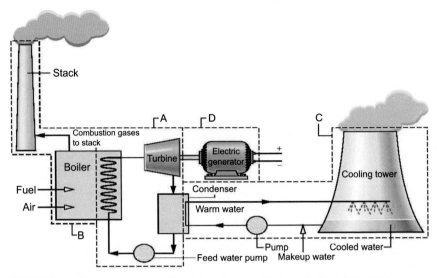

FIGURE 11.4　Schematic diagram of the Clausius–Rankine cycle (Bobok, 2014).

begins. As heat is added to the wet steam, its dryness fraction rises to the value of $x = 1$. The saturated steam becomes superheated steam, heated further in special equipment of the boiler; the so-called steam superheater (3). The temperature of the superheated steam increases near to 565°C, which is the creep limit of stainless steel. The superheated steam flows across nozzles to the impeller of the turbine (4), where it expands, and its high kinetic energy is converted into mechanical energy of rotation, driving the electric generator (5). At the turbine's exit, pressure and temperature of the steam substantially decrease. The steam then flows to the condenser (6), which is a heat exchanger to remove remaining heat from the steam. The steam condenses in the condenser, and its dryness fraction tends to zero. Thus, the condensed water flows to the intake of the boiler feed pump and the cycle will be repeated. The T–s diagram is especially suitable to represent the above mentioned power cycle as it is shown in Fig. 11.5.

The boiler feed pump increases the pressure of the water, while its temperature also increases adiabatically. The entropy increase through the pump is neglected (1-2). As the water is heated in the boiler, its temperature increases at a constant pressure. The pressure loss of the flowing water is also neglected. The temperature rises until attains the saturation temperature belonging the given pressure (2-3). As the water begins to boil, pressure and temperature will be constant, while its dryness fraction rises until the upper boundary curve (3-4). Attaining the upper boundary curve, the temperature of the saturated steam increases again as the steam

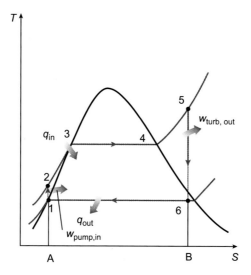

FIGURE 11.5 Clausius–Rankine cycle in a T–s diagram (Bobok, 2014).

becomes superheated (4-5). The superheated steam expands through flowing the turbine (5-6), its pressure and temperature decreases substantially about to 30 °C and 0.02 bar. Thus, the expanded superheated steam arrives the wet steam region while its dryness fraction is close to 1.0 yet. Along the isobar (6-1), the heat content of the wet steam is rejected and finally the condensed liquid flows to the suction side of the boiler feed pump and the cycle begins again.

The Clausius–Rankine cycle with superheating is the basic cycle of thermopower plants. For geothermal power plants, many details of the actual power cycle may be different, but the essential nature remains the same.

The amount of heat added to the working fluid is represented on the T–s diagram as the area under the actual section of the curve. The added heat during the Clausius–Rankine cycle is represented by the area A-1-2-3-4-5-B-A. The rejected heat in the cycle is equivalent to the area A-1-6-B-A. The work done during the cycle is obviously the difference of the added and the rejected heat, corresponding to the area of 1-2-3-4-5-6-1. Since both the heat addition and the rejection happens at constant pressure, the added and rejected heats are equal to the enthalpy differences between the beginning and the end point is the process:

$$Q_a = i_5 - i_2 \tag{11.8}$$

and

$$Q_r = i_6 - i_1 \tag{11.9}$$

The thermal efficiency of the cycle is obviously:

$$\eta = \frac{Q_a - Q_r}{Q_a} \tag{11.10}$$

or expressing by the enthalpies is:

$$\eta = \frac{(i_5 - i_2) - (i_6 - i_1)}{i_5 - i_2} \tag{11.11}$$

This expression can be written as:

$$\eta = \frac{(i_5 - i_6) - (i_2 - i_1)}{i_5 - i_2} \tag{11.12}$$

It means that $i_5 - i_6$ is the enthalpy decrease converted into the kinetic energy of the flowing steam and then into mechanical work in the turbine. Accordingly the enthalpy difference $i_2 - i_1$ equals to the mechanical work of the boiler feed pump.

It must be noted that the unique shape of the boundary curve is different for any material. Especially for hydrocarbons or other complex molecule fluids, the upper boundary curve has a positive slope. These fluids are used as secondary working fluids in binary power cycles.

The secondary working fluids have a low boiling point and high vapor pressure at low temperature relative to steam. The use of an organic fluid with the low temperature Rankine cycle has many advantages over using water. There is no substantial difference of the cycle efficiency between organic working fluids and water. The main advantage of an organic fluid is that it is able to extract more heat from the geothermal heat source than water. This results from so-called pinch point heat exchanger limitations, which are primarily a consequence of the organic fluid having a far lower ratio of latent heat of vaporization (at these lower boiling temperatures) versus specific heat capacity than water. The consequence is that the overall efficiency higher for the organic fluid even though the cycle efficiency is about the same for both fluids. The overall efficiency as it is known is the cycle efficiency times the ratio of thermal power extracted to thermal power available from the heat source. The available thermal power is obtained using an arbitrary minimum temperature 10°C higher than the lower temperature of the cycle. There are some further advantages of the use of organic fluids.

The expansion process takes place in the superheated region outside the upper boundary curve. Thus, in the absence of liquid droplets the blade erosion can be avoided.

The enthalpy drop is small, and it is possible to apply single stage turbines having better efficiency. For water, the enthalpy decrease is too high for expansion of a single stage. This requires a multiphase, complicated, and expensive turbine.

For a given power, the mass flow rate of organic fluids is proportionally higher, but the size of the turbine is not as large due to the high density of the vapor.

The density of an organic fluid at the exhaust is high. The flow rate of steam is about 16 times higher, thus the steam turbine size is considerably higher. Turbines using organic working fluids are more economical.

11.3 STEAM TURBINES

The energy conversion from heat to work is realized in a power cycle as the high-enthalpy steam expands flowing through the steam turbine. A steam turbine is a device that extracts thermal energy from the steam and converts it to mechanical work on a rotating output shaft. The rotary motion generated by the turbine is particularly suitable to drive the electrical generator. There are several ways to classify steam turbines.

In accordance to the principle of operation, steam turbines can be impulse or reaction turbines. According to their steam supply and exhaust condition, they can be distinguished as condensing or back-pressure types.

The basis of the distinction may be the number of the stages or the arrangements of the casing and the rotor.

The expansion of an ideal steam turbine is considered to be isentropic. Actual steam turbines are close to this approximation; their isentropic efficiency can be even more than 90%.

The isentropic efficiency is defined as the ratio of the actual work to the ideal work obtained by an isentropic expansion:

$$\eta = \frac{i_5 - i_6}{i_5 - i_6^*}, \tag{11.13}$$

where i_5 is the specific enthalpy at the beginning of the expansion, i_6 is the enthalpy at the end of the expansion for the actual turbine, and i_6^* is the enthalpy at the end of an isentropic expansion.

The impulse turbine has fixed nozzles in which the high enthalpy of the high temperature and pressure steam is converted into kinetic energy of high-speed jets. These jets flow to the bucket-shaped rotor blades changing the direction of the flow. The pressure drop occurs in the nozzle with an increase of velocity. As the steam flows thorough the rotating buckets, its pressure is unchanged. The absolute value of the velocity is also unchanged, but its direction is changed. The steam leaving the rotating buckets has a high amount of kinetic energy. This high exit velocity is the cause of the so-called leaving loss.

In the reaction turbines, the rotor blades themselves are designed to form convergent nozzles. The steam flowing through the blades accelerates and produces a reaction force acting on the rotor. Steam is directed to the rotor by the fixed blades of the stator. The steam then changes its direction, and its velocity increases relative to the rotation speed of the blades. The pressure drop occurs both in the stator and the rotor blades.

The steam leaving the turbine may flow to the atmosphere or even against a higher back-pressure. This is the practice for process steam applications. Condensation turbines are most frequently applied in electrical power generation. The condenser's pressure is 0.02 bar, and the temperature of the exchanged steam is only about 30°C.

The arrangement may be single casing, tandem compound, or cross-compound turbines. Single casing units are the most common type, where a single casing and the shaft are coupled to the generator. Tandem compound is used where two casings are directly coupled together to drive the generator. The cross-compound turbine arrangement contains two shafts not in line, driving two generators, which may operate at different speeds.

The flowing steam exerts not only rotating torque on the rotor, but also an axial thrust.

To balance this axial force, two-flow rotors are used. The steam inflows at the middle of the rotor, and outflows at both ends. The cross-section of the flow channel increases in the direction of the flow. The length of the blades increases also as it is shown in Fig. 11.6. Thus, axial forces are balanced while tangential forces act together.

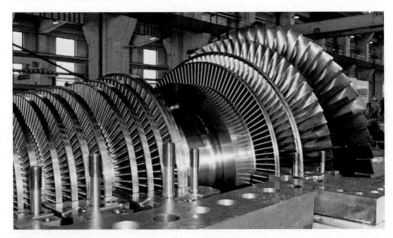

FIGURE 11.6 Steam turbine rotor. *http://www.energy.siemens.com/co/en/fossil-power-generation/steam-turbines/sst-800.htm.*

11.4 GEOTHERMAL POWER PLANTS

11.4.1 Single Flash Steam Power Plants

The Rankine cycle is the archetype of all working cycles operating with steam as working fluid. Geothermal energy conversion systems are different in lesser or greater degrees. The main difference is that the heat addition to the working fluid isn't done by a boiler, it is a natural underground heating. The pressure level of the heating is not produced by a boiler feed pump, it is induced by the hydrostatic pressure of the water column above the reservoir. In the Rankine cycle, the steam is superheated, while in a geothermal reservoir the geofluid can be at least saturated steam. Generally, the pressure and temperature of the Rankine cycle is higher than in geothermal working cycles. In consequence of the lower temperatures, the thermal efficiency of a geothermal working cycle is substantially lower than in the case of fossil-heated Rankine cycle (Tester et al., 2006).

The single flash steam power plants are the most widespread type of geothermal power plants. Except for the few dry steam reservoirs, the high temperature geofluids are in liquid state, since the formation pressure is substantially higher than the saturation pressure belonging to the given temperature. The thermal state of the reservoir fluid is represented by the point on the entropy—temperature diagram in the pressurized liquid region, close to the lower saturation curve as it is shown in Fig. 11.7. The pressure of the upflowing hot water decreases continuously in the tubing because of the increase of potential energy and fluid friction. The flow in the tubing is steady, incompressible, and turbulent. Where the reducing pressure attains the saturation pressure, a phase change begins; the liquid flashes into vapor. The temperature drop of the upflowing liquid is negligible between the reservoir and the saturation state. In spite of this, the pressure decrease of the isenthalpic two-phase flow induces substantial temperature drop while the part of the steam continuously increases to the wellhead. Thus, a two-phase water-steam mixture, the so-called wet steam is produced at the wellhead. The wet steam must be separated into steam and liquid phases with a minimum pressure loss. This is done in cylindrical cyclone separators, where steam and water disengage, owing to their substantially large density difference. The thermal state of the separated steam is represented by the far right point on the upper saturation curve, while the state of the separated water falls on the far left point at the lower saturation curve.

The steam mass fraction or dryness fraction x_c can be determined by the equation:

$$x_c = \frac{i_c - i_E}{i_D - i_E} \qquad (11.14)$$

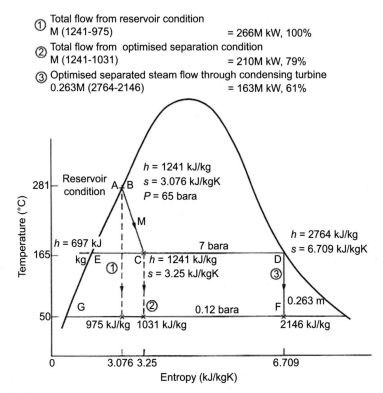

① Total flow from reservoir condition
   M (1241-975)   = 266M kW, 100%
② Total flow from optimised separation condition
   M (1241-1031)   = 210M kW, 79%
③ Optimised separated steam flow through condensing turbine
   0.263M (2764-2146)   = 163M kW, 61%

FIGURE 11.7   Flash process in T−S diagram (Bobok, 2014).

If $\dot{m}$ is the total mass flow rate upflowing in the tubing, then $x_C \cdot \dot{m}$ is the steam mass flow rate flowing to the turbine. The steam expands flowing through the turbine, while its pressure, temperature, and enthalpy decrease along the vertical line 3 as it is shown in Fig. 11.6. The work produced in the rotor is equal to the enthalpy drop.

The work produced by the turbine per unit mass flow rate of steam:

$$W_t = i_E - i_F \tag{11.15}$$

assuming no heat loss from the turbine, and neglecting the kinetic energy change between the inlet and the outlet of the turbine. The maximum possible work would be obtained if the turbine operated isentropically. This ideal expansion is done between points E and F.

It can be defined the isentropic turbine efficiency $\eta$is as the ratio of the actual work to the isentropic work:

$$\eta_{is} = \frac{i_E - i_F}{i_E - i_F} \tag{11.16}$$

The power produced by the turbine can be obtained:

$$P_t = x_c \cdot \dot{m}(i_E - i_F) \tag{11.17}$$

The gross electrical power is obtained as the product of the turbine power and the generator efficiency:

$$P_t = x_c \cdot \dot{m}(i_E - i_F) \tag{11.18}$$

The performance of the entire plant can be characterized by the utilization efficiency, or efficiency of the plant. This is the ratio the actual electric power output and the maximum theoretical thermal power that could be produced from the given geothermal fluid. The latter is the product of the total mass flow rate and the specific exergy:

$$P_{max} = \dot{m}e \tag{11.19}$$

The specific exergy can be determined as:

$$e = i(p, T) - i(p_0, T_0) - T_0[s(p, T) - s(p_0, T_0)], \tag{11.20}$$

where p and T is the pressure and temperature of the geofluid at the wellhead, p_0 and T_0 is the pressure and temperature of the surrounding. The i, specific enthalpy, and the s, specific entropy, values can be taken from the steam table. Finally, the utilization efficiency is:

$$\eta_u = \frac{P_e}{P_{max}}. \tag{11.21}$$

Naturally, many auxiliary units make complete the entire system. A schematic drawing can be seen in Fig. 11.8, where the main components of a single-flashed geothermal power plant are shown.

The production well (PW) provides the geofluid for the power plant. It is obvious that the reservoir is tapped by many PWs. As it is known, the maximum theoretical thermal power that could be produced from a single production well can be determined as:

$$P_{max} = \dot{m}[i_{WH} - i_0 - T_0(s_{WH} - s_0)]. \tag{11.22}$$

That is the total mass flow rate of the production well times the specific energy, referring to the thermal parameters of the produced geofluid at the wellhead and the surroundings. This value isn't satisfied by the energy demand of a power plant. A typical single flash power plant produces 25–30 MW electric powers, and needs at least five to seven production and four to five injection wells. The largest geothermal power plants are implemented to large reservoirs of many production wells. Some examples are found in Table 11.1 as follows.

The electric power obtained from a single well is 3.92 MW at the Geysers and 0.7 MW at Larderello. It can be taken as a preliminary estimation that the electric power of a geothermal well is about 2 MW.

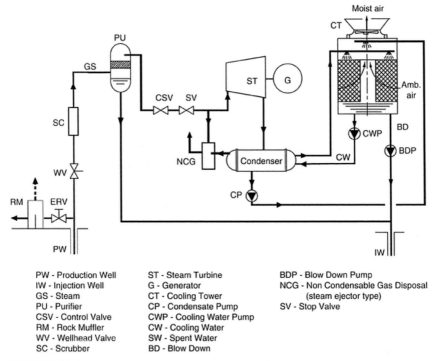

FIGURE 11.8   Schematic drawing of a single-flashed power plant (Bobok, 2014).

PW - Production Well
IW - Injection Well
GS - Steam
PU - Purifier
CSV - Control Valve
RM - Rock Muffler
WV - Wellhead Valve
SC - Scrubber

ST - Steam Turbine
G - Generator
CT - Cooling Tower
CP - Condensate Pump
CWP - Cooling Water Pump
CW - Cooling Water
SW - Spent Water
BD - Blow Down

BDP - Blow Down Pump
NCG - Non Condensable Gas Disposal
    (steam ejector type)
SV - Stop Valve

TABLE 11.1   The Largest Geothermal Power Plants (Bertrani, 2015)

Power Plant	Country	Electric Power (MW)
Geysers	USA	1584
Larderello-Travale	Italy	795
Tongonan/Leyte	Philippines	726
Cerro Prieto	Mexico	720
Olkaria	Kenya	591
Mak-Ban/Laguna	Philippines	458
Wairaki	New Zealand	399
Salton sea	USA	388
Coso	USA	292
Darajat	Indonesia	260

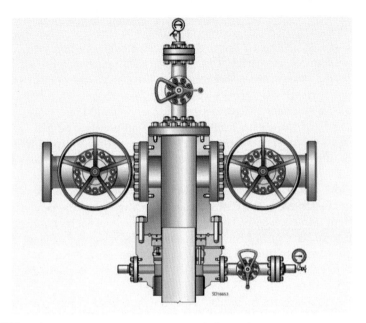

FIGURE 11.9   Wellhead equipment system (Bobok, 2014).

At each production well, there is an assemblage of equipment to control and monitor the flow of the produced steam-water mixture. The equipment includes the blowout preventer, master valves, bleed line, and a silencer, which is a cyclone separator for emergency venting. Pressure and temperature gauges also are equipped at the wellhead. A typical wellhead equipment system is shown in Fig. 11.9.

Production wells generally are placed on a wide area around the power plant.

The produced two-phase mixture flows from the wellhead to the centrifugal separator. It may be designed with different configurations. The separator can be placed directly at each wellhead. In the centrifugal separator, the mixture is separated into two distinct phases, steam and water. In this case, individual steam transporting pipes connect the separators to a steam collector at the plant, while the separated water flows directly to the injection well through individual water lines.

It is another arrangement, when the production wells supply with two-phase geofluid a large central cyclone separator placed directly to the powerhouse.

Production wells may be connected to intermediate satellite separators in the field. The separated steam flows from the satellite separators to the steam collector, while the water flows from the separators to the injection wells.

The type of the flow in the gathering pipe can be different depending on the arrangement of the separator. If the centrifugal separator is placed directly at the wellhead, homogeneous steam phase flows toward the steam collector. Since the length of the gathering pipe may be rather long, the occurring pressure losses of the steam flow cannot be neglected. Steam then flows through a ball check valve. To achieve a practically dry state of the steam, a further cyclone separator, the so-called moisture remover (MR) is built in the steam-transporting pipeline (SP). The MR is usually placed directly near the powerhouse. General design specifications of the cyclone separator and moisture remover are shown in Figs. 11.10 and 11.11.

The flow rate, the wellhead pressure, and the elevation are obviously different at each well. The distance between a well and the powerhouse is also different in each case. Thus the pressure loss of each gathering line is also different. In front of the powerhouse, the pressures of each arriving steam flow are not uniform. Control and stop valves are built in to adjust a uniform pressure level in the steam collector to provide a steady pressure level at the turbine inlet.

Geothermal steam turbines (T) differ markedly from the turbines usually found at fossil-fired power plants. In fossil plants, highly super-heated steam is used with high pressure and temperature (Thorhallsson and Ragnarsson, 1992). The entrancing steam at the inlet of a geothermal

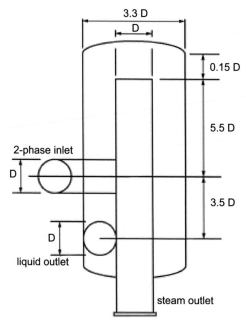

FIGURE 11.10 Cyclone separator (DiPippo, 2008).

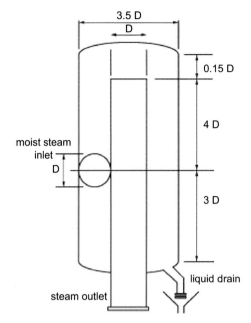

FIGURE 11.11 Moisture remover (DiPippo, 2008).

FIGURE 11.12 Impulse-reaction blade rows of a steam turbine (Bobok, 2014).

steam turbine is saturated with pressures that range 5–10 bar. Turbines for single-flash power plants consist of four to five stages of impulse-reaction blade rows. It is shown in Fig. 11.12. Especially at the last stage, a significant amount of condensed water appears. These small droplets strike the leading edge of the turbine blades causing erosion. The turbines must be made of corrosion-resistant materials because of the corrosive nature of the geofluid.

Turbines of single flash power plants are typically rated at 25–55 MW. Single-flow and double-flow arrangements are usually designed. The isentropic efficiency can be attained at the 80% value.

The steam exhausted from the turbine enters to the steam condenser. Steam condensers may be classified as surface type and direct-contact steam condensers. In the surface-type condensers, the steam and the cooling fluid (water or air) is separated. The condenser really is a shell-and-tube heat exchanger, which converts steam to its liquid state at a pressure below atmospheric pressure. During the condensation process the pressure and temperature is constant as it is shown in Fig. 11.7 between points F and G, while the latent heat of the steam is liberated and is carried away by the cooling water.

It can be expressed by the equation:

$$\dot{m}_{cw} \cdot c_w(T_{out} - T_{in}) = x_c \cdot \dot{m}(i_F - i_G), \tag{11.23}$$

in which $\dot{m}_{cw}$ is the mass flow rate of the cooling water, c_w is the specific heat capacity of the water, T_{out} and T_{in} is the outflowing and the inflowing cooling water temperature as it flows through the condenser, $\dot{x}_c\dot{m}$ is the mass flow rate of steam as it enters to the condenser, it is the specific enthalpy at the end of the expansion, while i_G is at the end of condensation process.

In this case, the Eq. (11.23) is modified as:

$$\dot{m}_{cw} \cdot c_w(T_G - T_{in}) = x_c\dot{m}(i_F - i_G). \tag{11.24}$$

In most surface condensers, the cooling water flows through the tube bundle, while the exhausted steam flows through the shell. The necessary internal vacuum in the shell is maintained by an external steam ejector (SE) or a vacuum pump. An SE uses steam as the motive fluid removing non-condensable gases, mostly carbon dioxide. A steam jet ejector is sketched in Fig. 11.13.

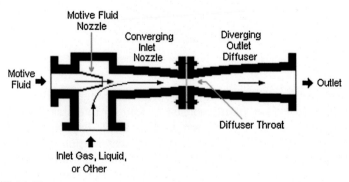

FIGURE 11.13   Steam jet ejector (Bobok, 2014).

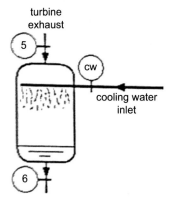

FIGURE 11.14   Direct contact heat exchanger.

A direct-contact steam condenser is shown in Fig. 11.14. Its main advantage is the simple construction. It is little more than a pressure vessel with an entrance opening for the exhausted steam, nozzles for the cooling water and an exit port for the condensed liquid. The steam and the cooling water flows are in direct contact and will mix. Since the heat transfer area is the surface of the cooling water droplets, it is larger than the area of the tube bundle, thus the heat transfer between the two flows is more effective and lower temperature difference is necessary between them. Another advantage is that there are no separating surfaces, and their corrosion and fouling are omitted, thus the heat transfer performance can be improved. The frictional pressure drop of direct contact condensers is lower than the surface type units. A final advantage is the lower capital cost.

The temperature of the cooling water rises as it flows through the condenser. The cooling tower is used to accommodate the heat load from the condensing steam (see Fig. 11.15).

At the end of the condensing process, a low-pressure condensate flows out from the condenser. The low-pressure steam condensate is pumped into the cooling tower. The cooling tower is essentially also a direct-contact heat exchanger where the mixture of the condensate and the cooling water is sprayed into a counter-current airflow.

An axial flow ventilator at the top of the cooling tower draws the upflowing air, as it is shown in Fig. 11.17. The ambient air enters with a certain amount of water vapor, determined by its relative humidity. The airflow picks up more water vapor as the condensate partially evaporates. The evaporation process requires heat that is provided from the water itself, thereby dropping its temperature. The process involves simultaneously both heat and mass transfer between the water and the air.

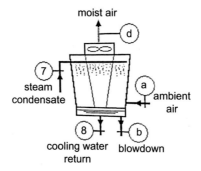

FIGURE 11.15   Cooling tower (DiPippo, 2008).

Assuming that the tower is an adiabatic system, neglecting the power of the ventilator, the balance equations of the enthalpy and the mass can be written obviously:

$$\dot{m}_7 i_7 - \dot{m}_6 i_8 = \dot{m}_d i_d - \dot{m}_a i_a + \dot{m}_b i_b \qquad (11.25)$$

The other two equations are the mass balance of water and the air:

$$\dot{m}_7 + \dot{m}_{wa} = \dot{m}_8 + \dot{m}_b + \dot{m}_{wd} \qquad (11.26)$$

and

$$\dot{m}_{ad} = \dot{m}_{aa} \qquad (11.27)$$

Note that both the inflowing and outflowing air flows contain water in the vapor phase. The terms $\dot{m}_{wa}$ and $\dot{m}_{wd}$ represent the vapor content of the entering and leaving air flow, respectively. These can be expressed by the relative humidity ω of the air flow:

$$\dot{m}_{wa} = \omega_a \cdot \dot{m}_a, \qquad (11.28)$$

and

$$\dot{m}_{wd} = \omega_d \cdot \dot{m}_d \qquad (11.29)$$

These equations can be used to determine the various mass flow rates. The properties of steam, water, and humid air are given in steam tables and graphically in psychometric charts referring to the given conditions.

It is remarkable that cooling towers for geothermal power plants have a much larger cooling capacity than a traditional fossil power plants of the same capacity. It is known that the net amount of work obtained by a power cycle is equal to the difference of the added and rejected heat:

$$Q_a - Q_r = W \qquad (11.30)$$

The thermal efficiency is obtained as:

$$\eta_{th} = \frac{W}{Q_a} \qquad (11.31)$$

The rejected heat can be expressed as:

$$Q_r = W\left(\frac{1}{\eta_{th}} - 1\right) \tag{11.32}$$

Fossil-fired, combined steam-and-gas turbine plants have a typical thermal efficiency of 50%, while binary geothermal plants have at least 14%. Thus, the rejected heat of a binary geothermal plant is seven times greater than the waste heat of a combined fossil-fired plant. Accordingly, the binary plant must have a cooling tower seven times larger in cooling capacity than a modern fossil fuel plant of the same power capacity.

A flash-steam power plant operates as an open system, thus Eq. (11.31) is not applicable. The waste heat can be calculated by the equation:

$$\dot{Q}_r = x_2 \dot{m}(i_5 - i_6) \tag{11.33}$$

This can be compared to the assaults obtained by the expression (11.32). The general qualitative conclusion is very similar to that obtained for a binary plant.

11.4.2 Dry-Steam Power Plants

Dry-steam geothermal reservoirs can be developed only where a series of necessary conditions are fulfilled simultaneously. The most important condition amongst them a high-strength heat source, which is able to raise the temperature of connate water to the boiling point. This heat source can be a shallow magma intrusion, not deeper than 5—6 km. Highly permeable formations must be located above the intrusion to form a reservoir. It is necessary that some vertical permeability between the reservoir and the surface allows the steam to escape to the surface during a geologically long duration of time. Meanwhile, the liquid level lowers substantially. It is necessary also the sufficient large vertical extension of the reservoir to allow the thermal convection currents of the reservoir fluid. There must be sufficient impermeable lateral boundaries of the reservoir, to avoid flowing the cooler groundwater into the steam reservoir. An impermeable uppermost formation, the so-called cap rock is also necessary to hold the steam in the reservoir. The fortunate coincidence of these various conditions makes very rare the existence of dry-steam reservoirs. Large dry-steam reservoirs are known only in two areas of the world: Larderello in Italy and The Geysers in California, United States. Smaller dry-steam fields are found in Indonesia, Japan, New Zealand, and in Utah, United States. There are more than 500 production wells both in The Geysers and Larderello. These two huge dry-steam reservoirs are the base of more than 60 electric power production unit, producing more than 25% the total geothermal power worldwide.

The produced steam is transported from the well to the powerhouse through the gathering pipeline system. It consists of one or more large diameter main pipelines to which smaller diameter branch pipes connect coming from the wellheads. The mass flow rate is increasing in the main pipeline as approaches the powerhouse. The main consideration in the design of the gathering system is the choice of the pipe diameters. The reservoirs, the wells, and the gathering pipelines form a synergetic flow system. The change of performance of any elements influences the performance of the whole system. The wellhead pressure, the mass flow rate, and the temperature are generally different, together with the length of the branch line and the elevation. All branch pipes are in parallel connection; their mass flow rates are added. Thus, the diameter of the main gathering pipe is increased in the direction of the flow. For steam transporting pipelines there is an accepted allowable velocity maximum based on the experience of everyday industrial practice. It is obtained to 50−60 m/s for large-diameter (>300 mm) pipes while for smaller diameter pipes 20−25 m/s is recommended. These velocity limits are necessary since the water droplets and fine solid particles carried by the high velocity steam flow cause erosion of elbows, valve seats, or other exposed parts.

At the wells, there are the usual valves and mainly an in-line axial centrifugal separator, to remove the droplets and particles before entering the pipe system. The steam pipes are thermally insulated and mounted on stanchions. Standard piping practice provides for axial thermal expansion to take place between stanchions or anchors to which the pipe is fixed and which transfer thrust to the ground. Main anchors are provided generally at pipe ends, at changes of pipeline direction, at shut off valves, and at manifolds where pipes are interconnected. Intermediate anchors are provided to divide the pipeline into separate expanding sections and to bear any unbalanced thrust. Pipe movement is accommodated by supporting the pipe on rollers, or by use of flexible hangers. Connections of the branch pipelines are suitable be made at or near anchor points of the main pipeline. Expansion bends of U shape, rectangular loops, or other configurations may be built in into straight sections of pipe. Steam traps are sited strategically along the pipes to remove condensate, which is than flow through separate lines for reinjection.

As the main pipeline approaches the powerhouse, there is an emergency pressure relief station. This allows for the temporary release of steam in the case of a turbine trip. The steam generally passes through a silencer before it enters the atmosphere. It is experienced that it is better to maintain the wells in a steady open state, than changing the wells through open and closed positions.

There are at the powerhouse a steam header, a final moisture remover, a vertical cyclone separator, and a Venturi meter for accurate measurement of the steam flow rate.

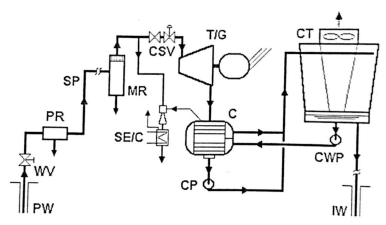

FIGURE 11.16 Schematic drawing of a dry-steam power plant (DiPippo, 2008).

The energy conversion system of a dry-steam power plant is rather similar to the single-flash system. Its block diagram is shown in Fig. 11.16. After the wellhead valve (WV), the particle remover (PR), and the MR, the steam flow toward the powerhouse. At the powerhouse the emergency pressure relief unit: control and stop valves are built in. An auxiliary pipe leads steam to the steam ejector (SE/C) to provide the vacuum in the condenser (C). The turbines may be single-flow for smaller units or double-flow for larger units (>60 MW). The condensers can be both direct-contact and surface-type.

The energy conversion process is shown in Fig. 11.17. It consists of two sections. Since the produced steam is in saturated state, the starting point

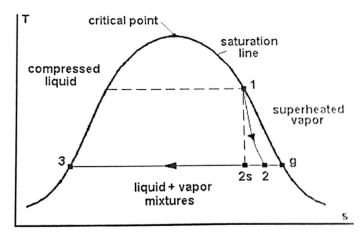

FIGURE 11.17 Dry-steam energy conversion process (DiPippo, 2008).

can be found at the upper saturation curve (1). The expansion process in the turbine is represented by the curve 1-2. The isentropic expansion would be the 1-2 s straight line. The condensation process goes on between the 2-3 section at constant temperature and pressure.

The work produced by the turbine while the unit mass steam flows through it can be obtained as:

$$W_t = i_1 - i_2 \tag{11.34}$$

The maximum possible work would be obtained if the turbine operated isentropically. The isentropic turbine efficiency can be calculated as the ratio of the effective work to the isentropic work, so:

$$\eta_t = \frac{i_1 - i_2}{i_1 - i_{2s}} \tag{11.35}$$

The power produced by the turbine can be obtained as:

$$P = \dot{m}_s w_t = \dot{m}_s (i_1 - i_2) \tag{11.36}$$

The produced electric power is obviously lower, it can be obtained as the product of the turbine power output and the generator efficiency:

$$P_e = \eta_g \cdot P \tag{11.37}$$

Considering the expansion process in the T−s diagram it can be recognized that the entire process occurs in the two-phase region. Adopting the Baumann rule to calculate the decrease in performance of a wet-steam expansion, it is obtained:

$$\eta_{tw} = \eta_{td} \frac{1 + x_2}{2} \tag{11.38}$$

in which η_{tw} is the efficiency of the wet-steam expansion, while η_{td} is the efficiency of the expansion in the dry-steam region. A conservative approximation is: to take $\eta_{td} = 0.85$.

It was mentioned earlier that the value of the wellhead pressure influences the power output of a dry-steam plant. There is an optimum wellhead pressure, at which the turbine power output has a maximum.

11.4.3 Binary Cycle Power Plants

In conventional geothermal power plants, the produced gas fluid is also the working fluid of the energy conversion process. In the binary cycle power plants the produced reservoir fluid and the secondary working fluid flows in two separated flow systems. The conventional geothermal energy conversion systems one open: the produced steam flows through the turbine, then it is rejected. The binary energy conversion system is a closed cycle: the same working fluid flows through the turbine repetitively.

A binary cycle is the closest to the classical Clausius—Rankine cycle. The working fluid receives heat from the geofluid in a heat exchanger, where it evaporates, expands flowing through the turbine, condenses, and the process repeats again.

The efficiency of the energy conversion cycle depends primarily on the steam temperature at the turbine inlet. The temperature of the wet stream decreases substantially in the flashing process. Because of this, flash technology is handily efficient and economic in the temperature region of the heat source below 170°C. For the utilization of these low temperature resources, binary cycle power plants can be applied advantageously.

A very important objective of the binary technology is to select a suitable working fluid. The suitable working fluids for binary plants have critical temperature and pressure far lower than water. These working fluids include the hydrocarbons, especially the aromatic-structure types, halogen-substituted hydrocarbons, steers, and simple-molecule materials such as ammonia, carbon dioxide, or even water.

Another important characteristic of binary cycle working fluids is the shape of the saturation curves: especially the upper boundary curve as viewed in the T—s diagram. This curve for simple molecule fluids up to four to five atoms has negative slope everywhere as it is shown in Fig. 11.18. When the number of the atoms increases the upper boundary curve becomes almost vertical for 6-10 atom molecules. For complex molecules having more than 10 atoms the slope of the saturated vapor curve is positive. In this case expansion from the saturated vapor curve happens in the superheated region, reducing blade erosion.

Since the critical pressure is reasonably low, it is feasible to consider supercritical cycles for these organic fluids. Considering Fig. 11.19, it can

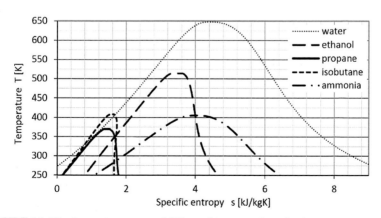

FIGURE 11.18 Saturation curves of different binary working fluids.

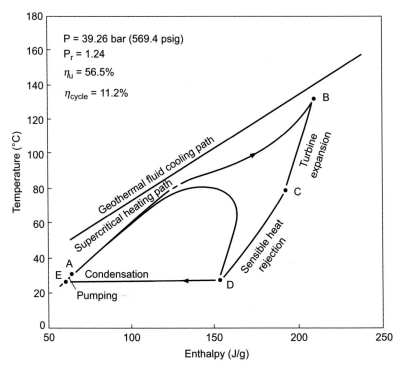

FIGURE 11.19 Supercritical binary cycle.

be recognized that the temperature difference between the geofluid cooling curve and the supercritical heating–boiling curve of the working fluid is not significant. Thus, the thermodynamical losses of the heat exchanger can be reduced. Some properties of organic fluids are tabulated in Table 11.2.

It is important to assess the chemical stability of the fluid, the aggressively towards metals, and thermodynamically and transport properties of them.

A simplified schematic drawing of a binary power plant is shown in Fig. 11.20.

The PW operates with a submersible pump (P) installed below the flash depth so as to prevent two-phase flow. The sand remover prevents erosion in the pipeline and the tubes. The heating and boiling process of the working fluid happens in two steps: first, the condensed working fluid is heated to its boiling temperature in the preheater (PH); it then turns to saturated or supercritical steam in the evaporator (E). The geofluid is always kept at a pressure above its flash point as flows through the evaporator and the preheater. An injection pump (IP) produces the

TABLE 11.2 Chemical and Physical Properties of Some Organic Fluids.

Name	Molar Weight	T_{crit} (K)	P_{crit} (bar)	Latent Heat (KJ/kg)
R125 Pentafluormethane	120	339	36.2	81.5
R218 Octafluorpropane	180	345	48.7	223
R290 Propane	44	369.8	42.5	292
R600 Butane	58	425	38.0	337
R601 Pentane	72	469.7	33.7	347
R717 Ammonia	17	405	113.3	1064
R718 Water	18	647	220.6	2392
R744 Carbon dioxide	44	304	738	167.53

necessary pressure to make the fluid flow through a fine-particle filter (FF) before finally reaching the injection well (IW).

The working fluid circulates in a secondary loop: through the pre-heater, the evaporator, it expands in the turbine (T) then condenses in the condenser (C). A condensate-feed pump (CP) maintains the flow in the loop.

A third, auxiliary loop is added to the system. The cooling water circulates through the condenser, and the cooling tower (CT).

The use of organic fluids with the low temperature Rankine cycle has many advantages over using water. The organic Rankine cycle efficiency is little different from that steam cycle between the same two top and bottom cycle temperature. Organic cycle deficiency is often less than water steam cycle.

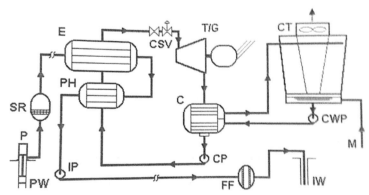

FIGURE 11.20 Scheme of a binary power plant (DiPippo, 2008).

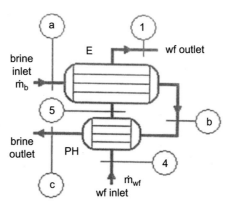

FIGURE 11.21   Heat transfer in the heat exchanger and pre-heater (DiPippo, 2008).

The main advantage of an organic working fluid is that it can extract more heat from the primary geothermal heat source than water. This is the consequence of the organic fluid having a far lower ratio of latent heat of vaporization at lower boiling temperature v3 specific heat capacity than water.

The only definitive difference between a binary and a fossil fuel power plant is the preheater and the evaporator, where the produced hot geofluid transfers heat to the working fluid. Analyzing this phenomenon we assume that the shells of the heat exchangers are perfectly insulated, the whole heat transfer is occurred between the geofluid and the working fluid. Further assumptions are the steady flow, and the differences of kinetic and potential energy are negligible between the inlet and the outlet.

Applying the notations of Fig. 11.21, the simplified energy equation can be written as:

$$\dot{m}_b(i_a - i_c) = \dot{m}_{wf}(i_1 - i_4),\tag{11.39}$$

If the geofluid is incompressible, having very low amount of dissolved gases, the enthalpy can be written as the product of its specific heat and the temperature:

$$\dot{m}_b \cdot c_b(T_a - T_c) = \dot{m}_{wf}(i_1 - i_4)\tag{11.40}$$

The required geofluid mass flow rate for a given cycle is obtained as:

$$\dot{m}_b = \dot{m}_{wf}\frac{i_1 - i_4}{c_b(T_a - T_c)}\tag{11.41}$$

The heat transfer process can be understood easier, using the so-called T−q, or temperature−heat transfer diagram. The abscissa represents to total amount of heat, transferred from the geothermal brain to the

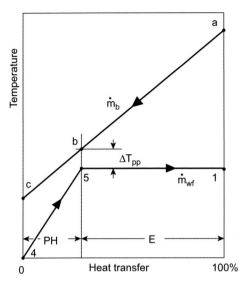

FIGURE 11.22 Temperature versus heat transfer diagram (DiPippo, 2008).

working fluid, as it can be seen in Fig. 11.22. The transferred heat can be given either in energy units (KJ/kg) or in percent.

The flows of the geofluid and the working fluid are counter-current. The geofluid flows at first through the evaporator (section of a-b) then the preheater (b-c). The preheater provides sensible heat for the working fluid rising its temperature to the boiling point 5. The evaporation occurs between the points 5 and 1, where the temperature is constant while the evaporator provides latent heat for the phase charge of the working fluid. The place in the process, where the temperature difference is the minimum between the brine and the working fluid is the so-called pinchpoint. The value of that difference is designated the pinch-point temperature difference: ΔT_{pp}.

In the state four the working fluid is a compressed liquid, at the outlet of the feed pump. In the state five the working fluid is a saturated liquid at the boiling point, while state 1. is saturated vapor at the turbine inlet. Thus the two heat exchangers: the preheater and the evaporator can be investigated separately as:

$$\dot{Q}_{PH} = \dot{m}_b c_b (T_b - T_c) = \dot{m}_{wf} (i_5 - i_1) \tag{11.42}$$

and

$$\dot{Q}_E = \dot{m}_b c_b (T_a - T_b) = \dot{m}_{wf} (i_1 - i_5) \tag{11.43}$$

The geofluid inlet temperature T_a is always known. The pinch-point temperature difference is given from manufacturer's data. Thus T_b can be determined knowing the turbine inlet temperature T_1. In the ideal case the pinch-point occurs at the outlet had of the preheater. The evaporator heat transfer surface area between the two fluids A_E can be determined from the expression:

$$\dot{Q}_E = U_E A_E \Delta T_{lnE} \tag{11.44}$$

where U_E is the overall heat transfer coefficient, ΔT_{lnE} is the logarithmic mean temperature difference of the evaporator

$$\Delta T_{lnE} = \frac{(T_a - T_1) - (T_b - T_5)}{\ln\left|\frac{T_a - T_1}{T_b - T_5}\right|} \tag{11.45}$$

The corresponding equations for the preheater are:

$$\dot{Q}_{PH} = U_{PH} \cdot A_{PH} \cdot \Delta T_{lnPH} \tag{11.46}$$

and

$$\Delta T_{lnPH} = \frac{(T_b - T_5) - (T_c - T_4)}{\ln\left|\frac{T_b - T_5}{T_c - T_4}\right|} \tag{11.47}$$

Since heat exchangers can be made in a great variety of geometrical arrangements, it must be applied different correction factors for shell and tube, pure counterflow, crossflow and plate units. These valves can be found in the handbook of Roshenow and Hartnett.

The thermal efficiency of the binary plant can be obtained as

$$\eta_{th} = \frac{\dot{W}}{\dot{Q}} \tag{11.48}$$

Since the net power of the cycle is the difference of the input and the rejected thermal power, it can be written:

$$\eta_{th} = \frac{\dot{Q}_{PH} + \dot{Q}_E - \dot{Q}_c}{\dot{Q}_{PH} + \dot{Q}_E} \tag{11.49}$$

This formula is valid to the cycle, not to the whole plant. In the latter case the auxiliary power needs (pumps, cooling power fan etc.) must be subtracted from the not cycle power W. The binary power plant efficiencies depending on the heat source temperatures are plotted in.

References

Bertrani, R., 2015. Geothermal Power Generation in the World 2010-2014 Update Report.
Bobok, E., 2014. Geothermal Power Plants. E-Book, Tankönyvtár. http://www.tankonyvtar.hu/hu/tartalom/tamop412A/2011_0059_SCORM_MFKGT5064-EN/sco_03_01.scorm.

DiPippo, R., 2008. Geothermal Power Plants, Principles, Applications, Case Studies and Environmental Impact, second ed. Elsevier, ISBN 978-0-7506-8620-4.

Tester, J.N., et al., 2006. The Future of Geothermal Energy: Impact of Enhanced Geothermal Systems on the United States in the 21st Century. MIT, Cambridge, MA.

Thorhallsson, S., Ragnarsson, A., 1992. What is geothermal steam worth. Geothermics 21, 901—915.

Propagation of the Cooled Region in a Small Fractured Geothermal Reservoir

12.1 INTRODUCTION

The implementation of the first Hungarian geothermal pilot power plant occurred in 2004. After a comprehensive site investigation, a fractured limestone reservoir was selected in southwestern Hungary, close to the Slovenian border. It is located at a depth of 3000 m. The reservoir temperature is 142°C. There are two unsuccessful petroleum prospecting boreholes with good mechanical integrity. The distance between the wells

Copyright © 2017 Elsevier Inc. All rights reserved.

is 1000 m. After some work a doublet was found to be a production and an injection well. A prefeasibility study investigated the hydraulic and thermal behavior of the reservoir within production and injection. The most important questions of this study were:

- What kind of flow system will be developed between and around the wells?
- How will injection affect the temperature distribution in the reservoir and the adjacent rock mass?
- How will the cooled region propagate from the injection well toward the production well?
- How will the produced water temperature decrease with time?

Armed with the answers to these questions we can predict the lifetime of the system.

This problem was investigated first by Bodvarsson (1975) and later Bodvarsson et al. (1985), and Ghassemi and Tarasov (2004).

12.2 THE CONCEPTUAL MODEL

The existing large, horizontal, fractured Triassic limestone reservoir is replaced by a single equivalent fracture bounded by parallel plane walls. The primary reason for this simplification is to get a preliminary result without suitable or reliable input data. The fracture is filled with hot water which is considered incompressible. Thus mechanical and thermal processes can be treated separately. There is no overpressure in the reservoir. The pressure distribution is hydrostatic with depth. The horizontal extent of the equivalent fracture is much greater than the distance between the two boreholes. The injection well occurs in the plane fracture as a source, the production well is a sink. This model is valid for the flow pattern developed in a large equivalent fracture. The flow net in this case extends over the whole plane. A relative smaller equivalent fracture has a closer boundary curve around the doublet. The closed boundary is modeled by superimposing a fictitious parallel flow to the source and the sink. The flow net remains inside this closed curve. The flow in the fracture is steady, laminar, and two-dimensional. This is the so-called Hele−Shaw flow. The Hele−Shaw flow is a quasipotential motion having a complex potential function. The bottom-hole temperature in the injection well is constant. The injected cold water occurs in the fracture as an abrupt thermal inhomogeneity, thus, near the fracture transient heat conduction is generated within the adjacent rock. Thus, the hot rock heats the injected cold water. The water temperature increases as it flows toward the production well, while the rock temperature decreases. As the result of this heat transfer, the produced water temperature decreases. If the produced water temperature drops below a given limit the doublet will not operate efficiently and its operation will be terminated.

12.3 THE MATHEMATICAL MODEL

An orthogonal coordinate system is chosen. The $x-y$ plane is parallel to the fracture walls at halfway between them, z is the transverse direction; $2b$ is the gap between the fracture surfaces. The x-axis passes through to the source and the sink. It is directed toward the sink. Because of the incompressibility of water, the flow and the heat transfer can be determined separately.

The governing equations of the Hele–Shaw flow (Polubarinova-Kotschina, 1952) are

$$\frac{\partial p}{\partial x} = \rho v \frac{\partial^2 v_x}{\partial z^2}; \quad \frac{\partial p}{\partial y} = \rho v \frac{\partial^2 v_y}{\partial z^2}; \quad \frac{\partial p}{\partial z} = 0 \tag{12.1}$$

It can be proven that

$$\frac{\partial^2 p}{\partial x^2} + \frac{\partial^2 p}{\partial^2 y} = 0 \tag{12.2}$$

Solving Eq. (12.1) using the no-slip boundary condition, we have

$$v_x = \frac{1}{2\rho v}\left(z^2 - b^2\right)\frac{\partial p}{\partial x}; \quad v_y = \frac{1}{2\rho v}\left(z^2 - b^2\right)\frac{\partial p}{\partial y} \tag{12.3}$$

Their integral means between the planes

$$c_x = \frac{b^2}{3v\rho}\frac{\partial p}{\partial x}; \quad c_y = \frac{b^2}{3v\rho}\frac{\partial p}{\partial y} \tag{12.4}$$

The pressure p is a harmonic function fulfilling Eq. (12.2), and c_x and c_y can be derived from a scalar potential.

$$c_x = \frac{\partial \Phi}{\partial x}; \quad c_y = \frac{\partial \Phi}{\partial y} \tag{12.5}$$

where

$$\Phi = \frac{b^2 p}{3\rho v} \tag{12.6}$$

Because the fluid is incompressible

$$\frac{\partial c_x}{\partial x} + \frac{\partial c_y}{\partial y} = 0 \tag{12.7}$$

which becomes an identity, if

$$c_x = \frac{\partial \Psi}{\partial y} \quad c_y = -\frac{\partial \Psi}{\partial x} \tag{12.8}$$

Eqs. (12.5) and (12.8) are Cauchy–Riemann equations

$$\frac{\partial \Phi}{\partial x} = \frac{\partial \Psi}{\partial y}; \quad \frac{\partial \Phi}{\partial y} = -\frac{\partial \Psi}{\partial x} \tag{12.9}$$

Fulfillment of Eq. (12.9) is equivalent to the existence of an analytic complex variable function $W(\xi)$, the so-called complex potential

$$W(\xi) = \Phi(x, y) + i \cdot \Psi(x, y) \tag{12.10}$$

of which real part is the velocity potential Φ and the imaginary part is the stream function Ψ. The $\Psi = const$ curves are the streamlines.

The complex potential of the Hele–Shaw flow between the source and the sink can be written applying the method of hydrodynamic singularities:

$$W = \frac{Q}{2\pi} \ln \frac{\xi + a}{\xi - a} \tag{12.11}$$

Fig. 12.1 shows the two singularities at the point $x = -a$, the source of Q and at the point $x = a$, the sink of capacity of $-Q$. Using the exponential form of ξ, the real and imaginary parts of W can be separated easily.

$$W = \Phi + i \cdot \Psi = \frac{Q}{2\pi} \ln \frac{r_1}{r_2} + i \cdot \frac{Q}{2\pi} (\varphi_1 - \varphi_2) \tag{12.12}$$

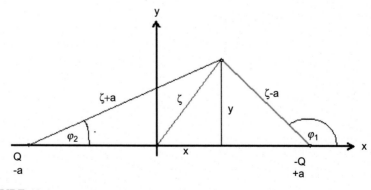

FIGURE 12.1 The two singularities.

Thus the equation of the streamlines is

$$\varphi_1 - \varphi_2 = \frac{2\pi k}{Q} = \text{const} \qquad (12.13)$$

Considering Fig. 12.1, we see that

$$\text{tg}(\varphi_1 - \varphi_2) = \frac{\frac{y}{x-a} - \frac{y}{x+a}}{1 + \frac{y}{x-a} - \frac{y}{x+a}} = C \qquad (12.14)$$

After some manipulation we obtain

$$x^2 + \left(y - \frac{a}{C}\right)^2 = a^2\left(1 + \frac{1}{C^2}\right) \qquad (12.15)$$

The shape of the streamlines is a family of circles between the source and the sink, with centers at $x = 0$ and $y = \frac{a}{C}$ as is shown in Fig. 12.2.

In the case of $C = 0$ the circle becomes a straight line along the x axis. If $C = \infty$, $\text{tg}\frac{2\pi}{Q}k = \infty$, $k = \frac{Q}{4}$ and the radius of the circle is a. Thus, half of the flow rate runs inside an origin-centered circle of radius a.

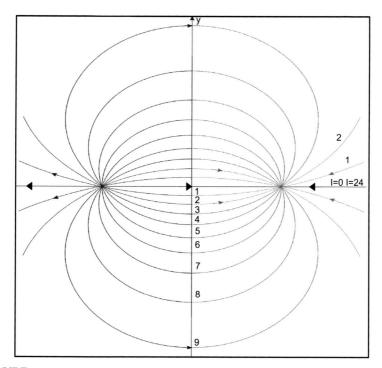

FIGURE 12.2 Subdivision of the fracture into part-channels.

In a small fractured reservoir the lateral extent of the equivalent fracture has the same order of magnitude as the distance between the two wells. The effect of the boundary proximity is the change of the streamline pattern. The circular-shaped streamlines are deformed, they are compressed in the direction of the y-axis.

This effect can be modeled by the method of singularities. The impermeable contour line coinciding with the fracture edge around the doublet can be replaced by an equivalent closed streamline. This closed streamline can be obtained by superimposing a fictitious apparent homogeneous parallel flow of x-direction into the doublet flow pattern. The complex potential in this case can be written as

$$W = c_{\infty x} + \frac{Q}{2\pi} \ln \frac{\xi + a}{\xi - a} \tag{12.16}$$

Its imaginary part is the stream function

$$\Psi = c_{\infty x} y - \frac{Q}{2\pi} (\varphi_2 - \varphi_1) \tag{12.17}$$

After some manipulation we obtain

$$\frac{2ay}{x^2 + y^2 - a^2} = tg\left[\frac{2\pi}{Q}(c_{\infty x} \cdot y - \varphi)\right] \tag{12.18}$$

The streamline equation is obtained by the substitution of $\Psi = \text{const.}$ The $\Psi = 0$ curve is the closed contour-streamline separating an "inner" and an "outer" flow. This streamline pattern is shown in Fig. 12.3.

As the apparent parallel flow velocity $c_{\infty x}$ increases, the compression of the contour streamline is greater. Thus $c_{\infty x}$ influences the size of the contour line in both directions. Substituting $y = 0$, and $\Psi = 0$ into Eq. (12.18) we got the length of the contour in the x-direction:

$$x_L = 2\sqrt{a^2 + \frac{aQ}{\pi \cdot c_{\infty x}}} \tag{12.19}$$

Similarly, substituting $x = 0$ and $\Psi = 0$ into Eq. (12.18) we obtained the width of the contour in the y-direction:

$$\frac{2y}{y^2 - a^2} = tg\frac{2\pi}{Q} c_{\infty x} \cdot y \tag{12.20}$$

This equation can only be solved numerically. These expressions are suitable to estimate the extent of the fracture based on tracer test data.

The conjugate of the velocity is the derivative of the complex potential:

$$\bar{c} = c_{\infty x} + \frac{Q}{2\pi}\left(\frac{1}{\xi + a} - \frac{1}{\xi - a}\right) \tag{12.21}$$

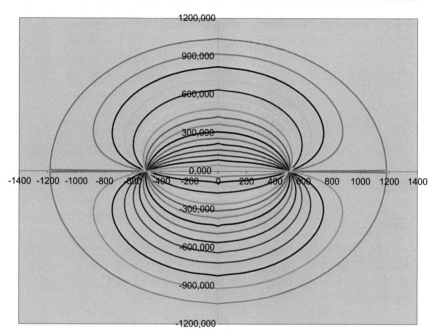

FIGURE 12.3 Streamline pattern inside the closed boundary curve.

The velocity along the x-axis has an x-component only:

$$c_x = c_{\infty x} + \frac{Q}{2\pi}\left(\frac{1}{x+a} - \frac{1}{x-a}\right) \qquad (12.22)$$

Its average is

$$\overset{\approx}{c}_x = c_{\infty x} + \frac{Q}{2\pi}\int_{x=0}^{a-x}\left(\frac{1}{x+a} - \frac{1}{x-a}\right)dx \qquad (12.23)$$

This average can be obtained from tracer tests, as the ratio of the distance between the two wells and the measured average residence time of the tracer.

$$\overset{\approx}{c} = \frac{2a}{t_{av}} \qquad (12.24)$$

In this way we can determine $c_{\infty x}$ and the length and width of the fracture.

The plane of the flow can be decomposed into part-channels, in each of which the flow rates are the same. In these part-channels the flow is one-dimensional in a curvilinear coordinate system. Thus the two-dimensional plane flow is replaced by a finite set of one-dimensional flows. In this way ordinary differential equations are obtained along the streamlines, while finite differences are obtained perpendicular to them. The heat transfer in the fracture can be solved by this complex method, simultaneously using finite differences and ordinary differential equations.

12.4 <u>HEAT TRANSFER IN THE FRACTURE</u>

At the beginning of the injection the fracture is filled with hot water. Its temperature is the same as the natural geothermal temperature at the given depth. As the injected water flows along the streamlines, it will warm up, while the rock temperature decreases. The whole heat transfer process can be separated into two subprocesses: advection in the water and transient one-dimensional heat conduction toward the fracture in the rock mass (Toth and Bobok, 2008). Since the injected water mass is much smaller than the rock, the slow transient heat transfer is followed by the fluid instantaneously. The internal energy balance for an infinitesimal volume element (as shown in Fig. 12.4) is written as:

$$\dot{m}\, c[(T + dT) - T] = 2UL(T_\infty - T)ds \qquad (12.25)$$

where $\dot{m}$ is the injected mass flow rate, c is the heat capacity, U is the transient overall heat transfer coefficient, T_∞ is the initial geothermal rock

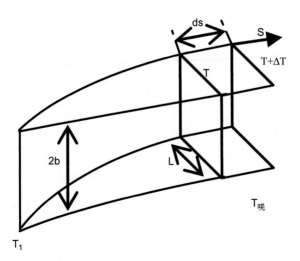

FIGURE 12.4 Infinitesimal volume element schema.

temperature, and L is the width of the part-channel. Solving Eq. (12.16) we can get

$$T = T_\infty - (T_\infty - T_1) \cdot e^{\frac{2ULs}{mc}} \tag{12.26}$$

where T_1 is the temperature of the injected water at the bottom-hole, and s is the actual length along the streamline in question.

The transient overall heat transfer coefficient can be calculated as

$$\frac{1}{U} = \frac{1}{h} + \sqrt{\frac{\pi \cdot t}{(\rho c k)_r}} \tag{12.27}$$

In which ρ is the density, c is the heat capacity, and k is the heat conductivity of the rock. From the point of view of computation, a very favorable condition exists in the Nusselt number which is constant for the heat transfer on a flat plate between the solid and the laminar flow. The experimentally determined value is $Nu = 5.12$ (Lundberg et al., 1963). Thus the heat transfer coefficient is obtained as

$$h = \frac{5 \cdot 12 \cdot k_w}{2b} \tag{12.28}$$

where k_w is the heat conductivity of the water. Consequently h is independent of the changing velocity along the streamlines, heat transfer coefficient can be determined without knowledge of the velocities.

Knowing the value of U, the water temperature can be calculated in any part-channel as a function with length and time. Note that Eq. (12.19) is valid only for that region of the fracture which is filled the injected water.

Propagation of the cooled region along the streamlines lags behind the motion of the injected fluid. This is an important difference in comparison to oil displacement by water. The boundary surface between the water and the oil phase moves together with the flowing fluids, and the material properties experience an abrupt jump on this strong material singular surface. In the geothermal reservoir the injected water temperature increases gradually as there is no sharp contour of the cooled region.

At the bottom-hole of the production well the cooled region will arrive first along the straight streamline between the two wells. All other part-channels still carry hot water of undisturbed reservoir temperature. The parallel connected part-channels carry water of different temperatures. This homogenized temperature is shown in Fig. 12.5 depending on time. These temperatures characterize the sustainability of the system.

Increase of the mass flow rate results in a temperature drop of the injected water and in effect, a smaller temperature decrease. Choosing the best and most economic temperature limit of this produced water enables one to estimate the lifetime of the doublet.

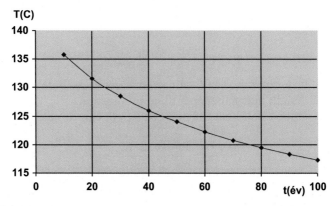

FIGURE 12.5   Temperature is dependent on time.

12.5 SUMMARY

This chapter is part of a prefeasibility study, made for the implementation of the first Hungarian geothermal pilot power plant. The site chosen is a fractured limestone reservoir in the southwestern part of Hungary. A doublet was planned to supply geothermal energy for the plant. For this preliminary investigation the fracture system is replaced by an equivalent fracture in which a Hele–Show flow is observed. The flow pattern and the streamline system are treated by methods of hydrodynamic singularities. The heat transfer in the fracture is advection and conduction in the adjacent rock. The cooled region propagates from the injection well towards the production well and it lags behind the motion of the fluid. The decrease in the produced water temperature is not a sudden drop, but it suffers from a gradual change. This is determined as the function of time. If we use appropriate heat temperature limits, an accurate, effective operational lifetime of the doublet can be determined as outlined in this work.

Acknowledgment

Portions of this chapter were previously published by the author as paper at PROCEEDINGS, Thirty-Fourth Workshop on Geothermal Reservoir Engineering Stanford University, Stanford, California, February 9–11, 2009 SGP-TR-187 titled "Cold Front Propagation in a Fractured Geothermal Reservoir" and are reprinted with permission of Stanford University.

References

Bodvarsson, G., 1975. Thermoelastic phenomena in geothermal system. In: Proceedings of 2nd UN Symposium on Geothermal Resource San Francisco, USA, vol. 1, pp. 903–906.

Bodvarsson, G., Preuss, K., O'Sullivan, M., 1985. Injection and energy recovery in fractured geothermal reservoirs. SPE J. 303–312.

Ghassemi, A., Tarasov, S., 2004. Three-dimensional modelling of injection induced thermal stresses. In: Proceedings of 29th Workshop, Stanford, CA, USA.

Lundberg, R.E., Mc Cuen, P.A., Reynolds, W.S., 1963. Heat transfer in annular passages. Int. J. Heat Mass Transfer 6, 495–501.

Polubarina-Kotshina, P.J., 1952. Teorija dvizsennyija gruntovih vod. Nauka, Moscow.

Toth, A., Bobok, E., 2008. Cold front propagation in a fractured geothermal reservoir. In: 33rd International Geological Congress. Norway, Oslo.

Borehole Heat Exchangers

13.1 INTRODUCTION

The heat content of rocks near the surface of the Earth is a huge resource of geothermal energy. The top few hundred meters of the Earth's crust is not a geothermal reservoir in its classical sense. The temperature of this region is too low for immediate utilization. Only adaption of heat pumps makes this vast resource accessible, increasing the temperature of the heat-carrying fluid. The most widespread technology to utilize the shallow geothermal resources is borehole heat exchangers (BHEs) equipped with geothermal heat pumps. The borehole heat exchanger is a device to extract geothermal heat from the shallow rocks without geofluid production. It is a heat exchanger installed inside a borehole, circulating any heat-carrying fluid through it. The heat exchanger inside the borehole can be a double U tube or two coaxial pipes. The borehole round the pipes is backfilled with a material of high thermal conductivity. The energy

Copyright © 2017 Elsevier Inc. All rights reserved.

supply of the BHE is transferred by conduction which is rather weak to reach at least a medium output temperature. The reason for this is the relatively shallow depth of the BHE, the low heat transfer area, the low surrounding temperature, and the weak heat conductivity of the rock around it. The output temperature of the circulating fluid can be increased to the required level by the operating electrical heat pump.

Higher output temperature can be attained by the so-called deep borehole heat exchangers (DBHEs), which are transformed from dry unsuccessful boreholes. In this case the depth is the same as thermal water or hydrocarbon wells, the heat transfer area is larger, the temperature of the surrounding rock is higher, and the heat conductivity of the deep compacted rocks is also higher than the shallow region. Naturally a DBHE has a lower thermal power capacity than a similar thermal water well.

To solve this problem some useful ideas were suggested by Horne (1980), Armstead (1983), and Morita et al. (1985, 2005). Their recommendations are to circulate water in a closed casing well. The water flows downward through the annulus between the casing and the tubing while it warms up, and it returns at the bottom-hole and flows upward through the tubing. The upward-flowing water cools to a certain extent because of the heat transfer across the tubing wall.

Such an experimental production unit was installed in 1989 in Szolnok, central Hungary. The results, as expected, were rather modest because of the insufficient heat transfer area around the well and the low heat conductivity of the surrounding rocks. The circumstances were analyzed by Bobok et al. (1991) and Bobok and Tóth (2000, 2002, 2008). In the following we shall introduce a more sophisticated mathematical model to describe the heat transfer mechanics of such a system, to predict its thermal behavior in order to avoid further inefficient and expensive experiments, and to show the range of the dry hole geothermal utilization. Our attention is focused on the annular heat transfer phenomenon.

13.2 THE MATHEMATICAL MODEL

The simplified model of a closed geothermal well is shown in the following. The casing is closed at the bottom without any perforations. The water flows downward through the annulus between the coaxial casing and tubing. Since the adjacent rock is warmer than the circulating water, the water temperature increases in the direction of the flow. An axisymmetric thermal inhomogeneity is developed around the well, together with radial heat conduction toward the well. This is the heat supply of the system. The warmed-up water flows upward through the tubing while its temperature slightly decreases, depending mainly on the heat conduction coefficient of the tubing. The system is analogous to a

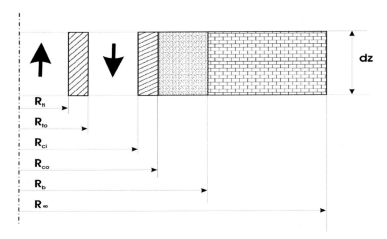

FIGURE 13.1 Schematic drawing of control volume.

countercurrent heat exchanger. The main difference is the increasing adjacent rock temperature distribution with the depth. Thus the familiar methods for design of heat exchangers are not sufficient for this case.

Let's consider the schematic drawing of the system in Fig. 13.1. The geometric parameters are defined as shown in the figure. It is convenient to separate the system into two subsystems. One is the flowing fluid, in which the convective heat transfer is dominant. The other is the adjacent rock mass around the well, with a radial conductive heat flux. Thus the internal energy balance can be written for the two subsystems in a simplified form. Cylindrical coordinates are chosen. The radial coordinate r is measured from the axis of the coaxial cylinders, while z lies along the axis directed downward. The steady, axisymmetric turbulent flow is taken to be uniform at a cross-section; the velocity v and temperature T are cross-sectional average values. The t, c, and a indices refer to the tubing, the casing, and the annulus, respectively. Thus the balance equation of the internal energy for the flow across the tubing is:

$$R_{ti}^2 \pi \rho c v_t dT_t = 2R_{ti}\pi U_{ti}(T_t - T_a)dz \tag{13.1}$$

in which ρ is the density, c is the heat capacity of the fluid, and U_{ti} is the overall heat transfer coefficient referring the inner radius of the tubing. For the annular flow we get:

$$\left(R_{ci}^2 - R_{t0}^2\right)\pi \rho c v_a dT_a = 2R_{ci}\pi U_{ci}(T_b - T_a)dz + 2R_{ti}\pi U_{ti}(T_t - T_a)dz \tag{13.2}$$

where T_b is the temperature at the borehole radius R_b, and U_{ci} is the overall heat transfer coefficient referring to the radius R_{ci}.

The unsteady axisymmetric heat flux around the well is equal to the heat flux through the casing. It can be expressed as:

$$\dot{Q} = 2\pi k_R \frac{T_\infty - T_b}{f(t)} = 2\pi R_{ci} U_{ci}(T_b - T_a) \tag{13.3}$$

in which $\dot{Q}$ is the heat flux over the unit length cylinder, and k_R is the heat conductivity of the rock. The undisturbed natural rock temperature is T_∞, its distribution linear with depth

$$T_\infty = T_s + \gamma z \tag{13.4}$$

where T_s is the annual mean temperature at the surface, and γ is the geothermal gradient. The parameter $f(t)$ is the transient heat conduction time function (Ramey, 1962).

13.3 SOLUTION

To solve the differential equation system it is necessary to know the overall heat transfer coefficients U_{ti} and U_{ai}. The determination of U_{1B} and U_{ci} needs the knowledge of the heat transfer coefficients h_{to} and h_{ci} referring to the inner and outer surface of the annulus.

The down-flowing fluid in the annulus of the closed-loop geothermal system is heated from two directions independently: across the tubing and across the casing. These heat fluxes can be varied independently. It is obvious that two independent heat transfer coefficients and two different Nusselt numbers are obtained on the inner and outer walls of the annulus. To determine this heat transfer mechanism is more difficult than in a simple circular tube.

Lundberg et al. (1963) have shown that it is possible to reduce the problem to four fundamental solutions in accordance with the different boundary conditions. These can be combined using superposition techniques to yield a solution for any desired boundary conditions. The present case can be interpreted as the superposition of two fundamental solutions. One is perfectly insulated. Another particular solution is obtained interchanging the two surfaces. The two particular solutions can be superimposed providing solution for the two-sided heating.

Following the familiar semianalytical treatment, we will employ the subscript t1 to designate conditions on the tubing surface when this surface alone is heated. The subscript c1 designates conditions on the casing surface when this surface alone is heated. The opposite surface in either case is insulated. The single subscript t or c refers to the conditions on the tubing or casing surfaces, respectively, under any conditions of simultaneous heating at both surfaces.

For the case of constant heat rate per unit tube length it is possible to express $\mathrm{Nu_T}$ and $\mathrm{Nu_c}$ for any heat flux ratio on the two surfaces in terms of $\mathrm{Nu_{t1}}$ and $\mathrm{Nu_{c1}}$. Similarly two influence coefficients can be derived, θ_t^* and θ_c^*. Finally the following expressions are obtained:

$$\mathrm{Nu_t} = \frac{\mathrm{Nu_{t1}}}{1 - \frac{q_{c1}}{q_{t1}}\theta_t^*} \tag{13.5}$$

$$\mathrm{Nu_c} = \frac{\mathrm{Nu_{c1}}}{1 - \frac{q_{t1}}{q_{c1}}\theta_c^*} \tag{13.6}$$

Kays and Leung (1963) carried out experiments in annuli using air for various values of radius ratio with constant heat rate per unit tube length but with various heat flux ratios at the inner and outer surfaces including the two limited cases of only one side being heated. They then obtained a solution in tabulated form for constant heat rate under fully developed turbulent flow based upon empirical data. The results are presented in the form $\mathrm{Nu_{t1}}$ and $\mathrm{Nu_{c1}}$ and the two influence coefficients θ_t^* and θ_c^*. These are given in tables for a wide range of Reynolds and Prandtl numbers and for radius ratios. These results are then directly applicable to Eqs. (13.5) and (13.6). Data are widened by some of our experimental results obtaining with water for $4\frac{1}{2}''$ and $7''$ tubing and casing diameters. The heat flux q_{c1} can be obtained from the temperature distribution of an injection well. As it is known (Ramey, 1962), the bottom-hole temperature of the injected water is

$$T_{bh} = T_s + \gamma(H - A) + (T_i - T_s + \gamma A)\cdot e^{\frac{-H}{A}} \tag{13.7}$$

The overall heat flow into the down-flowing water is

$$\dot{Q} = \dot{m}c(T_{bh} - T_i) \tag{13.8}$$

Thus the integral main of the heat flux per unit length of casing is

$$q_{c1} = \frac{\dot{m}c(T_{bh} - T_i)}{2R_{ci}\pi H} \tag{13.9}$$

In the second case the casing is perfectly insulated, the annular flow is heated across the tubing only. The mass flow rates in the tubing and in the annulus are the same. Eqs. (13.1) and (13.2) lead to the relation

$$\frac{dT_t}{dz} = \frac{dT_a}{dz} \tag{13.10}$$

Derivating Eq. (13.1) by z, we obtain

$$\dot{m}c\frac{d^2T_t}{dz^2} = 2\pi R_{ti}U_{ti}\left(\frac{dT_t}{dz} - \frac{dT_a}{dz}\right) \tag{13.11}$$

comparing Eqs. (13.1) and (13.2) it is obtained that

$$\frac{d^2T_t}{dz^2} = 0 \quad \text{and} \quad \frac{d^2T_a}{dz^2} = 0 \tag{13.12}$$

Thus the general solutions of Eq. (13.12)

$$T_t = K_1 z + K_2 \tag{13.13}$$

$$T_a = K_1 z + K_3 \tag{13.14}$$

The boundary conditions are the following:

if $z = 0$ $T_a = T_i$,   if $z = H$ $T_t = T_a$ and $K_1 = 2\pi R_{ti} U_{ti}(T_t - T_a)$

Solving the obtained equation system, finally the heat flux across the tubing wall

$$q_{T1} = \frac{\dot{m}c(T_{bh} - T_i)}{B + H} \tag{13.15}$$

Eqs. (13.9) and (13.15) can be substituted into Eqs. (13.5) and (13.6). Based on experimental data the particular Nusselt numbers can be calculated by the following formulas:

$$Nu_{t1} = 0.016 \cdot Re^{0.8} Pr^{0.5} \quad Nu_{c1} = 0.018 \cdot Re^{0.8} \cdot Pr^{0.5} \tag{13.16}$$

$$\theta_t^* = 0.410 \cdot Re^{-0.078} \cdot Pr^{-0.58} \quad \theta_c^* = 0.325 \cdot Re^{-0.078} \cdot Pr^{-0.58} \tag{13.17}$$

Determining the Nusselt numbers on both surfaces the heat transfer coefficients on the walls of the annulus are

$$h_t = \frac{k \cdot Nu_t}{2R_{t0}} \quad \text{and} \quad h_c = \frac{kNu_c}{2R_{ci}} \tag{13.18}$$

Knowing the h_{t0} and h_{ci} values, the overall heat transfer coefficients U_{ti} and U_{ci} can be determined as

$$\frac{1}{U_{ti}} = \frac{1}{h_{ti}} + \frac{R_{ti}}{k_{ins}} \cdot \ln\frac{R_{t0}}{R_{ti}} + \frac{R_{ti}}{R_{t0}h_{t0}} \tag{13.19}$$

and

$$\frac{1}{U_{ci}} = \frac{1}{h_{ci}} + \frac{R_{ci}}{k_s} \cdot \ln\frac{R_{c0}}{R_{ci}} + \frac{R_{ci}}{k_c}\ln\frac{R_b}{R_{co}} \tag{13.20}$$

Combining the Eqs. (13.1)–(13.3) we obtain two simple differential equations:

$$A\frac{d(T_t - T_a)}{dz} = T_\infty - T_a \tag{13.21}$$

in which

$$A = \frac{\dot{m} \cdot c \cdot (k_r + R_{ci}U_{ci}f)}{2\pi R_{ci}U_{ci}k_r} \tag{13.22}$$

and

$$B = \frac{dT_t}{dz} = T_t - T_a \tag{13.23}$$

where

$$B = \frac{\dot{m} \cdot c}{2\pi R_{ti}U_{ti}} \tag{13.24}$$

Combining Eqs. (13.21) and (13.23), a second-order inhomogeneous differential equation is obtained:

$$AB\frac{d^2T_a}{dz^2} + B\frac{dT_a}{dz} - T_a + T_s + \gamma(z - B) = 0 \tag{13.25}$$

In a similar way we can obtain for the flow through the tubing:

$$AB\frac{d^2T_t}{dz^2} + B\frac{dT_t}{dz} - T_t + T_s - \gamma z = 0 \tag{13.26}$$

These equations can be solved easily in the form

$$T_t = T_s + \gamma(z + B) + K_1 e^{x_1 z} + K_2 e^{x_2 z} \tag{13.27}$$

and

$$T_a = T_s + \gamma z - T_i + C_1 e^{x_1 z} + C_2 e^{x_2 z} \tag{13.28}$$

where x_1 and x_2 are the roots of the characteristic Eqs. (13.25) and (13.26), i.e.,

$$x_1 = -\frac{1}{2A}\left(1 - \sqrt{1 + \frac{4A}{B}}\right) \quad \text{and} \quad x_2 = -\frac{1}{2A}\left(1 + \sqrt{1 + \frac{4A}{B}}\right) \tag{13.29}$$

The constants of integration in Eqs. (13.27) and (13.28) can be determined satisfying the following boundary conditions.

1. At $z = 0$, $T_a = T_i$, where T_i is the temperature of the cooled injected water;
2. At $z = H$, $T_a = T_t$, the bottom-hole temperatures in the annulus and in the tubing are the same;
3. At $z = H$, $\frac{dT_t}{dz} = 0$, the depth derivative of the tubing temperature at the bottom-hole is zero. This is the consequence of Eq. (13.23);

4. The energy increase of the circulating fluid is equal to the integral of the heat flux across the borehole wall between the bottom and the surface:

$$\dot{m}c(T_{out} - T_i) = \int_0^H q(z)dz \tag{13.30}$$

where T_{out} is the temperature of the outflowing water at the wellhead. The obtained equations from the boundary conditions are

$$T_i - T_s = C_1 + C_2 \tag{13.31}$$

$$K_1 e^{x_1 H} + K_2 e^{x_2 H} + \gamma B = C_1 e^{x,H} + C_2 e^{x_2 H} \tag{13.32}$$

$$K_1 x_1 e^{x_1 H} + K_2 x_2 e^{x_2 H} = -\gamma \tag{13.33}$$

$$A(T_i - T_x + \gamma B + K_1 K_2) = -\frac{C_1}{x_1}\left(e^{x_1 H} - 1\right) - \frac{C_2}{x_2}\left(e^{x_2 H} - 1\right) \tag{13.34}$$

After solving the equation system for the constants C_1, C_2, K_2, and K_2, the temperature distributions in the annulus and in the tubing can be determined by Eqs. (13.27) and (13.28).

The thermal power can be calculated using the equation

$$P = \dot{m} - c(T_{wh} - T_i) \tag{13.35}$$

where T_{wh} is the temperature of the outflowing water.

13.4 RESULTS

The temperature distribution, both of the annulus and the tubing along the depth, is determined. The solution makes it possible to take into consideration the parameters influencing the temperature distribution and the thermal power of the system. The first example is the temperature distribution of a closed-loop well. The depth of the well is 2000 m, the casing is 7″, and the tubing is 4½″. The tubing is a steel pipe with polypropylene heat insulation. Three different mass flow rates are taken: 5, 10, and 15 kg/s. Geothermal gradient is 0.05°C/m, the heat conductivity of the polypropylene is 0.2 W/m °C. The average heat conductivity of the rock is 2.5 W/m °C.

It can be recognized that the bottom-hole temperature depends strongly on the mass flow rate. The temperature difference between the produced and the injected water is decreasing as the mass flow rate is increasing. These can be shown in Fig. 13.2.

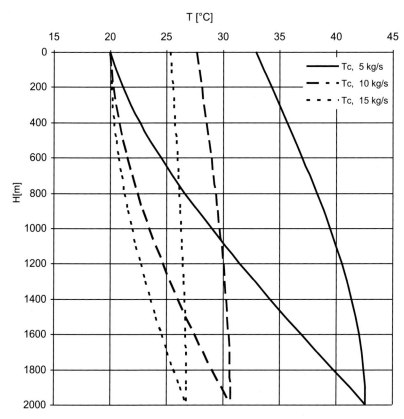

FIGURE 13.2   Temperature distributions along the depth.

The quality of the heat insulation of the tubing has a very important role. Applying a vacuum-insulated tubing (VIT) of an extreme low heat conductivity ($k = 0.006$ W/m °C) the up-flowing water temperature is almost constant. In this case the casing diameter is 9 5/8″, the inner tube diameter is 4½″, the outer is 5½″. This is shown in Fig. 13.3.

The temperature distribution both in the annulus and in the tubing depends strongly on the operation time of the system. There is a short initial period of important temperature decrease. Later the rate of change will be smaller. Finally the solution tends to a steady temperature, as is shown in Fig. 13.4.

The influence of the mass flow rate and the operation time on the thermal power of the system can be seen in Fig. 13.5. The thermal power is plotted against the production time. The mass flow rate is the parameter of the family of the curves. For small mass flow rates the effect of the time is very small. For higher flow rates the change with time is important.

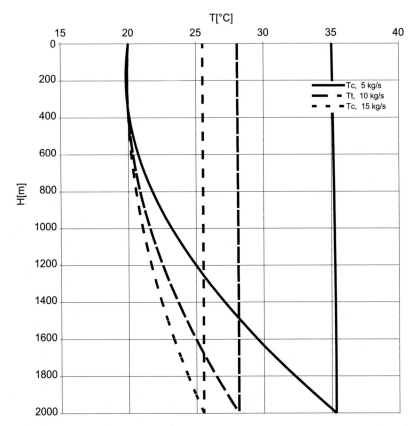

FIGURE 13.3 Temperature distributions obtained with vacuum-insulated tubing, $k_{VIT} = 0.006$ W/m °C.

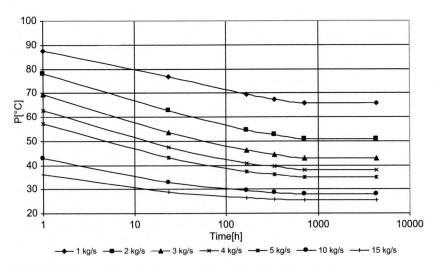

FIGURE 13.4 Outflowing water temperature decrease versus time.

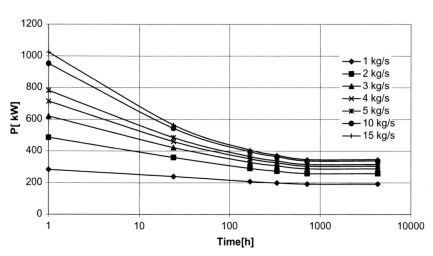

FIGURE 13.5 Thermal power decrease versus time.

While the growth of the mass flow rate has a substantial influence on the thermal power in the initial stage of production, as the operation time increases the differences of the power curves caused by the different mass flow rates decrease. The curves converge, especially for higher flow rates as they tend to an equilibrium state. It can be recognized that there exists an upper limit of the flow rate over which the equilibrium thermal power is not increasable. This thermal power determines the sustainability of the system. Heat conduction toward the well cannot carry more heat than this upper limit. Heat conductivity of the adjacent rock mass restricts the exploitable thermal power by a single closed-loop geothermal well. The sustainable power production of such a system can be determined knowing the depth and the completion of the well, the way of heat insulation of the tubing, the local geothermal gradient, and the material properties of the rocks around the well.

It can be seen that the temperature and the thermal power of such a closed-loop system is rather moderate. The cause of this is the small heat transfer area and the low heat conductivity of the rocks. It seems only small-scale utilizations can be based on this clean technology, even applying heat pumps as well.

Acknowledgment

Portions of this chapter were previously published by the author as paper at PROCEEDINGS, Thirty-Third Workshop on Geothermal Reservoir Engineering Stanford University, Stanford, California, January 28–30, 2008 SGP-TR-185 titled "Limits of Heat Extraction from Dry Hole" and are reprinted with permission of Stanford University.

References

Armstead, H.C.H., 1983. Geothermal Energy, second ed. EFN Spon, London.

Bobok, E., Mating, B., Navratil, L., Turzó, Z., 1991. Heat mining without water production (in Hungarian). Kőolaj és Földgáz 21 (6), 161−169.

Bobok, E., Tóth, A., 2000. In: Temperature Distribution in a Double_Function Production_ Reinjection Geothermal Well, vol. 24. Geothermal Resource Council Transactions, San Francisco, USA, pp. 555−559.

Bobok, E., Tóth, A., 2002. In: Geothermal Energy from Dry Holes: a Feasibility Study, vol. 26. Geothermal Resource Council Transactions, Reno, USA, pp. 275−278.

Bobok, E., Toth, A., 2008. Limits of heat extraction from dry hole. In: Proceedings, Thirty-third Workshop on Geothermal Reservoir Engineering, SGP-TR-185. Stanford University, USA, pp. 417−423.

Horne, R.N., 1980. Design considerations of a downhole coaxial geothermal heat exchanger. Geothermal Resource Council Transactions 4, 569−572.

Kays, W.M., Leung, E.Y., 1963. Heat transfer in annular passages − hydrodynamically developed turbulent flow with arbitrarily prescribed heat flux. Int. J. Heat Mass Transfer 6, 537.

Lundberg, R.E., McCuen, P.A., Reynolds, W.S., 1963. Heat transfer in annular passages. Int. J. Heat Mass Transfer 6, 495.

Morita, K., Tago, M., Ehara, S., 2005. Case Studies an Small-Scale Power Generation with the Downhole Coaxial Heat Exchanger, World Geothermal Congress, Paper 1622.

Morita, K., Matsubayashi, O., Kusunoki, I.C., 1985. Downhole coaxial heat exchanger using insulated inner pipe for maximum heat extraction. Geothermal Resource Council Transactions 9.

Ramey, H.J., 1962. Wellbore Heat Transmission Transactions of the AIME, pp. 427−435.

Flow and Heat Transfer During Drilling Operations

14.1 INTRODUCTION

On a drilling rig, the drilling fluid is pumped from the mud pit through the drill pipe down to the drilling bit. The fluid transmits down the necessary mechanical energy for cleaning the drill bit and the bottom hole, then carries the cuttings up to the surface through the annulus. On the surface, cuttings are filtered out and the fluid returns to the mud pit. The fluid is then pumped down to the drill pipe and re-circulated.

Copyright © 2017 Elsevier Inc. All rights reserved.

14.2 RHEOLOGY OF THE DRILLING FLUIDS

Drilling fluids must be suitable for several requirements during drilling operations. It is necessary to keep the drill bit clean and cool. Another demand is to remove the cuttings from the bottom hole and to transport them to the surface through the annular space between the drilling pipe and the casing. It is suitable to raise density of the fluid, since the greater density produces greater Archimedean and hydrodynamic lifting forces. The raised density provides a higher hydrostatic pressure to prevent formation fluids from entering into the bare wellbore. The greater hydrostatic pressure maintains the wellbore stability, too.

The most frequently used drilling fluids are water-based bentonite suspensions, the so-called drilling muds. Most drilling muds are shear-thinning, in other words pseudoplastic. Simultaneously, the thixotropic behavior also occurs. In the following, these rheological behaviors are taken with a glance.

Purely viscous fluids may be completely characterized by their shear stress-shear rate relationship under conditions of one-dimensional laminar flow. Thus the shear stress-shear rate function can be determined by measuring only one component of the stress tensor and the deformation rate tensor. In view of the wide range over which measurements need to be made for these variables, it is convenient to plot the shear rate-shear stress relationship. The curve thus obtained, the so-called flow curve, is suitable to characterize the rheological behavior of the fluid and forms an important aid in the classification of the different types of non-Newtonian fluids.

The flow curve of a Newtonian fluid is a straight line. Flow curves of Newtonian fluids of different viscosities pass through the origin; the slope of the lines representing viscosity of the fluid. Logarithmically plotted flow curves are parallel straight lines with slopes equal to unity. They have intercepts at $dv/dr = 1$ equal to their viscosities. Both types of flow curve are shown in Figs. 14.1 and 14.2.

The terms pseudoplastic and dilatant refer to fluids for which the logarithmically plotted flow curves have slopes of less than unity and greater than unity, respectively. The apparent viscosity decreases with increasing shear rate for pseudoplastic fluids, while it increases as the shear rate increases for dilatant material, as shown in Fig. 14.3. These flow curves also show that pseudoplastic and dilatant behavior may exist only over a limited range of the shear rate.

$$\mu_a = \frac{\tau}{\frac{dv}{dr}} \tag{14.1}$$

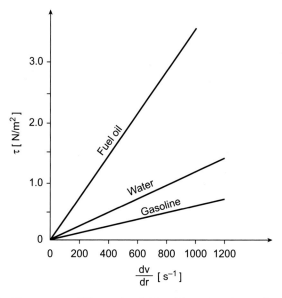

FIGURE 14.1   Flow curves of Newtonian fluids with arithmetic coordinates.

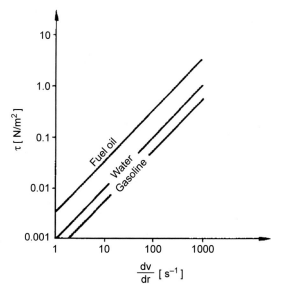

FIGURE 14.2   Flow curves of Newtonian fluids with logarithmic coordinates.

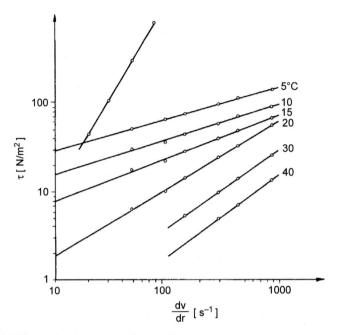

FIGURE 14.3 Pseudoplastic and dilatant flow curves.

There is no single simple form of the constitutive relation that adequately expresses the rheological behavior of pseudoplastic materials. Widely used in petroleum engineering is the so-called power law equation, which is valid over a limited range of the shear rate, where the logarithmic flow curve is a straight line. This law is obtained empirically as:

$$|\tau| = K\left|\frac{dv}{dr}\right|^n \tag{14.2}$$

in this equation, K is the so-called consistency index, and n is the behavior index. Both K and n depend on the temperature of the fluid. Fig. 14.3 shows the flow curves at different temperatures. The variation of the consistency index and the behavior index as a function of temperature is shown in Fig. 14.4.

A more precise expression for the power law can be written using cylindrical coordinates as:

$$\tau_{rz} = K\left|\frac{dv}{dr}\right|^{n-1}\frac{dv}{dr} \tag{14.3}$$

It is obvious that the apparent viscosity:

$$\mu_a = K\left|\frac{dv}{dr}\right|^{n-1} > 0 \tag{14.4}$$

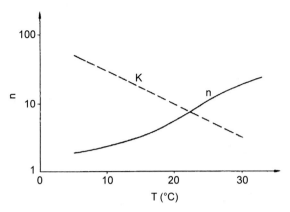

FIGURE 14.4 Consistency index and behavior index varying with temperature.

Within its validity range, the power law equation is in excellent agreement with experimental data. The accuracy of the power law equation decreases at very low and very high shear rates. For pseudoplastic fluids, the behavior index is always less than unity. Most crude oils, fine clay suspensions, certain types of drilling muds, and washing fluid exhibit pseudoplastic behavior.

The power law equation may also be applied to dilatant fluids with appropriately different values of the consistency index and the behavior index. The behavior index of dilatant fluids is always greater than unity. Dilatancy can be observed in dense suspensions of irregularly shaped solids in liquids. The dilatancy can vary rapidly with the concentration. The same suspension at lower a concentration may be pseudoplastic, while at a higher concentration the behavior is dilatant.

The so-called Bingham plastic materials are fluids for which a finite shearing stress is required to initiate motion, and for which there is a linear relationship between the shear rate and the shear stress once the initiating stress has been exceeded. This behavior characterizes asphalt, bitumen, certain drilling muds, fly ash suspensions, sewage sludge, etc. The main advantage of such a drilling mud is that it the cuttings are suspended while drilling is paused or the drilling assembly is brought in and out of the borehole.

The constitutive relation of a Bingham fluid is of the form:

$$|\tau| = \tau_y + \mu \left| \frac{dv}{dr} \right| \tag{14.5}$$

where τ_y is the yield stress. Fig. 14.5 shows the flow curves of a Bingham drilling mud at different concentrations.

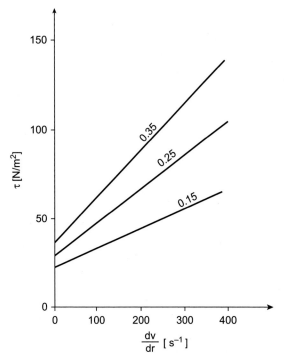

FIGURE 14.5   Flow curve of a Bingham plastic fluid: a drilling mud at different concentrations.

Certain materials exhibit a yield stress like a Bingham fluid, but the relationship between the shear rate and the shear stress after the yield stress has been exceeded is not linear. This category comprises mainly drilling muds, but there are some crude oils which also display this yield-pseudoplastic character. Fig. 14.6 shows the flow curves of a fluid which exhibits such behavior. The constitutive relation for these fluids can be written as:

$$\tau = \tau_y + K \left| \frac{dv}{dr} \right|^{n-1} \frac{dv}{dr} \tag{14.6}$$

It is an important observation that the yield stress decreases as the temperature increases.

Newtonian, pseudoplastic, and dilatant fluids respond instantaneously to a change in the shear stress. Their rheological behavior is influenced by structural changes in the system. Their equilibrium structure depends on the shear rate, and follows any change in the shear rate without the slightest delay. This structural change can be considered as instantaneous and reversible.

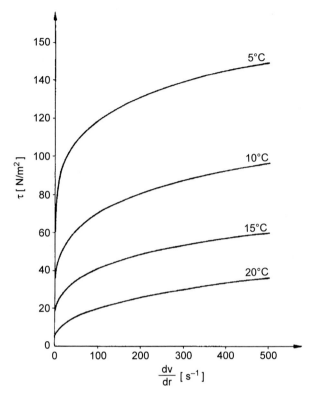

FIGURE 14.6 Flow curves of a yield-pseudoplastic fluid at different temperatures.

In contrast to this, certain other fluids exhibit slow structural changes, which follow changes in the shear rate with a considerable delay. The reconstruction of the changed structure may be so extremely slow that the process can be considered irreversible. This time-dependent rheological behavior is rather common. Certain drilling fluids and crude oils exhibit time-dependent rheological properties. Fig. 14.7 shows the decrease in the shear stress at a constant shear rate as a function of the duration of shear, as measured for a Kiskunhalas (Hungary) crude oil.

If the shear stress at a constant shear rate decreases with the duration of shear, the fluid is called thixotropic. As Fig. 14.7 shows, the shear stress decreases asymptotically to a stabilized value τ_s. The duration of shear, necessary to reach the stabilized value of shear stress, decreases as the shear rate increases. The value of the stabilized shear stress depends on the shear rate as:

$$\tau_s = K_s \left(\frac{dv}{dr}\right)^m \tag{14.7}$$

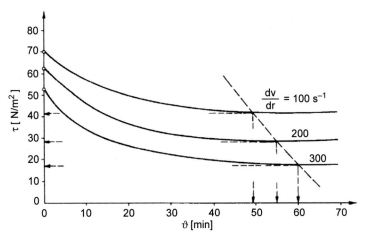

FIGURE 14.7   Thixotropic shear stress decrease depending on the duration of shear.

This power law equation has a behavior index m of less than unity, thus the relationship is of a pseudoplastic type. It characterizes an ultimate structural state which is independent of time, i.e., stable. Eq. (14.7) can be used as an adequate constitutive relation for a flow of crude oil after the first few kilometers of a pipeline.

The thixotropic character of a drilling mud may be important for relatively short pipes, especially for the drill string.

A great number of constitutive relations have been proposed for thixotropic fluids, which out of necessity are more complex than the power law equation. A relatively simple constitutive equation was developed by Bobok and Navratil (1982) for thixotropic fluids on the basis of a wide range of experimental results. In this model, the thixotropic material is considered to be a pseudoplastic fluid with changing rheological properties.

The shear stress in this model depends on two variables; the shear rate and a dimensionless structural parameter δ:

$$\tau = \left(\frac{dv}{dr}, \delta\right) \tag{14.8}$$

It is assumed that the structure of a thixotropic fluid can be completely characterized by the structural parameter, which depends on the shear rate and the duration of shear. The function given by Eq. (14.8) can be represented by a surface in the coordinate system dv/dr, δ, and τ as shown in Fig. 14.8. Any changes in the shear condition occur along a curve on this surface of the shear state. The path on this surface between two arbitrary points can be broken down along the orthogonal parameter coordinates $\left(\frac{dv}{dr} \text{ and } \delta\right)$, into a constant shear rate, and a constant

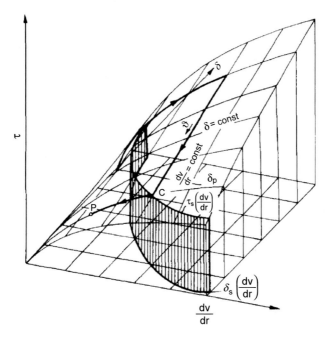

FIGURE 14.8 Shear state surface of thixotropic fluids.

δ section. The processes of constant shear rate, and of constant structural parameter have a merely different nature.

The constant-shear-rate process is considered irreversible. It can proceed along a $dv/dr =$ const. curve in one direction only, namely that in which the structural parameter decreases, i.e., the thixotropic structure is broken down. Since the breakdown of the structure is much faster than its regeneration (less than an hour as opposed to a few days), and regeneration may already have started at rest, the irreversibility of this process seems to be acceptable. Another important feature of the constant-shear-rate process is that its ultimate state is on the stabilized flow curve. At any constant shear rate, both δ and τ decrease as the duration of shear increases, until the stabilized values δ_s and τ_s, are reached. Values of δ and τ that are smaller than δ_s and τ_s cannot be achieved by the further increase in the duration of shear only. The stabilized-flow curve cannot be crossed along a constant-shear-rate line.

Along the constant structural–parameter curves, the consistency index and the flow-behavior index are constant, thus the parameter curves $\delta =$ const. are real pseudoplastic flow curves, i.e.:

$$\tau = K\left(\frac{dv}{dr}\right)^n \tag{14.9}$$

Experimental observations show that the consistency index changes with δ, but that the behavior index may be considered constant. Along the parameter lines $\delta = $ const., the stabilized flow curve can be crossed in both directions. The shear rate of the crossing point C is an important quantity in evaluating the consistency index K of an arbitrary flow curve of constant δ.

As is shown in Fig. 14.9, at the point C, the stabilized flow curve and an arbitrary flow curve of constant δ intersect each other. It is obvious that:

$$\tau_{sc} = K_s \left(\frac{dv}{dr}\right)_c^m = K\left(\frac{dv}{dr}\right)_c^n = \tau_c \tag{14.10}$$

Thus, K can be expressed as:

$$K = K_s \left(\frac{dv}{dr}\right)_c^{m-n} \tag{14.11}$$

Substituting this into Eq. (14.9), we get:

$$\tau = K_s \left(\frac{dv}{dr}\right)_c^{m-n} \left(\frac{dv}{dr}\right)^n \tag{14.12}$$

In the interval where:

$$\frac{dv}{dr} < \frac{dv}{dr_c}$$

it is possible, at a given dv/dr, to achieve smaller shear stresses than the stabilized shear stress at this dv/dr.

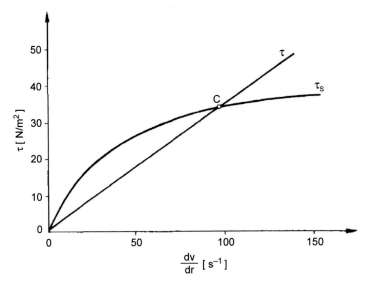

FIGURE 14.9 Flow curves of thixotropic fluid.

The flow curve in the interval:

$$0 \leq \frac{dv}{dr} \leq \frac{dv}{dr_c}$$

is reversible, changing the shear rate whenever the measured shear stress falls to the curve 0C. Increasing the shear rate to a greater value $\frac{dv}{dr_p}$, the immediately obtained shear stress τ_p will be on the same flow curve of constant δ, since the thixotropic structure remains the same. But from this instant the breakdown of the structure begins, and the state of shear will reach stabilized flow curve, along a constant-shear-rate line at point C'.

The shear stress for this constant-shear-rate breakdown process can be expressed as:

$$\tau = K_s \left(\frac{dv}{dr}\right)^m \left\{ 1 + \left[\frac{\left(\frac{dv}{dr}\right)^{n-m}}{\left(\frac{dv}{dr}\right)_c^{n-m}} - 1 \right] e^{-\left(\alpha\frac{dv}{dr}+\beta\right)\vartheta} \right\} \tag{14.13}$$

in which $\frac{dv}{dr}$ is the actual shear rate, $\left(\frac{dv}{dr}\right)_c$ is the earlier shear rate for the stabilized state of shear, α and β are material constants, ϑ is the duration of shear at the present shear rate.

The equation contains five material constants (K_s, m, n, α, β), which can be determined using a rotational viscometer. Thus we have piecewise, valid constitutive relations for certain restricted changes of the shear state of a thixotropic fluid.

For the entrance region of a one-dimensional steady thixotropic flow in a pipe, Eq. (14.13) can be used. Naturally, the "entrance region" is considered in a thixotropic sense. After the necessary duration of shear, the stabilized shear state will be attained for which Eq. (14.9) is valid. For an abrupt change in the shear rate (change in pipe diameter or flow rate) the new shear stress can be determined using Eq. (14.12). For a decreased shear rate (increase in pipe diameter or decrease in flow rate) the developed shear stress is stable, and smaller than the stabilized shear stress τ_s. For an increased shear rate (decrease in pipe diameter or increase in flow rate), the obtained shear stress represents the initial value of a beginning structure breakdown process, which can be determined using Eq. (14.13).

An important consequence of Eq. (14.12) is that a considerable shear stress decrease can be achieved by applying an initially large shear rate. The flow through a centrifugal pump is not sufficient for this purpose because of the very short duration of shear. An entrance pipe section of smaller diameter may be satisfactory provided it is of a suitable length to produce the necessary duration of shear.

If the shear stress at a constant shear rate increases with the duration of shear, the fluid is called rheopectic. The best-known example of this type of behavior is that of egg-white. Rheopectic behavior is less commonly encountered than thixotropy, although a rheopectic material would be useful as a fracturing fluid.

14.3 LAMINAR FLOW OF PSEUDOPLASTIC FLUIDS IN PIPES

Laminar flow occurs only on rare occasions in the drillpipe. Nevertheless, it is discussed briefly for the sake of completeness. Consider a straight cylindrical pipe of constant cross-section. The orientation of the pipe is arbitrary w.r.t. the gravity field. A steady, one-dimensional flow of an incompressible, pseudoplastic fluid is investigated. It is convenient to choose a cylindrical coordinate system with the notation as shown in Fig. 14.10. The momentum equation for the depicted control volume, similarly to the Newtonian case, can be written as:

$$\frac{\rho gJr}{2} - K\left|\frac{dv}{dr}\right|^n = 0 \qquad (14.14)$$

Rearranging, the velocity gradient can be expressed as:

$$-\frac{dv}{dr} = \left(\frac{\rho gJr}{2K}\right)^{\frac{1}{n}} \qquad (14.15)$$

Integrating this equation yields:

$$v = -\left(\frac{\rho gJ}{2K}\right)^{\frac{1}{n}} \frac{n}{n+1} r^{\frac{n+1}{n}} + k \qquad (14.16)$$

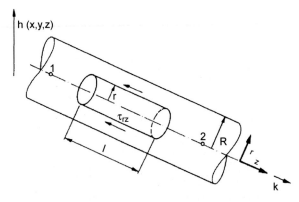

FIGURE 14.10 Cylindrical coordinate system and control volume for pseudoplastic pipe flow.

where k is a constant of integration, which can be obtained from the usual boundary conditions:

$$r = R; \quad v = 0$$

Substituting these values into Eq. (14.16), we get:

$$k = \frac{n}{n+1} \left(\frac{\rho g J}{2K} \right)^{\frac{1}{n}} R^{\frac{n+1}{n}}$$

which leads to the final form of the velocity distribution along the pipe radius:

$$v = \frac{n}{n+1} \left(\frac{\rho g J R}{2K} \right)^{\frac{1}{n}} R \left[1 - \left(\frac{r}{R} \right)^{\frac{n+1}{n}} \right] \tag{14.17}$$

The flow rate is obtained as:

$$Q = 2\pi \int_0^R rv\,dr = \frac{n}{3n+1} \left(\frac{\rho g J R}{2K} \right)^{\frac{1}{n}} R^3 \pi \tag{14.18}$$

Consequently the cross-sectional average velocity is:

$$c = \frac{Q}{R^2 \pi} = \frac{n}{3n+1} \left(\frac{\rho g J R}{2K} \right)^{\frac{1}{n}} R \tag{14.19}$$

It is convenient to express the velocity distribution in terms of the averaged velocity:

$$v = \frac{3n+1}{n+1} c \left[1 - \left(\frac{r}{R} \right)^{\frac{n+1}{n}} \right] \tag{14.20}$$

The velocity maximum is obtained at the pipe axis:

$$v_{max} = \frac{3n+1}{n+1} c \tag{14.21}$$

The maximum shear rate occurs at the pipe wall:

$$\left(\frac{dv}{dr} \right)_R = - \left(\frac{\rho g J R}{2K} \right)^{\frac{1}{n}} = - \frac{c}{R} \frac{1+3n}{n} \tag{14.22}$$

The shear stress at the wall is:

$$\tau_R = - \frac{\rho g J R}{2} \tag{14.23}$$

The shear-stress distribution along the radius is obviously linear:

$$\tau = \frac{\rho g J R}{2} = \frac{\tau_R}{R} r \tag{14.24}$$

The relation between the wall-shear stress and the wall-shear rate is obtained by combining Eqs. (14.22) and (14.23):

$$\tau_R = K\left(\frac{dv}{dr}\right)_R^n \tag{14.25}$$

The dimensionless velocity distributions are plotted in Fig. 14.11 for pseudoplastic fluids of different behavior indexes, as:

$$\frac{v}{c} = \frac{3n+1}{n+1}\left[1 - \left(\frac{r}{R}\right)^{\frac{n+1}{n}}\right] \tag{14.26}$$

This figure illustrates the effect of the behavior index on the velocity profile. In the special case of a Newtonian fluid, $n = 1$, and the usual parabola is obtained. It can be seen that as n approaches zero, the velocity profiles flattens (resembling a plug flow), whereas as n increases in the dilatant region the profile steepens, tending towards an inclined straight line.

The head loss of a pipe section of length L and diameter D can be also determined. The mechanical energy equation in its general form can be written as:

$$\frac{c_1^2}{2g} + \frac{p_1}{g} + h_1 = \frac{c_2^2}{2g} + \frac{p_2}{g} + h_2 + h'_{1-2} \tag{14.27}$$

Since the cross-section of the pipe is constant:

$$c_1 = c_2$$

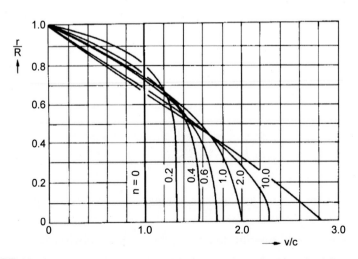

FIGURE 14.11   Dimensionless velocity distribution of pseudoplastic fluid flows.

thus we get:

$$h'_{1-2} = \frac{p_1 - p_2}{\rho g} + h_1 - h_2 = JL \tag{14.28}$$

Using Eq. (14.19) to express the hydraulic gradient J, and after substitution into Eq. (14.28), the head loss is obtained as:

$$h'_{1-2} = \frac{2K}{\rho g} \left(\frac{3n+1}{n}\right)^n \frac{L}{R^{n+1}} c^n \tag{14.29}$$

It is conspicuous that the head loss is not a linear function of the cross-sectional average velocity, not even for laminar flow. The head loss for pseudoplastic flow increases with the averaged velocity to a lesser degree than in the Newtonian case. The head loss is linearly proportional to the length of the pipe. For pseudoplastic flow, a decrease in the diameter causes a smaller increase in the head loss than for a Newtonian fluid. Naturally, for $n = 1$, Eq. (14.29) reduces to the special case of the Hagen—Poiseuille equation.

Metzner and Reed (1955) proposed extending the Weisbach equation:

$$h'_{1-2} = \lambda \frac{L}{D} \frac{c^2}{2g} \tag{14.30}$$

to pseudoplastic fluid flow, as well.

Comparing Eqs. (14.29) and (14.30), the friction factor is obtained as:

$$\lambda = \frac{8K}{\rho} \left(\frac{6n+2}{n}\right)^n \frac{c^{n-2}}{D^n} \tag{14.31}$$

In the manner usual for Newtonian fluids, we shall express the friction factor in terms of the Reynolds number:

$$\lambda = \frac{64}{Re_p^*} \tag{14.32}$$

Using this, we obtain the expression:

$$Re_p^* = \frac{c^{n-2}D^n \rho}{\frac{K}{8}\left(\frac{6n+2}{n}\right)^n} \tag{14.33}$$

If the friction factor is plotted against Re_p^* all points fall along the same curve, irrespective of the value of n, as it is shown in Fig. 14.12.

It has already been noted that the power law equation is not applicable at very low shear rates. Near to the pipe axis, the shear rate approaches zero, thus in this region the accuracy of the power law approximation decreases. Fortunately, the shear stress similarly tends to zero in this inner region so that the contribution of this zone to the energy dissipation is

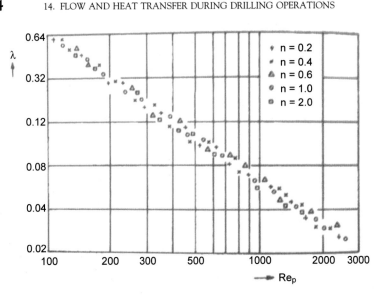

FIGURE 14.12   Friction factor and Reynolds number.

negligible. Thus friction factor and head loss calculations based on the power law equation are sufficiently accurate for practical purposes.

14.4 PSEUDOPLASTIC FLUID FLOW IN ANNULI

Flow in annuli is encountered in drilling and well-completion technology, in which drilling mud or cement paste flows between the borehole wall and the drilling pipe, or between two concentric pipes of casing. Since most drilling fluids are pseudoplastic, or thixotropic–pseudoplastic, the importance of this problem is obvious.

We consider a steady laminar flow of an incompressible, pseudoplastic fluid in a straight, infinitely long, concentric annulus. The position of the symmetry axis is arbitrary w.r.t. the gravity field. A cylindrical coordinate system is chosen; its z-axis directed parallel with the flow. The velocity field is axisymmetrical, having only one non-zero component:

$$v_z = v; \quad v_r = 0; \quad v_\varphi = 0$$

The momentum equation for this case can be written as:

$$\rho g J r = -\frac{d}{dr}\left(r\frac{d\tau_{rz}}{dr}\right) \tag{14.34}$$

Considering Fig. 14.13, it is seen that at a radial position $r = R_0$ both dv/dr and τ_{rz} change signs.

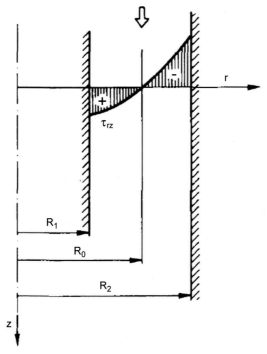

FIGURE 14.13   Pseudoplastic fluid flow through an annulus.

The sign of shear stress is determined by the sign of the velocity gradient; where:

$$r < R_0; \quad \frac{dv}{dr} > 0, \text{ and thus } \tau_{rz} > 0$$

In the region:

$$r > R_0; \quad \frac{dv}{dr} < 0, \text{ and } \tau_{rz} < 0$$

The differential equation for this problem is:

$$-\rho gJr = \frac{d}{dr}\left[rK \left|\frac{dv}{dr}\right|^{n-1} \frac{dv}{dr} \right] \qquad (14.35)$$

Its first integral is readily obtained:

$$rK \left|\frac{dv}{dr}\right|^{n-1} \frac{dv}{dr} = A - \frac{\rho gJr^2}{2} \qquad (14.36)$$

Using the fact that at:

$$r = R_0 \quad \frac{dv}{dr} = 0$$

we obtain the constant of integration as:

$$A = \frac{\rho g J R_0^2}{2}$$

Substituting into Eq. (14.36), we obtain:

$$\left| \frac{dv}{dr} \right|^{n-1} \frac{dv}{dr} = \frac{\rho g J}{2K} \left(\frac{R_0^2}{r} - r \right) \tag{14.37}$$

It is clear that the differential Eq. (14.37) has two domains of solution which interconnect at $r = R_0$.

When $r < R_0$, the right-hand side of the equation has a positive sign, since $\frac{R_0^2}{r} > r$, accordingly we have the condition $dv/dr > 0$.

On the other hand, when $r > R_0$, the right-hand side of the equation has a negative sign, since $\frac{R_0^2}{r} < r$, so that we now have the condition $\frac{dv}{dr} < 0$.

We will first solve the differential equation in the domain $r < R_0$. The sign of the nth root is of course positive:

$$\frac{dv}{dr} = \left(\frac{\rho g J}{2K} \frac{R_0^2 - r^2}{r} \right)^{\frac{1}{n}} \tag{14.38}$$

The function:

$$\frac{R_0^2 - r^2}{r}$$

cannot be integrated in quadratic form, since it does not satisfy Tschebischew's conditions. Since the function is monotonic in the region of $R_1 \leq r \leq R_0$, it can be replaced by a linear function of interpolation.

Since the velocity must be zero at $r = R_1$ the constant of integration can be obtained. Thus the velocity distribution in the interval $R_1 < r < R_0$ is given by:

$$v = \left(\frac{\rho g J}{2K} \right)^{\frac{1}{n}} \frac{\left(\frac{R_0^2 - R_1^2}{R_1} \right)}{R_1 - R_0} \left(\frac{r^2}{2} - R_0 r + R_0 R_1 - \frac{R_1^2}{2} \right) \tag{14.39}$$

This equation contains the temporarily unknown parameter R_0. This can be eliminated by relating the two velocity distributions obtained from the two solutions. To do this, it is necessary to make use of the maximum velocity. Substituting the condition $r = R_0$; $v = v_{max}$ we get:

$$v_{max} = \left(\frac{\rho g J}{2K} \right)^{\frac{1}{n}} \frac{\left(\frac{R_0^2 - R_1^2}{R_1} \right)}{R_1 - R_0} \left[\frac{R_0^2 - R_1^2}{2} - R_0(R_0 - R_1) \right] \tag{14.40}$$

After simplification the maximum velocity is obtained as:

$$v_{max} = \left(\frac{\rho g J}{2K} \frac{R_0^2 - R_1^2}{R_1}\right)^{\frac{1}{n}} \frac{R_0 - R_1}{2} \tag{14.41}$$

The second part of the solution is that for the region of $R_0 \leq r \leq R_2$, where:

$$r > R_0, \quad \frac{dv}{dr} < 0, \text{ and } \left(\frac{R_0^2}{r} - r\right) < 0$$

The shear stress is also negative in this region. To take into account the signs, Eq. (14.37) may be written as:

$$\left|\frac{dv}{dr}\right|^n \text{sign}\left(\frac{dv}{dr}\right) = -\frac{\rho g J}{2K}\left(\frac{R_0^2}{r} - r\right) \tag{14.42}$$

For the velocity gradient we get:

$$\frac{dv}{dr} = \left[\frac{\rho g J}{2K}\left(\frac{R_0^2}{r} - r\right)\right]^{\frac{1}{n}} \text{sign}\left(\frac{R_0^2}{r} - r\right) \tag{14.43}$$

Since:

$$\left(\frac{R_0^2}{r} - r\right) < 0, \quad \text{we have } \left|\frac{R_0^2}{r} - r\right| = r - \frac{R_0^2}{r}$$

and thus sign $\left(\frac{R_0^2}{r} - r\right) = -1$.

Thus the differential equation which we have to solve is:

$$\frac{dv}{dr} = -\left[\frac{\rho g J}{2K} \frac{r^2 - R_0^2}{r}\right]^{\frac{1}{n}} \tag{14.44}$$

The integral of this expression cannot be obtained in quadratic form; we must again interpolate using the linear function:

$$f_2 = a_2 r + b_2$$

The function is fitted to the points R_0 and R_2. The coefficients are found to be:

$$a_2 = \frac{\left(\dfrac{R_2^2 - R_0^2}{R_2}\right)^{\frac{1}{n}}}{R_2 - R_0}$$

$$b_2 = \frac{\left(\dfrac{R_2^2 - R_0^2}{R_2 - R_0}\right)^{\frac{1}{n}}}{R_2 - R_0} R_0$$

After substitution and integration we obtain:

$$v = -\left(\frac{\rho g J}{2K}\right)^{\frac{1}{n}} \frac{\left(\frac{R_2^2 - R_0^2}{R_2}\right)^{\frac{1}{n}}}{R_2 - R_0}\left(\frac{r^2}{2} - R_0 r\right) + C_2 \tag{14.45}$$

Since at $r = R_2$, and $v = 0$, expressing and substituting the constant of integration C_2, we obtain:

$$v = \left(\frac{\rho g J}{2K}\right)^{\frac{1}{n}} \frac{\left(\frac{R_2^2 - R_0^2}{R_2}\right)}{R_2 - R_0}\left[\frac{R_2^2 - r^2}{2} - R_0(R_2 - r)\right] \tag{14.46}$$

The maximum velocity at $r = R_0$ is obtained as:

$$v_{max} = \left[\frac{\rho g J}{2K} \frac{R_2^2 - R_0^2}{R_2}\right]^{\frac{1}{n}} \frac{R_2 - R_0}{2} \tag{14.47}$$

(See Fig. 14.14).

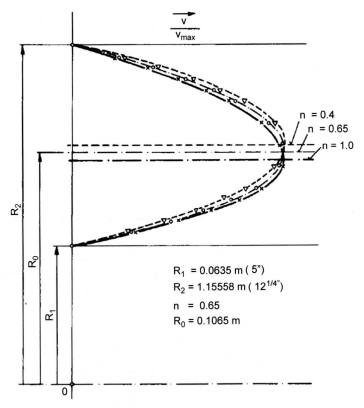

$R_1 = 0.0635 \text{ m } (5")$
$R_2 = 1.15558 \text{ m } (12^{1/4}")$
$n = 0.65$
$R_0 = 0.1065 \text{ m}$

FIGURE 14.14   Velocity distribution of the annular flow.

Since the v_{max} obtained by Eqs. (14.41) and (14.47) has to have the same value, we can equate Eqs. (14.41) and (14.47), and obtain the following equation for R_0:

$$\left(\frac{R_2}{R_1}\frac{R_0^2 - R_1^2}{R_2^2 - R_1^2}\right)^{\frac{1}{n}} = \frac{R_2 - R_0}{R_0 - R_1} \tag{14.48}$$

This implicit equation can be solved by iteration if R_1, R_2 and n are known. The results can be tabulated, for example for a borehole with a diameter of $D_2 = 12\,\frac{1}{4}$ in. and a drill-pipe diameter of $D_1 = 5$ in.

Knowing the two parts of the velocity distribution the flow rate can be determined. After determination of the integrals:

$$Q = 2\pi \left(\int_{R_1}^{R_0} vr\,dr + \int_{R_0}^{R_2} vr\,dr \right) \tag{14.49}$$

the flow rate is obtained as:

$$
\begin{aligned}
Q = 2\pi &\left(\frac{\rho gJ}{2K}\right)^{\frac{1}{n}} \left\{ \frac{\left(\frac{R_0^2 - R_1^2}{R_1}\right)^{\frac{1}{n}}}{R_1 - R_0} \left[\frac{R_0^4 - R_1^4}{8} - \frac{R_0(R_0^3 - R_1^3)}{3} + \frac{R_0 R_1(R_0^2 - R_1^2)}{2} \right.\right.\\
&\left. - \frac{R_1^2(R_0^2 - R_1^2)}{4}\right] + \frac{\left(\frac{R_2^2 - R_0^2}{R_2}\right)^{\frac{1}{n}}}{R_2 - R_0} \left[-\frac{R_2^4 - R_0^4}{8} + \frac{R_0(R_2^3 - R_0^3)}{3} \right.\\
&\left.\left. - \frac{R_0 R_2(R_2^2 - R_0^2)}{2} + \frac{R_2^2(R_2^2 - R_0^2)}{4}\right] - \left[\frac{R_0 R_2(R_2^2 - R_0^2)}{2} + \frac{R_2^2(R_2^2 - R_0^2)}{4}\right] \right\}
\end{aligned}
\tag{14.50}
$$

If we designate the quantity in the curl bracket by A, then:

$$Q = 2\pi A \left(\frac{\rho gJ}{2K}\right)^{\frac{1}{n}} \tag{14.51}$$

Representative values of A together with values of R_0 are listed in Table 14.1. The cross-sectional average velocity is:

$$c = \frac{2A}{R_2^2 - R_1^2} \left(\frac{\rho gJ}{2K}\right)^{\frac{1}{n}} \tag{14.52}$$

Using this equation the hydraulic gradient is found to be:

$$J = \frac{2K}{\rho g} \left(\frac{R_2^2 - R_1^2}{2A}\right)^{n} c^n \tag{14.53}$$

TABLE 14.1   Borehole Diameter 12 ¼″ Outer
Diameter Drilling Pipe 5ⁿ

n	$R_0(m)$	ϕ
0.30	0.1116	1.422274
0.35	0.1108	1.617641
0.40	0.1099	1.827188
0.45	0.1091	2.049351
0.50	0.1084	2.283278
0.55	0.1078	2.534876
0.60	0.1071	2.803840
0.65	0.1065	3.092167
0.70	0.1060	3.401992
0.75	0.1054	3.735550
0.80	0.1049	4.092524
0.85		
0.90	0.1040	4.903305
0.95	0.1036	5.357264
1.00	0.1032	5.847680

The head loss of an annulus of length L can thus be expressed as:

$$h' = \frac{2KL}{\rho g} \left(\frac{R_2^2 - R_1^2}{2A} \right)^n c^n \tag{14.54}$$

A so-called annulus shape parameter can be defined by the equation:

$$\phi = (R_2 - R_1)^{n+1} \left(\frac{R_2^2 - R_1^2}{2A} \right)^n \tag{14.55}$$

Applying this notation the head loss is obtained as:

$$h' = \frac{2K}{\rho g} \frac{\phi L}{(R_2 - R_1)^{n+1}} c^n \tag{14.56}$$

The Weisbach equation for non-circular channels can be written as:

$$h' = \lambda \frac{L}{4R_H} \frac{c^2}{2g} \tag{14.57}$$

where R_H is the hydraulic radius. By comparing Eqs. (14.56) and (14.57), we obtain the friction factor for the annulus:

$$\lambda_a = \frac{8K\phi}{\rho} \frac{c^{n-2}}{(R_2 - R_1)^n} \qquad (14.58)$$

A modified "annulus" Reynolds number can be defined in terms of the equation:

$$\lambda_a = \frac{64}{Re_a} \qquad (14.59)$$

From Eqs. (14.58) and (14.59), the annulus Reynolds number can be expressed as:

$$Re_a = \frac{8\rho c^{2-n}(R_2 - R_1)^n}{K\phi} \qquad (14.60)$$

Thus the recommended steps of calculation to determine the head loss for a laminar annular flow of a pseudoplastic fluid are the following:

1. Determination of the radial position R_0 of the maximum velocity by iteration from Eq. (14.48).
2. Knowing R_0, the constant A, and the annulus form, parameter ϕ can be calculated from Eqs. (14.50) and (14.55).
3. Determination of the modified annulus Reynolds number from Eq. (14.60).
4. Calculation of the friction factor using Eq. (14.59).
5. Finally we can determine the head loss for the annulus by applying Eq. (14.57) or, more directly, by omitting steps 3 and 4 and using Eq. (14.56).

14.5 TURBULENT FLOW OF NON-NEWTONIAN FLUIDS IN PIPES

For any kind of non-Newtonian fluid, laminar flow gives way to turbulent flow once a critical value of the Reynolds number is exceeded. Experimental data show that the value of the critical Reynolds number differs slightly from fluid to fluid. For the most common pseudoplastic fluids, the following relationship is found to hold:

$$\left(Re_p^*\right)_{cr} = \frac{6464n(n+2)^{\frac{n+2}{n+1}}}{(3n+1)^2} \qquad (14.61)$$

The friction factor values corresponding to the critical Reynolds number are shown in Fig. 14.15. The laminar–turbulent transition is strongly influenced by the elastic properties of the fluid. Methods of

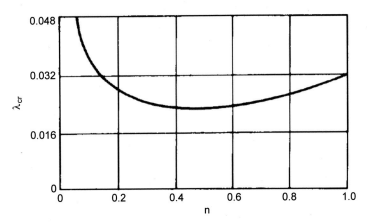

FIGURE 14.15   Critical friction factor values for different pseudoplastic fluids.

retarding this transition will be discussed later in connection with turbulent drag reduction.

In the following, our analysis will be restricted solely to pseudoplastic
turbulent flow. The analogous equations for other types of flow are obtained in a similar way.

We consider the case of a steady, one-dimensional axisymmetrical
turbulent flow in a circular, infinitely long pipe. A cylindrical coordinate
system is chosen. The momentum equation for this case is.

$$\frac{\rho g J r}{2} + K \left| \frac{dv}{dr} \right|^{n-1} \left(\frac{dv}{dr} \right) - \rho \overline{u'w'} = 0 \tag{14.62}$$

where $-\rho \overline{u'w'}$ is the τ'_{rz} component of the turbulent stress tensor.

The solution of this differential equation can be resolved into two parts
in accordance with the nature of the flow. The flow near the wall is
laminar within a very thin laminar sublayer. Turbulent momentum
transfer cannot be developed here. Since the thickness of the laminar
sublayer δ is very small relative to the radius R of the pipe, the velocity
distribution within the laminar sublayer may be taken as linear. This is
equivalent to the existence of a uniform shear stress in the sublayer equal
to the wall-shear stress τ_R. Thus, we can write:

$$\frac{\rho g J r}{2} - K \left| \frac{dv}{dr} \right|^n = 0 \tag{14.63}$$

The definition of the friction velocity:

$$v_* = \sqrt{\frac{g J R}{2}} \tag{14.64}$$

can be used again, so that we obtain:

$$-\frac{dv}{dr} = \left(\frac{v_*^2 \rho}{K}\right)^{\frac{1}{n}}$$

(14.65)

The linearized velocity profile within the laminar sublayer can be expressed in dimensionless form:

$$\frac{v}{v_*} = v_*^{\frac{2-n}{n}} \left(\frac{\rho}{K}\right)^{\frac{1}{n}} (R - r)$$

(14.66)

Outside the laminar sublayer, the viscous shear stresses are negligibly small compared to the turbulent stresses. Applying Kármán's (1930) equation for the mixing length, the momentum equation is obtained as:

$$\frac{\rho g J r}{2} = \rho \kappa^2 \frac{\left(\frac{dv}{dr}\right)^4}{\left(\frac{d^2v}{dr^2}\right)^2}$$

(14.67)

The integration of this differential equation is the same as for the Newtonian case. Thus, the dimensionless velocity distribution for the turbulent core flow is obtained as:

$$\frac{v}{v_*} = \frac{1}{\kappa} \left[\sqrt{\frac{r}{R}} + \ln\left(1 - \sqrt{\frac{r}{R}}\right) \right] + \frac{v_{max}}{v_*}$$

(14.68)

The velocity distribution of the laminar sublayer and that of the turbulent core flow have to yield the same value for the velocity at the boundary of the two regions, i.e., where $r = R - \delta$. Thus, equating the two expressions for the velocity distributions, we obtain:

$$\frac{v_{max}}{v_*} + \frac{1}{\kappa}\left[\sqrt{1 - \frac{\delta}{R}} + \ln\left(1 - \sqrt{1 - \frac{\delta}{R}}\right)\right] = \frac{v_*^{\frac{2-n}{n}} \delta}{\left(\frac{K}{\rho}\right)^{\frac{1}{n}}}$$

(14.69)

Since $\frac{\delta}{R} \ll 1$, the quantity $\left(1 - \sqrt{1 - \frac{\delta}{R}}\right)$ can be expanded into a binomial series. Neglecting the higher-order terms we get:

$$\sqrt{1 - \frac{\delta}{R}} \cong 1 - \frac{\delta}{2R} = 1$$

$$\ln\left(1 - \sqrt{1 - \frac{\delta}{R}}\right) \cong \ln\left[1 - \left(1 - \frac{\delta}{2R}\right)\right] = \ln\frac{\delta}{2R}$$

After substitution the maximum velocity is obtained as:

$$\frac{v_{max}}{v_*} = \left(\frac{v_*^{2-n}\rho}{K}\right)^{\frac{1}{n}} \delta - \frac{1}{\kappa}\left(1 + \ln\frac{\delta}{D}\right)$$

(14.70)

To calculate the maximum velocity it is necessary that δ be known. Prandtl assumed that for a Newtonian fluid:

$$Re_\delta = \frac{v_* \delta}{\nu} = const.$$

Analogously, we assume that for a pseudoplastic fluid:

$$Re_{p\delta} = \frac{v_*^{2-n} \delta^n \rho}{\frac{K}{8} \left(\frac{6n+2}{n}\right)^n} const. \tag{14.71}$$

Using Eqs. (14.71) and (14.33), the thickness of the laminar sublayer can be expressed as:

$$\delta = \left(\frac{Re_{p\delta}}{Re_p}\right)^{\frac{1}{n}} \left(\frac{c}{v_*}\right)^{\frac{2-n}{n}} D \tag{14.72}$$

Substituting this into Eq. (14.70), we get:

$$\frac{v_{max}}{v_*} = \frac{1}{m\kappa} \ln\left[Re_p \left(\frac{v_*}{c}\right)^{2-n}\right] - \frac{1}{\kappa}\left(1 + \frac{\ln Re_{p\delta}}{n}\right) + \frac{6n+2}{n}\left(\frac{Re_{p\delta}}{8}\right)^{\frac{1}{n}} \tag{14.73}$$

This expression contains two temporarily unknown constant $Re_{p\delta}$ and κ. The velocity distribution is obtained as:

$$\frac{v}{v_*} = \frac{1}{\kappa}\left[\sqrt{\frac{r}{R}} + \ln\left(1 - \sqrt{\frac{r}{R}}\right)\right] + \frac{1}{n\kappa} \ln\left[Re_p \left(\frac{v_*}{c}\right)^{2-n}\right]$$
$$- \frac{1}{\kappa}\left(1 + \frac{\ln Re_{p\delta}}{n}\right) + \frac{6n+2}{n}\left(\frac{Re_{p\delta}}{8}\right)^{\frac{1}{n}} \tag{14.74}$$

The cross-sectional average velocity can be calculated in the usual way, and is found to be:

$$\frac{c}{v_*} = \frac{v_{max}}{v_*} - \frac{1}{\kappa}\left(\frac{25}{12} - \frac{4}{5}\right) \tag{14.75}$$

thus:

$$\frac{c}{v_*} = \frac{1}{n\kappa} \ln\left[Re_p \left(\frac{v_*}{c}\right)^{2-n}\right] - \frac{1}{\kappa}\left(\frac{77}{60} + \frac{\ln(Re_{p\delta})}{n}\right) + \frac{6n+2}{n}\left(\frac{Re_{p\delta}}{8}\right)^{\frac{1}{n}} \tag{14.76}$$

The friction factor can be calculated in the same manner as for the Newtonian case:

$$\frac{1}{\sqrt{\lambda}} = \frac{0.8141}{n\kappa} lg\left(Re_p \lambda^{\left(1-\frac{n}{2}\right)}\right) + 0.7532 \frac{n-2}{2n}$$
$$+ 0.3535 \left[\frac{6n+2}{n}\left(\frac{Re_{p\delta}}{8}\right)^{\frac{1}{n}}\right] - \frac{1}{\kappa}\left(2.283 + \frac{\ln Re_{p\delta}}{n}\right) \tag{14.77}$$

Laboratory and in situ measurements by Bobok and Navratil (1982) both provided similar values of:

$$Re_{p\delta} = 12.087,$$

$$\kappa = 0.407$$

These values are the same as those obtained for the Newtonian case. Thus, finally we can write:

$$\frac{1}{\sqrt{\lambda}} = \frac{2}{n}\lg\left(Re_p\lambda^{\left(1-\frac{n}{2}\right)}\right) + 1.511^{\frac{1}{n}}\left(\frac{0.707}{n} + 2.121\right)$$
$$- \frac{4.015}{n} - 1.057 \tag{14.78}$$

or, in a simpler form:

$$\frac{1}{\sqrt{\lambda}} = \frac{2}{n}\lg\left(Re_p\lambda^{\left(1-\frac{n}{2}\right)}\right) + \beta \tag{14.79}$$

For a fully rough pipe, the friction factor is invariably:

$$\frac{1}{\sqrt{\lambda}} = 2\lg\left(3.715\frac{D}{K}\right) \tag{14.80}$$

In the transition region between the friction factor curve for a smooth pipe and that for a wholly rough region, an interpolation formula can be obtained analogous to the Colebrook equation:

$$\frac{1}{\sqrt{\lambda}} = -2\lg\left[\frac{10^{-\frac{\beta}{2}}}{Re_p^{\frac{1}{n}}\lambda^{\frac{2-n}{2n}}} + \frac{k}{3.715D}\right] \tag{14.81}$$

This so-called BNS-equation was elaborated by Bobok et al. (1981). Friction factor charts based on the BNS equation are shown in Figs. 14.16 and 14.17. The slope of the friction factor curves for hydraulically smooth pipes decreases as the behavior index decreases. The eventual laminar–turbulent transition is the other remarkable feature of the curves for a certain range of n. Experimental data obtained for crude oil pipelines with diameters ranging from 100 to 600 mm are in good agreement with the equation.

14.6 TURBULENT FLOW OF PSEUDOPLASTIC FLUIDS THROUGH ANNULI

The vertical upflow of the drilling mud through the annular space between the casing and the drill string is a complex mechanical and thermodynamical phenomenon. Considering the fluid to be incompressible, it is possible the substantial simplification of the mathematical model. It is

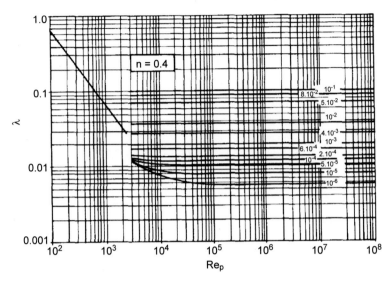

FIGURE 14.16   Friction factor chart for turbulent flow of a pseudoplastic fluid $n = 0.4$.

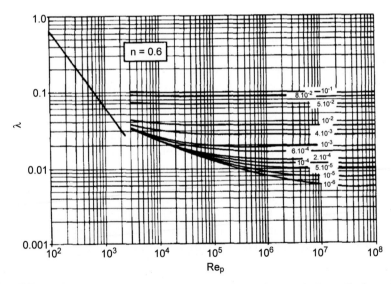

FIGURE 14.17   Friction factor chart for turbulent flow of a pseudoplastic fluid $n = 0.6$.

well-known fact that for incompressible fluids, the pressure is not a thermodynamic variable, thus it is independent of the temperature. Since there is no relation between the pressure and the temperature, the mechanical energy equation and the thermal energy equation can be solved independently. This fact makes possible an easy method of calculating the head loss of an incompressible, non-isothermal flow. It is obvious that the mathematical model refers to a simplified hypothetical phenomenon.

Consider the annular space between two coaxial cylindrical surfaces with radii of R_1 and R_2. Between the two cylindrical surfaces incompressible, pseudoplastic fluid flows in an axial direction. The flow is steady, axisymmetrical, and turbulent. An axisymmetrical, cylindrical control surface is chosen between the radii of r and r + dr. The momentum equation in this case can be written as:

$$\rho g J r + \frac{d}{dr}\left(rK\left|\frac{dv}{dr}\right|^{n-1}\frac{dv}{dr}\right) - \frac{d}{dr}\left[\rho r \kappa^2 \frac{\left(\frac{dv}{dr}\right)^4}{\left(\frac{d^2v}{dr^2}\right)^2}\right] = 0 \tag{14.82}$$

It can be recognized that the first term is the force that maintains the flow. The second term represents the viscous shear stress, while the third one is the apparent turbulent shear stress, based on the mixing length formula of Kármán's (1930). Both shear stress acts against the flow. A more brief form of Eq. (14.82) is:

$$\rho g J r = -\frac{d}{dr}\left[r\left(\tau_{rz} + \tau'_{rz}\right)\right] \tag{14.83}$$

The radial distribution of the sum of the stresses is obtained as:

$$\tau_{rz} + \tau'_{rz} = -\frac{\rho g J r}{2} + \frac{A}{r} \tag{14.84}$$

Since both the viscous and the turbulent shear stresses are zero in the location of the velocity maximum, the constant of integration A can be determined by satisfying the boundary condition; if $r = R_0$, then $\tau_{rz} + \tau'_{rz} = 0$.

Thus, we get:

$$\tau_{rz} + \tau'_{rz} = -\frac{\rho g J r}{2} + \left(\frac{R_0^2}{r} - r\right) \tag{14.85}$$

R_0 is the temporarily unknown radius of the location of the velocity maximum. This separates the two domain of the solution of Eq. (14.82). In the region of:

$$R_1 < r < R_0 \qquad \frac{dv}{dr} < 0,$$

while within the interval:

$$R_0 < r < R_2 \qquad \frac{dv}{dr} > 0,$$

Thus the momentum equation is solved separately in these two regions. Further, two regions of solution are obtained within the two laminar sublayers on the two cylindrical bounding surfaces. The laminar

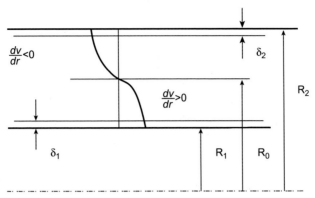

FIGURE 14.18   Different regions of solution of the momentum equation.

sublayers have different thicknesses of δ_1 and δ_2. All this is shown in Fig. 14.18.

The differential equation for the laminar sublayers is:

$$-\rho g J r = \frac{d}{dr}\left(rK\left|\frac{dv}{dr}\right|^{n-1}\frac{dv}{dr}\right) \qquad (14.86)$$

In the inner, turbulent region, the following equation is valid:

$$gJr = \frac{d}{dr}\left[r\kappa^2\frac{\left(\frac{dv}{dr}\right)^4}{\left(\frac{d^2v}{dr^2}\right)^2}\right] \qquad (14.87)$$

Thus in the whole annular cross-section, the velocity distribution is composed from four sections as it is shown in Fig. 14.19. These part-distributions are obtained by the following expressions:

The linearized velocity distribution refers to the laminar sublayer on the outer cylindrical surface is:

$$v_2 = \left[\frac{\rho g J}{2K}\cdot\frac{R_2^2 - R_0^2}{R_2}\right]^{\frac{1}{n}}(R_2 - r) \qquad (14.88)$$

The turbulent velocity profile, which is valid in the interval of $R_2 \geq r \geq R_0$, is obtained as:

$$\frac{v_2 - v_{max}}{v_{*2}} = -\frac{1}{\kappa}\left[\sqrt{\frac{r - R_0}{R_2 - R_0}} + \ln\left(1 - \sqrt{\frac{r - R_0}{R_2 - R_0}}\right)\right] \qquad (14.89)$$

The turbulent velocity profile belonging to the interval of $R_0 \geq r \geq R_1$ is:

$$\frac{v_1 - v_{max}}{v_{*1}} = -\frac{1}{\kappa}\left[\sqrt{\frac{R_0 - r}{R_0 - R_1}} + \ln\left(1 - \sqrt{\frac{R_0 - r}{R_0 - R_1}}\right)\right] \qquad (14.90)$$

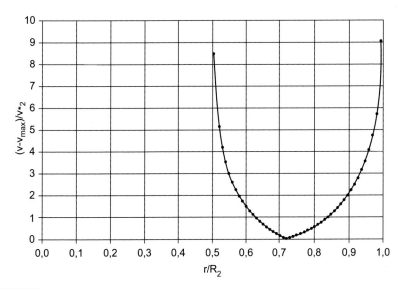

FIGURE 14.19 Turbulent velocity profile in the annulus.

The linearized laminar velocity distribution of the laminar sublayer on the inner cylindrical surface is:

$$v_1 = \left[\frac{\rho g J}{2K} \cdot \frac{R_0^2 - R_1^2}{R_1}\right]^{\frac{1}{n}} (r - R_1) \qquad (14.91)$$

The location of the velocity maximum depends on the relationship of R_1/R_2. The turbulent velocity distributions are fitted to the corresponding laminar sections. The velocity maximum is obtained the same from both equations. The radius belonging to the velocity maximum is determined by the equal values of the laminar and turbulent velocities obtained at the edge of the laminar sublayer. Finally, the obtained equation can be solved by the Newton–Raphson method. This result can be approximated for the everyday practice by the formula:

$$R_0 = \frac{R_1 + R_2 \left(\frac{R_1}{R_1}\right)^{\frac{7}{20}}}{1 + \left(\frac{R_1}{R_1}\right)^{\frac{7}{20}}} \qquad (14.92)$$

Based on Eq. (14.90), the cross-sectional mean velocity is obtained as:

$$\frac{c}{v_{*2}} = \frac{v_{max}}{v_{*2}} - \frac{4}{\kappa(R_2^2 - R_1^2)} \left\{ \frac{77}{240}(R_2 - R_0)^2 + \frac{19}{12} R_0(R_2 - R_0) \right.$$

$$\left. + \sqrt{\frac{R_2}{R_1} \cdot \frac{(R_0^2 - R_1^2)}{(R_2^2 - R_0^2)}} \left[\frac{77}{240}(R_0 - R_1)^2 + \frac{11}{12} R_0(R_0 - R_1)\right] \right\} \qquad (14.93)$$

The friction factor for the turbulent annular flow can be determined by the equation:

$$\frac{1}{\sqrt{\lambda}} = -2\lg\left\{\frac{\left[1+\left(\frac{2R_1}{R_2}\right)^2\right]10^{-\frac{\beta}{2}}}{Re_*^n \cdot \lambda^{\frac{2-n}{2n}}} + \frac{4+\frac{R_1}{R_2}}{30}\frac{k}{R_2-R_1}\right\}$$ (14.94)

In this equation, the modified Reynolds number for the annular flow is:

$$Re_a^* = \frac{c^{2-n}[2(R_1+R_2)]^n\rho}{\frac{K}{8}\left(\frac{6n+2}{n}\right)^n}$$ (14.95)

The power is:

$$\beta = 1,511^{\frac{1}{n}}\left(\frac{0,707}{n}+2,121\right)-\frac{4,015}{n}-1,057$$ (14.96)

As a special case, when $R_1 = 0$ and $R_0 = 0$, this formula obtains the BNS-equation for cylindrical pipes. The dependence of the friction factor of the annular flow on the modified Reynolds number and the relative roughness is shown in Fig. 14.20.
Knowing the friction factor, the head loss in the annular section is:

$$h' = \lambda\frac{L}{2(R_2+R_1)}\frac{c^2}{2g}$$ (14.97)

Thus the head losses for all elements of the closed-loop circulation system of the drilling mud are determined.

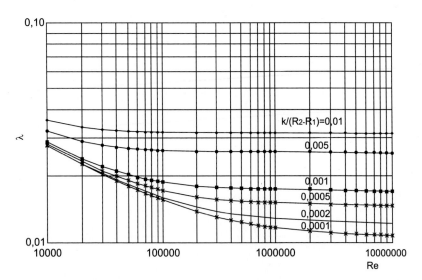

FIGURE 14.20   Friction factor chart of the annular flow.

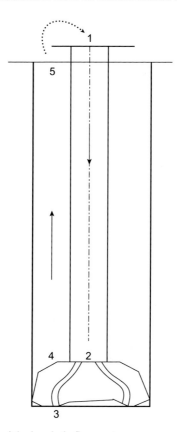

FIGURE 14.21 Sketch of the borehole flow system.

Consider the mud circulation system as it is shown in Fig. 14.21. The point 1 denotes the inlet cross-section of the drill pipe. The drill pipe is connected to the drill bit at the point 2. The point 3 belongs to the outlet cross-section of the drill bit nozzle. The point 4 denotes that cross-section of the annulus where, after the cleaning of the bottomhole, the drilling fluid and the removed drill cuttings are completely mixed. The point 5 is the outlet cross-section of the annulus, from where the mud flows to the shale shaker, which separates drill cuttings from the mud before it is pumped back down the drill pipe. The streamlines of the mud circulation system are closed curves, the flow can be considered to one-dimensional and steady. All elements of this system are in serial connection. It is obvious that the flow rate is constant while all arising head losses are added. Thus the mechanical energy equation can be written between the points 1 and 5 as:

$$\frac{p_1}{\rho g} + h_1 + \frac{c_1^2}{2g} = \frac{p_5}{\rho g} + h_5 + \frac{c_5^2}{2g} + h_{12}' + h_{23}' + h_{34}' + h_{45}' + W_{bh} \quad (14.98)$$

W is the specific work of the unit-weight fluid necessary to clear away the drill cuttings from the bottomhole. Some terms of this equation can be neglected. Since the inlet pressure is about 10 MP, the differences of the potential energies $h_1 - h_5$ and the kinetic energies $c_1^2/2g - c_5^2/2g$ can be taken negligible. The outlet pressure of the annulus is atmospheric: $p_5 = p_0$. The density difference of the downflowing drilling mud and the upflowing mixture of mud and cuttings can be also neglected. Thus the manometric head of the pump is obtained as:

$$H = \frac{p_1 - p_0}{\rho g} = h'_{12} + h'_{23} + h'_{34} + h'_{45} + W_{bh} \qquad (14.99)$$

The manometric head provides the necessary energy against the head losses and the work of for bottomhole cleaning. The useful energy consumption to accelerate the jets, to remove the drill cuttings from bottomhole and to mix the mud and cuttings is:

$$W_{bh} + h'_{23} + h'_{34} = H - h'_{12} - h'_{45} \qquad (14.100)$$

The efficiency of the circulation system can be improved by decreasing the head losses in the drill pipe and the annulus. It is a very effective method for this the addition of long-chain polymer materials to the drilling fluid. The addition a very small amount (a few ppm only) of polymers to a turbulent pipe flow can result a large reduction of the friction factor. This effect is not viscosity reduction, but the turbulent shear stress is suppressed. The head loss reduction can be even 70%.

14.7 DETERMINATION OF THE TEMPERATURE DISTRIBUTION IN THE CIRCULATING DRILLING FLUID

The knowledge of the temperature distribution in the circulating drilling fluid has a great importance to design drilling operations. Flow and heat transfer in drilling operations is a complex simultaneous interaction between the circulating fluid and the surrounding rock mass around the borehole. The temperature of the flowing fluid in the drill pipe and the annulus is lower than the undisturbed rock temperature. This temperature-inhomogeneity induces a radial heat flux toward the borehole. It must be noted that the upper section of the borehole the fluid is warmer than the rock, thus the heat flux is directed radially outward.

The drilling fluid is heated by the surrounding rock, which is cooled. The heat transfer process is a time-dependent phenomenon during the drilling history. An analytical mathematical model is elaborated to describe this transient heat transfer. It makes it possible to calculate temperature profiles for the entire wellbore and to investigate the influence of the main performance parameters.

Two types of mathematical models, analytical and numerical, have emerged for determination the circulating fluid temperature. Their theoretical bases are the same; the balance equations of mass, momentum, and energy. The initial and boundary conditions, and the material properties are obviously the same. The definitive difference between the two models is the different ways to solve the equations. Applying analytical solutions the obtained differential equations are integrated in closed form as mathematical analytic functions. In order to easy integration some simplifying assumptions can be made. Analytic functions can give information about the system behavior even for lack of detailed calculations. Numerical models use some time and space stepping procedure; the solution is obtained by a generated table or a graph. Numerical techniques are suitable to consider more realistic models of greater complexity, at the cost of extended amount of calculations.

Many early models assumed constant temperatures of the flowing fluid as Edvardson et al. (1962). Others used experimentally determined approximative correlations as Dowdle and Cobb (1975). A sophisticated analytical model was made by Boldizsár (1958), neglecting the thermal resistance of the multiple casing string. Raymond's first numerical model (1969) has included the transient response of the flow for the initial short-time period.

The present model is an analytical approach, in which the unnecessary simplifications are avoided, and a rigorous analytical treatment is applied as far as it is possible within reasonable limits. The temperature distribution is determined along the depth both in the drilling pipe and the annulus. Influencing factors include flow rate, well completion, elapsed time, and geothermal conditions of the surrounding rock are investigated.

The so-called forward circulation system is investigated, where the drilling fluid flows down in the drill pipe and back up in the annulus.

The thermal interaction between the drilling fluid and the formation is considered axisymmetrical. In accordance to the system's geometry a cylindrical coordinate system is chosen. It's z-axis is coaxial with the drillpipe, directing downward. The origin $z = 0$ is at the surface.

At the depth of z, a suitable chosen control volume is taken in order to write the balance equation of the internal energy. The control volume is coaxial with the borehole. Its boundaries consist of a cylindrical surface of radius R_∞ and two horizontal parallel planes of distance dz between them. The radius R_∞ belongs to the location of the undisturbed formation temperature T_∞. The temperature difference between the formation and the drilling fluid induces a radial heat flux through several elements of the wellbore. The control volume is shown in Fig. 14.22. It is convenient to separate it into four sub-systems: the flowing fluid in the drill pipe and the annulus, the well completion, and the formation.

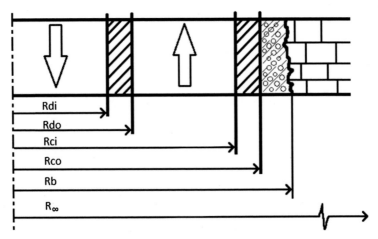

FIGURE 14.22   Elementary control surface.

In the downward flow through the drill pipe, the forced convection is the dominant mechanism of the heat transfer. In the upward flow through the annulus there is a twofold forced convection on both surfaces of the annulus. The well completion may be a completed section of the borehole, but it may be an open hole where the only thermal resistance is simply the forced convection between the formation and drilling fluid. In the formation radial conduction is the definitive phenomenon. These heat flux crossing the boundaries of the sub-systems must be the same on both sides.

Before balance equations are written, some simplifying assumptions can be taken. The drilling fluid is considered incompressible. The flow is steady and turbulent both through the drill pipe and the annulus. It is well-known that the velocity and temperature distribution becomes more uniform over the pipe cross-section as the Reynolds number increases, while the average velocity tends to the hypothetical velocity distribution of a perfect fluid. Thus the cross-sectional average velocities and temperatures can be used in the balance equations. The steep temperature change near the pipe wall in the thermal boundary layer is replaced by an abrupt temperature drop between the solid wall and the flowing fluid. The axial component of the conductive heat flux in the fluid is negligible. The temperature field of the formation is considered to axisymmetrical around the borehole. The rate of change of the temperature in the fluid is substantially greater than in the surrounding rock. Thus the transient heat transfer process can be considered to a slow temperature change of the huge heat capacity rock mass, while the thermal response of the tiny fluid filament in the borehole follows it instantaneously.

Thereafter the balance equation of the internal energy can be written for the drill pipe as follows:

$$\dot{m}cdT_D = 2\pi R_{Di}U_{Di}(T_A - T_D)dz \qquad (14.101)$$

where $\dot{m}$ is the mass flow rate of the fluid (kg/s), c is its specific heat capacity (J/kg°C). U_{iv} is the overall heat transfer coefficient referring to the inner surface of the drill pipe (W/m²°C), T_D and T_A are the flowing fluid temperatures in the drill pipe and the annulus. The other notations are clearly shown in Fig. 14.24.

The internal energy balance for the upflowing fluid in the annulus is:

$$\dot{m}cdT_A = 2\pi R_{Ci}U_{Ci}(T_B - T_A)dz - 2\pi R_{Di}U_{Di}(T_A - T_D)dz \qquad (14.102)$$

where U_{zi} is the overall heat transfer coefficient referring to the inner surface of the casing, T_B is the temperature of the cement sheet at the borehole radius R_B.

The heat fluxes at the boundary of the cement sheet and the surrounding rock are the same:

$$2\pi R_{Ci}U_{Ci}(T_B - T_A) = 2\pi k_R \frac{T_\infty - T_B}{f(t)} \qquad (14.103)$$

where k_R is the heat conductivity of the rock (W/m°C), T_∞ is the undisturbed temperature of the rock at the given depth:

$$T_\infty = T_0 + \gamma z \qquad (14.104)$$

T_0 is the surface temperature, γ is the geothermal gradient (°C/m), and f(t) is the so-called transient heat conduction function, depending on the Fourier number and the quantity $\frac{R_{Ci}U_{Ci}}{k_R}$.

The balance Eq. (14.102) can be slightly modified:

$$\frac{dT_D}{dz} = \frac{T_A - T_D}{B} \qquad (14.105)$$

where the quantity:

$$B = \frac{\dot{m}c}{2\pi R_{Di}U_{Di}} \qquad (14.106)$$

doesn't depend on the depth z.

Balance Eqs. (14.102) and (14.103) are added, thus we get:

$$\dot{m}c\left(\frac{dT_A}{dz} + \frac{dT_D}{dz}\right) = 2\pi R_{Ci}U_{Ci}(T_B - T_A) \qquad (14.107)$$

The temperature difference $T_B - T_A$ can be expressed from (14.107):

$$T_B - T_A = \frac{\dot{m}c}{2\pi R_{Ci}U_{Ci}}\left(\frac{dT_A}{dz} + \frac{dT_D}{dz}\right) \qquad (14.108)$$

Similarly, $T_\infty - T_B$ is obtained from Eq. (14.107):

$$T_\infty - T_B = \frac{R_{Ci}U_{Ci}f(t)}{k_R}(T_B - T_A) \qquad (14.109)$$

The sum of Eq. (14.108) and Eq. (14.109) obtains:

$$T_\infty - T_A = \left(\frac{\dot{m}c}{2\pi R_{Ci}U_{Ci}} + \frac{R_{Ci}U_{Ci}f(t)}{k_R} \cdot \frac{\dot{m}c}{2\pi R_{Ci}U_{Ci}}\right)\left(\frac{dT_A}{dz} + \frac{dT_D}{dz}\right) \quad (14.110)$$

After some simplification, we get the expression:

$$\frac{dT_A}{dz} + \frac{dT_D}{dz} = \frac{T_\infty - T_A}{A} \qquad (14.111)$$

in which the so-called performance state coefficient:

$$A = \frac{\dot{m}c(k_R + R_{Ci}U_{Ci}f)}{2\pi R_{Ci}U_{Ci}k_R} \qquad (14.112)$$

It seems to be independent of the depth z. Nevertheless, A depends linearly on the circulating mass flow rate and the specific heat capacity of the drilling fluid. Thus at a given state of performance, A can be considered constant. In the case of constant A, the equation system is linear and an analytic solution can be obtained relative easily. However, A has a weak dependence on depth, temperature, and time.

In accordance to the well completion, R_{Ci} and U_{Ci} changes in different depth intervals. Heat conductivity k_R can be replaced by its depth-averaged value. Heat transfer coefficient between flowing fluid and pipe wall depends on the viscosity; that is the temperature. In fluid-filled annular sections natural convection occurs. Its heat transfer coefficient depends on the temperature both directly, and because of the viscosity-dependence, indirectly, too. The transient heat conduction function f depends on the Fourier–number, that is the time and the well completion. At a given time it can be taken to constant, thus we obtain different f values for different time-step. However, Eq. (14.112) has a slightly non-linear character, simple depth-averaged material properties are suitable to determine temperature distributions.

Finally, the differential equation system from pure mathematical point of view is:

$$A\frac{dT_A}{dz} + A\frac{dT_D}{dz} = T_\infty - T_A \qquad (14.113)$$

and:

$$B\frac{dT_D}{dz} + T_D = T_A \qquad (14.114)$$

Derivating Eq. (14.114) by z, after substitution we obtain a second-order, linear, inhomogeneous differential equation with constant coefficient for T_s:

$$AB\frac{d^2T_D}{dz^2} - B\frac{dT_D}{dz} - T_D + T_0 + \gamma z = 0 \qquad (14.115)$$

The homogeneous differential equation belonging to it is:

$$AB\frac{d^2T_D}{dz^2} - B\frac{dT_D}{dz} - T_D = 0 \qquad (14.116)$$

Its characteristic equation is:

$$AB\lambda^2 - B\lambda - 1 = 0 \qquad (14.117)$$

The roots of the characteristic equation are:

$$\lambda_1 = \frac{1}{2A}\left(1 + \sqrt{1 + \frac{4A}{B}}\right) \qquad (14.118)$$

and:

$$\lambda_2 = \frac{1}{2A}\left(1 - \sqrt{1 + \frac{4A}{B}}\right) \qquad (14.119)$$

Since both A and B are real, the solution of the homogeneous differential equation is obtained as:

$$T_{Dhom} = C_1 e^{\lambda_1 z} + C_2 e^{\lambda_2 z} \qquad (14.120)$$

Since the right-hand side of Eq. (14.113) is linear, a particular solution of it is looking for also in linear form:

$$T_{Dinh} = \alpha + \beta z \qquad (14.121)$$

Substituting it into Eq. (14.114), we get:

$$T_{Dinh} = B\gamma - T_0 - \gamma z \qquad (14.122)$$

Thus, the general solution of Eq. (14.114) is the sum of Eqs. (14.121) and (14.120):

$$T_D = C_1 e^{\lambda_1 z} + C_2 e^{\lambda_2 z} + T_0 + \gamma z - B\gamma \qquad (14.123)$$

The temperature distribution of the annular flow can be determined substituting Eq. (14.123) into Eq. (14.114).

Thus we get:

$$T_A = C_1(1 + B\lambda_1)e^{\lambda_1 z} + C_2(1 + B\lambda_2)e^{\lambda_2 z} + T_0 + \gamma z \qquad (14.124)$$

The constant coefficients C_1 and C_2 can be determined satisfying the following boundary conditions; if $z = 0$, $T_D = T_e$. That is, the drilling fluid temperature at the entrance in the drill pipe on the surface is T_e. The other is that the bottomhole temperatures both in the drill pipe and the annulus are the same. If $z = H$, $T_D(H) = T_A(H)$. Thus, the following linear algebraic equation system is obtained:

$$C_1 + C_2 = T_e - T_0 + B\gamma \tag{14.125}$$

and:

$$C_1\lambda_1 e^{\lambda_1 H} + C_2\lambda_2 e^{\lambda_2 H} = -B\gamma \tag{14.126}$$

The roots of this equation system are:

$$C_1 = \frac{D_1}{D}; \quad C_2 = \frac{D_2}{D} \tag{14.127}$$

where:

$$D = \lambda_2 e^{\lambda_2 H} - \lambda_1 e^{\lambda_1 H}$$

$$D_1 = \lambda_2(T_e - T_0 + B\gamma)e^{\lambda_2 H} + \gamma \tag{14.128}$$

$$D_2 = -\lambda_1(T_e - T_0 + B\gamma)e^{\lambda_2 H} - \gamma$$

Temperature distributions $T_D(z)$ and $T_A(z)$ can be determined by Eqs. (14.123) and (14.124). In order to calculate these, it is necessary to evaluate the constants A and B. Both constants depend on the data of borehole geometry, drilling fluid properties, surrounding rock parameters, and the data of performance. These are the following:

- borehole depth, drill pipe outer and inner diameters, drill bit size, and the instantaneous completion of the borehole
- entrance mud temperature, density, viscosity, specific heat capacity, and heat conductivity of the mud
- average density, heat conductivity, specific heat capacity of the rock, surface earth temperature, and geothermal gradient
- circulation mass flow rate, mud temperature at the entrance, and elapsed time during drilling operation

Knowing these data, the overall heat transfer coefficients U_{Di} and U_{Ci}, and the transient heat conduction function $f(t)$ can be calculated in the well-known manner (Willhite, 1967). The transient heat transfer function has different values belonging successive steps of elapsed time.

Using the evaluated constants A and B, temperature distributions is obtained from Eqs. (14.123) and (14.124). It is obvious that many independent variables influence temperature distributions in the borehole. This can be followed in the calculated temperature distribution diagrams.

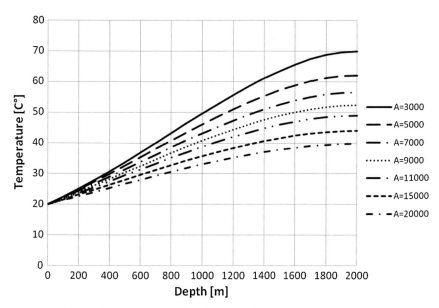

FIGURE 14.23   The temperature distribution in the drillpipe.

Considerable observations can be made examining the calculated temperature distribution functions.

The temperature of the downward flow in the drillpipe increases monotonically. Bottomhole temperatures decrease as the mass flow rates increase. Temperature gradient of the downward flow at the bottomhole equals zero. These observations are demonstrated in Fig. 14.23.

As the fluid turns upward, at the bottomhole, temperature increases until attains its maximum. The location of this maximum is obtained at hardly different depths as it is shown in Fig. 14.24. It is remarkable that the annular temperature distribution differs greatly from the undisturbed natural geothermic temperature.

Above this depth, the temperature of the upflowing fluid decreases, tending to the downflowing fluid temperature. It is shown in Fig. 14.25.

Both the downflowing and the upflowing fluid temperature depends on the mass flow rate and the specific heat capacity of the circulating drilling fluid. As mass flow rate increases, the fluid temperature decreases since the performance coefficients A and B are linear functions of the mass flow rate.

Bottomhole temperature is also influenced by the mass flow rate. Entrance temperature of the drilling fluid influences the bottomhole temperature especially at high mass flow rates. The change of the bottomhole temperature can be seen in Fig. 14.26.

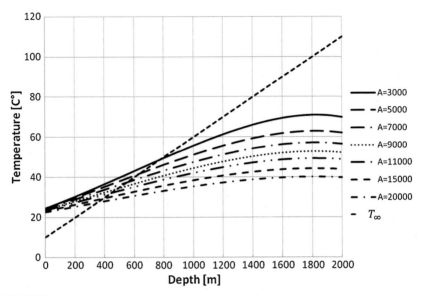

FIGURE 14.24   The temperature distribution in the annulus.

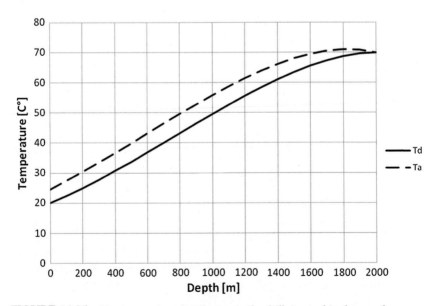

FIGURE 14.25   The temperature distribution in the drillpipe and in the annulus.

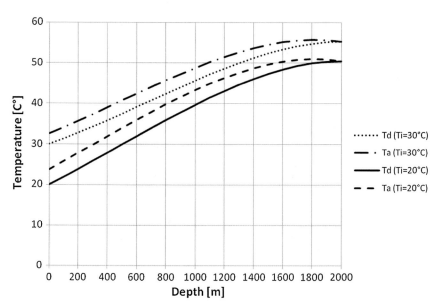

FIGURE 14.26 The effect of the injection temperature on the temperature distributions.

The findings, based on this mathematical model, seem to be general valid.

The above analytic mathematical model is convenient to predict the temperature distribution of the circulating drilling fluid both in the drill pipe and the annulus. Calculations are referred to for forward circulation. Many independent variables influence temperature distribution in the drill pipe and the annulus. Some of these are constant, like rock properties geothermal gradient, borehole geometry, etc. Others change as performance parameters vary, e.g., mass flow rate, entrance fluid temperature, and the elapsed time. The knowledge of the temperature distribution in the circulating drilling fluid can be applied in the design of different drilling operations.

References

Bobok, E., Navratil, L., 1982. Constitutive Equation for Thixotropic-Pseudoplastic Crude Oils. Kőolaj és Földgáz 15/115, 6, 161–167 (in Hungarian).

Bobok, E., Navratil, L., Szilas, A.P., 1981. Investigation of Turbulent Head Loss for Pseudoplastic Crude Oils. MTA X. Oszt. Közl. 14/1 75–93 (in Hungarian).

Boldizsár, T., 1958. The distribution of temperature in flowing wells. Am. J. Sci. 256.

Dowdle, W.L., Cobb, W.N., 1975. Static formation temperature from well log - an empirical method. J. Petroleum Technol. 1326–1330.

Edwardson, M.J., et al., 1962. Calculation of formation temperature disturbances caused by mud circulation. J. Petroleum Technol. 416–426.

Kármán, T., 1930. Mechanische Ahnlichkeit and Turbulenz. Math. Phys. Göttingen.

Metzner, A.B., Reed, J.C., 1955. Flow of non-Newtonian fluids- correlation of the laminar, transition and turbulent flow regions. AIChE J. 1 (434).

Szilas, A.P., Bobok, E., Navratil, L., 1981. Determination of turbulent pressure loss of non-Newtonian oil flow in rough pipes. Rheol. Acta 20, 487–496.

Willhite, G.P., 1967. Overall heat transfer coefficients in steam and hot water injection wells. J. Petroleum Technol. 607–615.

15

A Case Study About a Serious Industrial Accident

OUTLINE

Serious plant accidents can occur in geothermal energy generation plants. Primarily, the period of drilling, well preparation works, and the trial run is critical, and unforeseeable, hard to handle events can take place. A serious field accident happened in Hungary during the recovery of a deep, high temperature, high-pressure geothermal reservoir, causing severe environmental damage. The suppression of the steam blowout of Fábiánsebestyén's well Number 4, which occurred under unique conditions all over the world, was achieved by the exemplary cooperation of Hungarian and American professionals of the oil industry. The review of the case study is edifying, and provides a great opportunity to sum up what was learned so far by reconstructing the fluid dynamic and thermodynamic processes of the eruption.

Copyright © 2017 Elsevier Inc. All rights reserved.

15.1 THE BRIEF STORY OF THE BLOWOUT

The Hungarian Oil Research Company drilled several deep, prospecting wells on the southeastern part of the Great Plain of Hungary in the hope of hydrocarbon deposits in the 1980s. In the area of Nagyszénás and Fábiánsebestyén, after several 3200–3500 m deep wells, a deeper region was targeted with well Number 4 of Fábiánsebestyén. On the 16th of December, 1985, the actual depth of the borehole had been 4239 m, when the replacement of the drilling bit became necessary. While the drilling string was pulled out, the movement induced pressure waves in the mud column. This caused the pressure balance between the mud column and the over-pressured reservoir fluid to end, and a remarkable mud overflow occurred at the wellhead. The master valve was successfully shut off, but the choke line remained open yet. While the choke line was closed, the resulting pressure surge pushed the drill collar and the bit to the blowout preventer and the master valve was broken off. As the donuts were opened, the high overpressure (about 360 bar) threw out the drill collar and a piece of the drill pipe from the well. The drill collar was broken into three pieces and fell to the platform, killing the head driller. The over-pressured hot water displaced the mud from the well soon, and became a steam blowout produced from a mixture of hot water and steam. Since the inner shaft of the blowout preventer was also broken, the closing of the upper valve was not successful. Later a huge explosion happened, shooting the broken shaft laterally, and another lateral jet occurred in place of the shaft.

The eruption reproduced the legendary steam-cannon of Archimedes. According to the antique scripts, Archimedes fired one-pound stone balls by steam pressure on Roman ships assaulting Syracuse. Leonardo da Vinci planned the steam-cannon from the scripts, and it was assembled and tested by researchers of the Massachusetts Institute of Technology in 2006. The fired one-pound stone ball was accelerating to 300 m/s speed, and its energy exceeded the energy of the today's machine-gun bullet.

The hot steam cloud from the horizontal steam jet perfused the equipment creating almost zero visibility and making the suppression attempt impossible. After the blowing of the jet, different filling materials were applied to stop the eruption, unsuccessfully.

The Hungarian and American well control experts, and the engineering corps of the Army dismounted and transported the drilling rig between the 3rd and 21st of January 1986. Then a new blowout preventer was mounted in the place of the old one between the 22nd and 26th. From the 27th until the 30th, they tried to fit production tubing under pressure, unsuccessfully. Hence, on the 31st, traditional killing techniques were used and "killed" the well by pumping sludge into the 8 5/8 inch casing and closing of the choke line.

The destroyed blowout preventer, the desperate work environment, and the extreme conditions of the well did not allow the usual well analysis with proper measurements to the strict standards. Despite these conditions, many valuable data were recorded along with the main aim of suppressing the blowout. A pithy, accurate, and reliable summary of the incident was done by E. Buda (1996). His excellent case study is the basis of much research. The overview, systemization, evaluation, and reliability check are the aims of the following discussion.

15.2 THE HYDRODYNAMIC AND THERMODYNAMIC RECONSTRUCTION OF THE BLOWOUT

The suppression of the steam blowout of Fábiánsebestyén's well Number 4 was an internationally acknowledged, remarkable achievement of Hungarian and American professionals of the oil industry. The reconstruction of the fluid dynamic and thermodynamic processes of the blowout can be done by the reliable, accurate case study of E. Buda (1996), and the observations, notes, and photographs from the well control experts. This is a very useful tool in unveiling the parameters and behavior of the reservoir. Fig. 15.1 shows a typical photo of the initial phase of the blowout.

The reconstruction of the blowout pf well Number 4 in Fábiánsebestyén can be done through the analysis of its mathematical model. Before the mathematical model can be constructed, the conceptual model of the phenomenon should be determined.

FIGURE 15.1   Initial phase of the blowout (Buda, 1996).

The essence of the conceptual model is the following. A deep, over-pressured hot water reservoir was tapped by the borehole of Fábiánse-bestyén-4. The well blew out; the high pressure and temperature hot water flowed up from the depth of about 4 km to the surface through the damaged blowout preventer. The upflowing fluid pressure decreased substantially as the decrease of the depth and because of the friction losses. Nevertheless, the wellhead pressure attained the 360 bar level. This high wellhead pressure excluded the occurrence of continuous fluid flow at the outflow opening.

The high pressure at the wellhead disintegrated the continuous liquid phase into a set of discrete small particles generating a spray. It was a sudden change of state, at the outflow opening a finite pressure jump formed a discontinuity surface. The phenomenon is analogous to the injection process into the combustion chamber of diesel engines. In the case of modern diesel engines, the pressure of injection can be even 300 bar into the combustion chamber, having a lower pressure of 10–12 bar. Relative to this, the vaporization from 360 bar to the atmospheric pressure is not an essential difference. The vaporization terminates the continuous nature of the jet. Thus, the hot water flows through the well at a high-pressure level; at the outlet cross-section it is vaporized and moves away as a set of individual particles. The liquid–steam phase change happens as the over-pressured droplets arrive to the atmospheric pressure surroundings. This theory is confirmed by the irregular shape of the jet, as can be seen in Fig. 15.2.

The mathematical model is made within this conceptual framework. It is based upon three pillars. The first is the system of balance equations. The second is the conditions of uniqueness: the geometry of the system, the initial and boundary conditions, and the material properties. The third pillar is the method of the solution; it can be analytical, numerical, or experimental.

FIGURE 15.2   The irregular shape of the jet (Buda, 1996).

The basic equations are the balance equation of the mass, the momentum, and the energy. These can be simplified remarkably for the one-dimensional flow across the well. An adequate approximation is to consider the flowing water incompressible. Thus the mechanical and thermal processes can be treated separately, the equation system becomes substantially simple.

It is necessary to determine the boundaries of the system. It is convenient to subdivide the whole system into two subsystems. These are the reservoir and the well. The reservoir is considered two-dimensional. Its outer boundary is the cylindrical contour of the undisturbed region around it. The upper and lower plane boundaries surrounded the reservoir part of interest. The internal cylindrical boundary is the wellbore surface. The well is bounded by its cylindrical surface with the casing and the cement sheet. There is no mass transfer across this surface, but heat is transferred across it. Thus, the upflowing hot water heats the surrounding rock mass around the well, while its temperature decreases. The flow of the hot water ends at the outflowing cross-section of the blowout preventer. This cross-section is a so-called strong singular surface on which the disintegration of the continuous liquid phase occurs. The mathematical model can operate only if the adequate data of the system are known and used for calculations. In order to achieve this, it is necessary to account for the available data.

15.2.1 Available Data Measured During the Blowout

The geometry of the system is shown in Fig. 15.3, which is the sketch of the well completion. The wellhead equipment, and the damaged blowout preventer with the place of the outflow can be seen in Fig. 15.4. The actual bottomhole depth is 4239 m. The density of the mud used before the blowout was 2130 kg/m^3.

The pulling out of the drill string induced pressure waves. The pressure balance between the mud column and the formation was disturbed by the pressure minimum of the waves caused an intense inflow into the well. The depth of the inflow was about 3880 m, where the well is surrounded a fractured Triassic dolomite and dolomite breccia formation.

The last formation temperature and pressure measurements were made at the depth of 3684 m. The obtained formation temperature was 190.5°C, the pressure was 712.26 bar. The geothermal gradient can be calculated considering that the annual mean temperature is 10.5°C at the surface. Thus we get:

$$\gamma = \frac{T - T_0}{H} = \frac{190.5 - 10.5}{3684.5} = 0.04885°C/m \qquad (15.1)$$

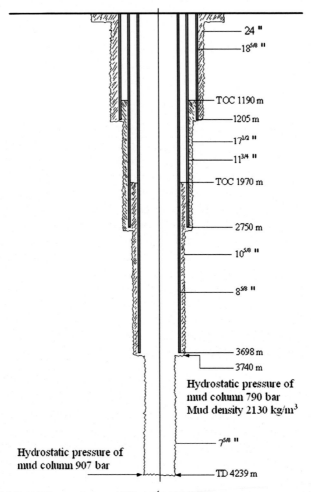

FIGURE 15.3   Well completion of Fáb-4 (Árpási, 1997).

The assumed depth of the sudden inflow into the well is 3880 m. The extrapolated formation temperature is here:

$$T = T_0 + \gamma H = 10,5 + 0,04885 \cdot 3880 = 199,6 \,^{\circ}C \qquad (15.2)$$

The extrapolated formation pressure in the over-pressured region at this depth is 731 bar.

During the blowout, the wellhead pressure had a stabilized value of 360 bar. This was measured by the manometer at the wellhead equipment. There were several attempts to block the damaged blowout preventer with different kinds of filling materials. When the outflow was temporarily stopped for a few minutes, the wellhead pressure increased to

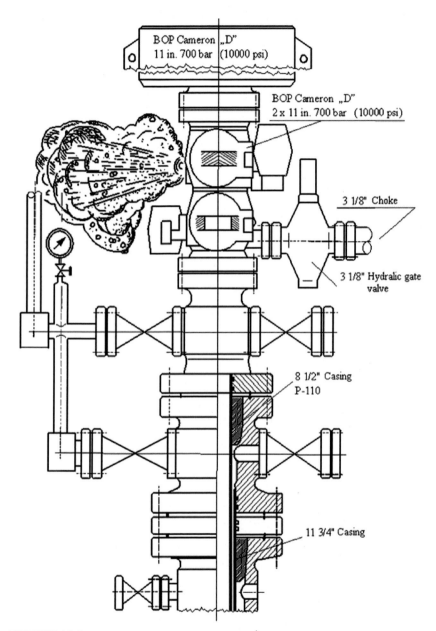

FIGURE 15.4   The damaged blowout preventer (Árpási, 1997).

410 bar. The temperature of the wellhead equipment at its outer surface was measured by a contact thermometer. The result was 150°C. The preceding listed measured parameters are certain. The characteristic pressure values in the well can be calculated based on these certainly measured values.

When the flow is stopped, transient pressure waves occur in the well. The amplitude of the pressure waves is obtained as:

$$\Delta p = \rho a c = 920 \cdot 1012 \cdot 3, 1 = 28, 86 \cdot 10^5 \; \frac{N}{m^2} \cong 29 \; \text{bar} \qquad (15.3)$$

where a is the speed of sound in the water-filled steel pipe, and c is the cross-sectional average velocity before the flow stopped. The static pressure at the temporarily closed wellhead is obviously the difference of the pressure maximum and the amplitude:

$$p_{st.k} = p_{max} - \Delta p = 410 - 29 = 381 \; \text{bar} \qquad (15.4)$$

When the flow is stopped, the well can be considered a piezometric tube. Naturally the flow is also stopped in the formation, thus the whole system is in hydrostatic state. The formation pressure is balanced by the wellhead pressure and the hydrostatic pressure of the hot water column. Consider the pressure distribution along the depth in the well as it is shown in Fig. 15.5. The straight line AB represents the natural hydrostatic pressure distribution of the hot water of 190.5°C. The section AC is the lithostatic pressure distribution of the rock. The line CD represents the

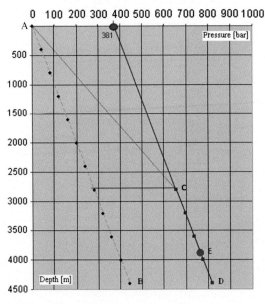

FIGURE 15.5   Static pressure distribution in the well.

pressure distribution in the over-pressured formation, where the fluid bears the weight of the overburden rock mass. The point E marks the depth of the latest pressure measurement (3684 m) where the static formation pressure is 712 bar. It is assumed that the well column is filled by hot water of density 920 kg/m^3. Thus, the pressure gradient is constant from the surface to the over-pressured region (AC) and in the over-pressured zone to the depth of the inflow (CD).

The static wellhead pressure was 381 bar when the flow had been temporarily stopped. The average rock density over the over-pressured region is 2340 kg/m^3. Two static equations can be written with the preceding data. The lithostatic pressure at the point C is:

$$P_c = \rho_k \cdot g \cdot h_c \tag{15.5}$$

The hydrostatic pressure at the point C is obtained as:

$$P_c = p_k + \rho g h_c \tag{15.6}$$

From these two equations the depth of the top of the over-pressured region can be calculated:

$$h_c = \frac{p_k}{g(\rho_k - \rho)} = \frac{381 \cdot 10^5}{9,81 \cdot (2340 - 920)} = 2735 \text{ m} \tag{15.7}$$

This depth is obtained to 2730 m from geophysical logging. The coincidence of these two values seems to be satisfactory. The static pressure at the depth of inflow (point D) is:

$$P_H = p_k + \rho g H = 3,81 \cdot 10^5 + 920 \cdot 9,81 \cdot 3880 = 731 \cdot 10^5 \ \frac{N}{m^2} = 731 \text{ bar} \tag{15.8}$$

These data form the cardinal points of the reconstruction of the blowout phenomenon. One of the most important basic parameters is the flow rate in the well. Correct flow rate measurements were not feasible in the time of the blowout, but it can be determined using the preceding available data. Thus the flow in the reservoir and the well can be determined.

15.2.2 The Flow in the Reservoir and the Well

As mentioned in the previous section, there were not adequate flow rate measurements at the time of blowout. All of the present experts were in complete agreement that the flow rate was uncommonly high; it was estimated to 5000–8000 m^3/day. The wellhead pressure remained constant, even if the remarkable scaling resulted a cross-sectional decrease in the upper section of the well. The flow rate of the well can be calculated reliably based on the measured pressure and temperature data.

As the hot water upflows in the well, the pressure decreases from 731 bar to 360 bar while its temperature also decreases from 199.6°C to 191.5°C. It is obvious that the water must be in liquid phase along its pathway. Since the saturation pressure of the hot water of 191.5°C is 13.3 bar, the occurrence of steam phase in the well is impossible. Consequently the frictional pressure loss can be determined exactly depending on the flow rate for the homogeneous liquid phase. On the other hand, the formula for this case is much more simple than in the case of two-phase flow. This is also true for the calculations of the temperature distribution.

The well is a very suitable diagnostic tool to determine the behavior of the reservoir. The flowing pressure and temperature at the depth of the inflow can be calculated reliably knowing the wellhead pressure and temperature.

Knowing the pressures at the inflow and the wellhead, the flow rate of the upflowing water can be calculated with high accuracy. The short initial transient period of the blowout is excluded in the calculations. It is assumed that during this initial period, the drilling mud was displaced from the well by the hot water, thus homogeneous, steady, turbulent hot water flow is considered. Since the temperature of the hot water changes only slightly, its temperature-dependent material properties such as density and viscosity are taken at an average temperature. The water is considered to be incompressible. The mechanical energy balance between the inflow depth and the wellhead can be written as:

$$p_{wf} = p_{vh} + \rho g H + \Delta p' \qquad (15.9)$$

in which p_{wf} is the pressure of the inflowing water, p_{wh} is the wellhead pressure, H is the depth of the inflow, and $\Delta p'$ is the frictional pressure drop in the well. The lower, uncased section of the well has a length of L_1, a diameter of D_1 and a friction factor λ_1. The length of the cased section is L_2, its diameter is D_2 and its friction factor is λ_2. It is assumed that the hydraulic behavior of the well is fully rough, depending on the relative roughness only. In this case:

$$\lambda = \frac{1}{\left(2 \lg 3,715 \frac{D}{k}\right)^2} \qquad (15.10)$$

where D/k is the relative roughness, the ratio of the diameter and the average height of the roughness of the pipe wall.

Knowing the friction factor, the pressure loss can be calculated by Weisbach's equation:

$$\Delta p' = \lambda \frac{L}{D} \rho \frac{c^2}{2} \qquad (15.11)$$

The cross-sectional average velocity is obtained with the mass flow rate as:

$$c = \frac{4\dot{m}}{D^2 \pi \cdot \rho} \qquad (15.12)$$

Substituting this into Eq. (15.11), we get that:

$$\Delta p' = \frac{8}{\pi^2 \rho} \cdot \lambda \frac{L}{D^5} \, \dot{m}^2 \qquad (15.13)$$

The inflowing pressure p_{wf} also depends on the mass flow rate because the frictional pressure loss of the flow toward the well in the reservoir:

$$p_{wf} = p_\infty - \frac{\dot{m}v}{2\pi hK} \ln \frac{R_\infty}{R_1} \qquad (15.14)$$

where R_∞ is the radius of the contour of the drained area, p_∞ is the undisturbed reservoir pressure there. The thickness of the reservoir is h, its permeability is K, and the kinematic viscosity is v.

The difference of the undisturbed reservoir pressure and the wellhead pressure is the supply of the potential energy increase of the upflowing water, and the frictional pressure loss in the formation and in the well:

$$p_\infty - p_k = \rho gH + \frac{\dot{m}v}{2\pi hK} \ln \frac{R_\infty}{R_1} + \frac{8}{\pi^2 \rho} \left(\lambda_1 \frac{L_1}{D_1^5} + \lambda_2 \frac{L_2}{D_2^5} \right) \dot{m}^2 \qquad (15.15)$$

Thus, we get for the mass flow rate a quadratic algebraic equation:

$$a\dot{m}^2 + b\dot{m} + c = 0 \qquad (15.16)$$

In which:

$$a = \frac{8}{\pi^2 \rho} \left(\lambda_1 \frac{L_1}{D_1^5} + \lambda_2 \frac{L_2}{D_2^5} \right) \qquad (15.17)$$

$$b = \frac{v}{2\pi h \cdot K} \ln \frac{R_\infty}{R_1} \qquad (15.18)$$

$$c = p_\infty - p_k - \rho gH \qquad (15.19)$$

To determine the constants a, b, and c, most of the necessary parameters are known accurately, while others are estimated only. The parameters used for calculations are the following:

The undisturbed reservoir pressure is 731 bar, while the wellhead pressure during the blowout is 360 bar. The density of the hot water is 920 kg/m^3, its kinematic viscosity is 1.3×10^{-7} m^2/s. The depth of the inflow is 3881 m. The relative roughness of the uncased borehole is 200, while of the cased section is 1000. The length of the uncased section is 197 m, its diameter is 194 mm, and the friction factor is obtained to 0.0303.

The length of the cased section is 3684 m, its diameter is194 mm, and the friction factor is 0.0197.

The estimated parameters are: The radius of the contour of the drained area is 500 m, the reservoir thickness is 24 m, the permeability is 1 darcy (10^{-12} m^2).

Substituting these values into Eqs. (15.16)–(15.19), the mass flow rate is obtained as 89.45 kg/s. This can be proven with easily accounting the pressure drops. The pressure loss in the reservoir is:

$$\Delta p_r' = p_\infty - p_{wf} = \frac{\dot{m}v}{2\pi h K} \ln \frac{R_\infty}{R_1} = \frac{89,45 \cdot 1,3 \cdot 10^{-7} \cdot 8,5172}{6,28 \cdot 24 \cdot 10^{-12}}$$

$$= 4,11 \cdot 10^5 \ \frac{N}{m^2} \cong 4,11 \text{ bar} \tag{15.20}$$

The friction pressure loss in the well is obtained:

$$\Delta p' = \left(\lambda_1 \frac{L_1}{D_1^5} + \lambda_2 \frac{L_2}{D_2^5} \right) \frac{8\dot{m}^2}{\pi^2 \cdot \rho} = \left(0,03 \cdot \frac{197}{0,194} + 0,02 \cdot \frac{3684}{0,2} \right) \frac{8 \cdot 89,5^2}{3,14^2 \cdot 920}$$

$$= 17,11 \cdot 10^5 \frac{N}{m^2} \cong 17,1 \text{ bar} \tag{15.21}$$

The hydrostatic pressure of the hot water column is $p = \rho g H$.
The static pressure of the undisturbed reservoir is:

$$p_\infty = \Delta p_r' + \rho g H + \Delta p' + p_k = 4,11 + 350 + 17,1 + 360 = 731 \text{ bar} \tag{15.22}$$

Thus the mass flow rate is obtained to 89.45 kg/s, that is7728.5 t/day. The volume flow rate is 8400 m^3/day, which is slightly higher than the estimated value of 8000 m^3/day by E. Buda. This value decreased somewhat, since the scaling caused the decreasing of the casing diameter during the blowout.

15.2.3 Temperature Distribution in the Flowing Well

The temperature of the inflowing hot water is the same as the formation temperature at the given depth:

$$T = T_0 + \gamma H \tag{15.23}$$

where T_0 is the annual mean temperature at the surface, γ is the geothermal gradient, and H is the depth of the inflow. The temperature of the surrounding rock around the well is:

$$T_\infty = T_0 + \gamma z \tag{15.24}$$

The temperature inhomogeneity $T - T_\infty$ induces a radially outward heat flow toward the far rock mass of undisturbed temperature. Thus the temperature and the thermal energy content of the upflowing water decreases continuously. As the consequence of the thermal energy loss of the water, the surrounding rock is warmed up, thus the inhomogeneity of the temperature field is decreased together with the radial heat flow. Thus the water temperature at the wellhead increases gradually until the whole system attains a steady state. A heated region of characteristic shape is developed around the well, an axially symmetric body of revolution (Tóth, 2005). This transient thermal interaction will be investigated in the following sections. The calculated temperature distribution is suitable to obtain another independent method for verification the calculated value of the inflow depth.

In accordance to the geometry of the well, a cylindrical coordinate system is chosen. Its downward directed z-axis coincides with the symmetry axis of the well, the $z = 0$ point is belongs to the surface. A suitable control volume is chosen as a limit for the balance equation of the internal energy. It is a cylinder, coaxial with the well. At the arbitrary depth of z, there are two parallel planes of an infinitesimal distance of dz. These are the upper and lower boundaries of the control volume. The outer cylindrical boundary has a radius of R_∞, which belongs to the undisturbed temperature of the formation. The control volume is shown in Fig. 15.6.

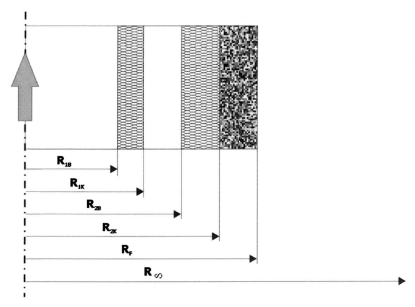

FIGURE 15.6   Control volume for the energy equation.

It is divided to two subsystems: one of them is the upflowing fluid confined by the tubing wall of the radius R_{1B}. The heat transfer between the upflowing hot water and the tubing is happened along this surface. The other is the well completion and the surrounding rock around the well. All radii necessary for the calculations are marked in the figure. The radially outward heat conduction is the dominant phenomenon across the well completion and the surrounding rock. Since the determination of the two different phenomena needs disparate kinds of differential equations, it is convenient to calculate the two subsystems separately. The joint condition between them is that the heat flux transferred from the water to the tubing is the same that is conducted across the well components toward the rock:

$$\dot{m}cdT = 2R_{1B}\pi U_{1B}(T - T_F)dz \tag{15.25}$$

The heat flux transferred across the well completion is the same as the conductive heat flux in the rock:

$$2R_{1B}\pi U_{1B}(T - T_F) = 2\pi k_K \frac{T_F - T_\infty}{f(t)} \tag{15.26}$$

where c is the specific heat capacity of the water, U_{1B} is the overall heat transfer coefficient, T_F is the temperature at the borehole wall, k_K is the heat conductivity of the rock, T_∞ is the undisturbed geothermal temperature, and $f(t)$ is the so-called transient heat conduction function depending on the Fourier number and the coefficient $R_{1B}U_{1B}/k$.

In order to calculate the overall heat transfer coefficient, consider the horizontal section of the well completion of unit thickness, as it can be seen in Fig. 15.6. The elements of the well completion: tubing, casings, and cement sheets have thermal resistances in serial connection. Across all elements, the temperature drops are added, while the heat fluxes are the same.

The mechanism of the heat transfer across the single elements is different. There is forced convection between the upflowing water and the tubing wall, conduction occurs across the tubing and casing walls and free convection develops in the fluid-filled annulus. The heat flux between the water and the tubing wall is:

$$Q = 2\pi R_{1B}h_{1B}(T - T_{1B}) \tag{15.27}$$

where h_{1B} is the heat transfer coefficient on the cylindrical surface of radius R_{1B}, while T_{1B} is the wall temperature there. The heat flux across the tubing wall is:

$$Q = 2\pi k_a \frac{T_{1B} - T_{1K}}{\ln \dfrac{R_{1K}}{R_{1B}}} \tag{15.28}$$

where k_a is the heat conductivity of the steel. If the annulus between the tubing and the casing is filled with water, the heat flux by free convection can be written as:

$$Q = 2\pi R_{1K} h_{GY}(T_{1K} - T_{2B}) \tag{15.29}$$

in which h_{GY} is the heat transfer coefficient of the free convection through the annulus.

The heat is transferred also by conduction across the casing wall. The heat flux is obtained as:

$$Q = 2\pi k_a \frac{T_{2B} - T_{2k}}{\ln \dfrac{R_{2K}}{R_{2B}}} \tag{15.30}$$

The heat is conducted through the cement sheet with a flux:

$$Q = 2\pi k_c \frac{T_{2K} - T_F}{\ln \dfrac{R_F}{R_{2K}}} \tag{15.31}$$

where k_c is the heat conductivity of the cement.

Expressing the temperature differences from Eqs. (15.27)–(15.31) and summing up them, we get the temperature difference between the water and the rock at the borehole wall:

$$T - T_F = \frac{Q}{2R_{1B}\pi} \left(\frac{1}{h_{1B}} + \frac{R_{1B}}{k_a} \ln \frac{R_{1k}}{R_{1B}} + \frac{R_{1B}}{R_{1K}} \frac{1}{h_{GY}} + \frac{R_{1B}}{k_a} \ln \frac{R_{2K}}{R_{2B}} \right. \\ \left. + \frac{R_{1B}}{k_c} \ln \frac{R_F}{R_{2K}} \right) \tag{15.32}$$

The so-called overall heat transfer coefficient can be defined based on the equation:

$$Q = 2R_{1B}\pi U_{1B}(T - T_F) \tag{15.33}$$

Thus, we obtain the expression:

$$\frac{1}{U_{1B}} = \frac{1}{h_{1B}} + \frac{R_{1B}}{k_a} \ln \frac{R_{1K}}{R_{1B}} + \frac{R_{1B}}{R_{1K}h_{GY}} + \frac{R_{1B}}{k_a} \ln \frac{R_{2K}}{R_{2B}} + \frac{R_{1B}}{k_c} \ln \frac{R_F}{R_{2K}} \tag{15.34}$$

to calculate U_{1B}.

Naturally the overall heat transfer coefficient is not constant along the whole depth of the well. In the uncased section, it is $1/h_{1B}$, and as the number of the casings increases as the terms of Eq. (15.34) increases also. Combining the Eqs. (15.25)–(15.34) we get the differential equation:

$$\frac{dT}{dz} = \frac{2\pi R_{1B} U_{1B} k_K (T - T_0 - \gamma z)}{\dot{m}c(k_K + R_{1B} U_{1B} f(t))} \tag{15.35}$$

Introducing a few approximations we obtain a first-order, linear, inhomogeneous differential equation of constant coefficients. The necessary assumptions are the following: the heat conductivity coefficient of the rock is replaced by its depth-averaged value, the overall heat transfer coefficient is similarly averaged along the depth and we assume that the transient heat conduction function does not depend on the depth. Thus the so-called well-performance coefficient can be introduced in which all parameters are constant along the depth:

$$A = \frac{\dot{m}c(k_K + R_{1B}U_{1B}f)}{2\pi R_{1B}U_{1B}k_K} \tag{15.36}$$

The dimension of A is m. Using it, Eq. (15.35) can be written in the simple form:

$$A\frac{dT}{dz} = T - T_0 - \gamma z \tag{15.37}$$

It can be solved easily by pure analytic means (Tóth, 2005). The boundary condition is that the temperature of the inflowing water is the same as the reservoir temperature. Thus we get:

$$T = T_0 + \gamma(z + A) - \gamma A e^{\frac{z-H}{A}} \tag{15.38}$$

This equation expresses the temperature distribution of the upflowing water along the depth. The time-dependence is considered implicitly, as the well-performance coefficient contains the transient heat conduction function. The time-dependent nature occurs remarkably in the initial period of the blowout, later it tends gradually to the steady state.

Calculate the temperature of the flowing water at the wellhead using the following actual data: the mass flow rate is 89.45 kg/s, and the specific heat capacity of the water is 4187 J/kgC. The average heat conductivity of the rock is 3.5 W/mC. The inner radius of the tubing is 0.1 m, and the overall heat transfer coefficient is 42 W/m²C. The value of the transient heat conductivity function f is 1.80. Thus, the well performance coefficient is obtained to:

$$A = \frac{\dot{m}c(k_K + R_{1B}U_{1B}f)}{2\pi R_{1B}U_{1B}k_K} = \frac{89.45 \cdot 4187(3.5 + 0.1 \cdot 42 \cdot 1.8)}{6.28 \cdot 0.1 \cdot 42 \cdot 3.5} = 44870 \text{ m} \tag{15.39}$$

Thus the temperature of the water at the wellhead, in the depth of $z = 0$ is:

$$T_{ki} = 10 + 0.04885 \cdot 44870\left(1 - e^{-\frac{3880}{44870}}\right) = 191.5\,^\circ C \tag{15.40}$$

The depth of the inflow H can be calculated from the measured wellhead temperature:

$$T_{ki} = T_0 + \gamma A - \gamma A e^{-\frac{H}{A}} \tag{15.41}$$

The wellhead temperature of the water was not measured directly. The temperature of the outer surface of the blowout preventer was measured only by a contact thermometer. Knowing this measured temperature, the water temperature can be calculated. The depth of inflow can be obtained as:

$$H = A \ln \frac{1}{1 + \dfrac{T_0 - T_{ki}}{\gamma A}} \tag{15.42}$$

Thus, the depth of the inflow can be calculated not only by the overpressure distribution, but it can be verified by another independent method based on the temperature measurement. As it was mentioned, the measured temperature is available on the outer cylindrical surface of the casing directly below the blowout preventer. Thus the temperature of the flowing water can be calculated knowing the overall heat transfer coefficient and the heat flux across the well completion. The former is obtained as:

$$\frac{1}{U_{1B}} = \frac{1}{h_{1B}} + \frac{R_{1B}}{k_a} \ln \frac{R_{1K}}{R_{1B}} + \frac{R_{1B}}{R_{1K} h_{GY1}} + \frac{R_{1B}}{k_a} \ln \frac{R_{2K}}{R_{2B}} + \frac{R_{1B}}{R_{2K} h_{a2}} + \frac{R_{1B}}{k_a} \ln \frac{R_{3K}}{R_{3B}} \tag{15.43}$$

The heat flux across the well completion is equal with the sum of the leaving heat flux due the free convection, and radiation on the outermost cylindrical surface of radius R3k. It can be expressed by the equation:

$$Q = 2R_{3K} \pi h_{3K} (T_{3K} - T_L) + 2R_{3K} \pi \varepsilon \phi \left(T_{3K}^4 - T_L^4 \right) \tag{15.44}$$

in which h_{3k} is the heat transfer coefficient of the free convection, T_L is the temperature of the surrounding air, ε is the emissivity of the cast steel surface, ϕ is the Stefan–Boltzmann coefficient. In order to get the heat transfer coefficient of the free convection, it is necessary to determine the Grashof number by the parameters of the air:

$$Gr = \frac{\beta \cdot \Delta T \cdot g \cdot D^3}{v^2} = \frac{3.2 \cdot 10^{-3} \cdot 100 \cdot 9.8 \cdot 10.5^3}{1.3^2 \cdot 10^{-12}} = 2.32 \cdot 10^7 \tag{15.45}$$

The product of the Grashof number and the Prandtl number is $1.16 \cdot 10^7$, thus the Nusselt number is obtained as:

$$Nu = 0.52 (GrPr)^{0.25} = 30.35 \tag{15.46}$$

The heat transfer coefficient can be calculated as:

$$h_{3K} = \frac{k_L \cdot Nu}{D} = \frac{0.025 \cdot 30.35}{0.5} = 1.518 \ \frac{W}{m^2 \, ^\circ C} \tag{15.47}$$

On the other hand, the heat flux is:

$$Q = 2R_{1B}\pi U_{1B}(T_{ki} - T_{3K}) \tag{15.48}$$

The temperature of the flowing water is:

$$T = T_{3K} + \frac{Q}{2\pi R_{1B}U_{1B}} \tag{15.49}$$

Calculate the heat flux using the following data: $R_{3k} = 0.25$ m, $T_L = 50$ C, $\varepsilon = 0.25$, $\phi = 5.67.108 \ W/m^2K^4$, $Gr = 2.32.10^7$, $Nu = 30.35$, $h_k = 1.518 \ W/m^2C$, and $h_{1B} = 29 \ W/m^2C$. Finally, we get:

$$\begin{aligned} Q = \ & 6.28 \cdot 0.25 \cdot 1.518(150 - 50) + 6.28 \cdot 0.25 \cdot 0.25 \cdot 5.67 \cdot 10^{-8}) \\ & \times (4.23 \cdot 10^8 - 3.24 \cdot 10^8) = 707 \ W/m \end{aligned} \tag{15.50}$$

Thus the water temperature at the wellhead is:

$$T_k = 150 + \frac{707}{6.28 \cdot 0.1 \cdot 29} = 188.8\,^\circ C \tag{15.51}$$

This temperature is in good agreement with the value obtained by Eq. (15.41).

The experts working on the suppression of the steam blowout at Fábiánsebestyén, despite the hard circumstances, recorded many valuable data, which gave the opportunity to discover the features of the revealed high pressure over-pressured reservoir. The verification of the given data was achieved by a coherent reconstruction based on fluid dynamic and thermodynamic calculations. The physical laws as governing principles were suitable to check the reliability of the recorded data and could fit in a non-contradictory system. The estimated value of output, which was one of the most doubtful data among the experts, proved the excellent sense of reality of the onsite petroleum engineers. From the calculations, without considering the effect of scaling the volumetric flow rate, was 8400 m^3/day, which strengthens the estimated 8000 m^3/day for the initial period of the eruption. Because of the suppressing effect of scaled casing and choke line, the volumetric flow rate could easily drop to the estimated 5000 m^3/day or even less.

The 360 bar wellhead pressure could hardly fit in the model, while the actions at the wellhead were considered as isenthalpic expansion and the jet was seen as continuous. The conception of the individual droplet set broken by vaporization made it possible to understand and fit the outlier pressure value in the model.

The wellhead temperature also caused some discrepancy. The water temperature was estimated much higher already by E. Buda (1996) than it was deduced from the measured temperature of the casing head's outer surface with a contact thermometer. Because of the extreme flow rate, the temperature decrease of the upwelling water was relatively low, even before the stationary state.

The conformity of the results of heat loss calculations of the well and deduced temperatures from the casing surface measurements are well-looking. This initial period temperature is increasing with time until a heated region evolves around the well. This final value can be well calculated, and they show the changes of temperature with time in the simulation model. Nevertheless, our current knowledge on the reservoir can only become fact if it is supported by a well analysis on a future production well.

References

Árpási, 1997 Árpási, M., 1997. Steam Blowout in Fábiánsebestyén. MOL, Budapest.

Buda, E., 1996. Case Study of the Steam Blowout in Fábiánsebestyén. MOL, Budapest [In Hungarian].

Miscellaneous Geothermal Applications

16.1 A PROSPECTIVE GEOTHERMAL POTENTIAL OF AN ABANDONED COPPER MINE

The Recsk copper mine is an unfortunate implementation of the Hungarian ore mining industry. Recsk, as can be seen in Fig. 16.1, is situated in the Mátra Mountains, Northern Hungary.

The Mátra Mountains belong to the Inner Carpathian volcanic arc. This is the highest (1014 m) and the largest Tertiary volcanic range of Hungary.

Copyright © 2017 Elsevier Inc. All rights reserved.

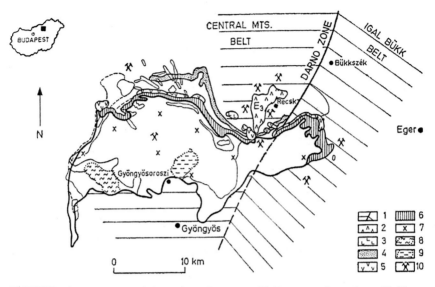

FIGURE 16.1 Geology of the Mátra Mountains. (1) Basement formations. (2) Upper Eocene biotite-hornblende andesites. (3) Eggenburgian andesites. (4) Lower Rhyolite Tuffs. (5) Carpathian pyroxene andesites. (6) Middle Rhyolite Tuffs. (7) Badenian pyroxene andesites. (8) Diatomites (Caldera stage sediments). (9) Hydroquartzies (Caldera stage). (10) Mining area. *After Földessy (1975).*

In the last century detailed geological and geophysical surveys have been made, providing a great number of data for both the surface and subsurface geology. The most informative contributions to our knowledge about this area are the works of Kubovics and Pantó (1950), Földessy (1975), and Zelenka (1973). More than 1200 ore exploratory drillings have been undertaken to find and evaluate the important copper deposits of Recsk.

16.1.1 Geological Background

The pre-Tertiary basement of the Mátra Mountains is separated by a regional scale deformation zone, the so-called Darnó line. Two essentially different basement structures can be recognized: the folded Mesozoic of the Eastern Mátra and the faulted Mesozoic structural belt of Western Mátra. The structural differences between the two units separated by this zone have been maintained throughout the Tertiary period.

The first Tertiary volcanic activity belonged to the vertical movements along the Darnó line, and four substages of volcanism resulted. The first was entirely subaqueous volcanism in the Upper Eocene. The rocks are typically biotite-hornblende andesites. The second substage has developed by step-by-step assimilation and contamination as well as the build up of a stratovolcanic character. The originally andesitic character was shifted toward the more acidic range, producing dacites. In the third substage the eruptive center was shifted northward, produced a stratovolcanic sequence of biotite partly overlapping the earlier volcanic sequences. The fourth stage was the development of a central explosive caldera of the stratovolcano, and resulted in the formation of radial and irregular dyke patterns. The quickly subsided volcanic area has filled with reef limestones.

This subsidence reached its maximum by the middle Oligocene, when the largest part of the Eocene volcanics were covered by marine sediments. The Upper Eocene volcanic activities have associated with very significant mineralizations in connection with shallow intrusive porphyric body and its skarn environment producing porphyric copper ores, skarnous copper ores in the intrusives, and altered country rocks. The third substage of volcanism has produced intensive hydrothermal alterations as well as formation of stockwork copper ores. In the caldera area exhalative-sedimentary copper mineralization developed during this stage.

The Neogene volcanism includes andesitic and rhyolitic phases. Through these phases the initial rhyolitic predominance has been changed toward the andesitic character.

The entire ore-forming process was restricted to the hydrothermal temperature range. Its complexity is due to its temporally multiphase nature and the variety in the environmental controls of localization. Two stages of mineralization can be distinguished. The main stage comprises mineralization related to the intrusive host rocks. A second, less important stage is coupled to the latest effusives. The ore formation began at 400°C, and ended at about 150°C.

The most important ore type is the porphyry copper mineralization, in the form of disseminations, microveinlets and veins throughout the inner alteration zones within intrusive bodies. The porphyry copper ore reserves total several hundred million tons at 0.4% copper cut off grade, with a 0.77% average copper grade. From the low-grade central core a gradual enrichment occurs, 0.4—0.6% values in the phyllic region and 0.9—1% Cu maxima in the propylitic zone.

The highest concentrations of copper can be found in the limestone skarns, with an average 1.5% Cu content. Two main localizations of these skarn ores have been recognized: one is stratabound and parallel to the original bedding of the skarnified sediments, the other is represented by cross-cutting steep lenses and veins. The ores related to the skarn zone represent 30% of the economic ore reserves of the deposit.

There are two main ore zones in the Recsk area. The so-called upper ore zone is situated at the depth interval between −490 and −690 m below sea level. The lower ore zone can be found between −690 and −890 m below sea level. The two ore zones are separated by a quartzit layer without any ore content.

16.1.2 The Story of the Mine

Recognizing the existence of the important copper ore reserves in 1969, it was decided to deepen a shaft directly instead of further exploratory drillings. The first shaft—Recsk I—was deepened with an internal diameter of 8 m, with a depth of 1202 m. It was completed in 1974. In the same year the deepening of the second shaft (Recsk-II) had been started. During this deepening, there was some serious water inrush. The largest happened at a depth of 770 m, with a flow rate of 0.95 m^3/min. The salinity of the inrushed water was very high, at 9000 g/m^3. The scale deposit was removed steadily from the wall of the shaft.

To connect the two shafts, two horizontal roadways were driven to a depth of −700 and −900 m below sea level. The cross-section of the roadway is 20 m^2. Generally the roadways have provd to be consistent, but mainly were supported by roofbolts of 1.8 m length. The roadway system of the mine is shown in Fig. 16.2.

There was a water inrush during the roadway driving too. This happened at the lower −700 m level with a flow rate of the 2 m^3/min. The inflowing cross-section was cemented with difficulty, because the pressure of the water was as high as 70 bar. In the lower roadway at the level

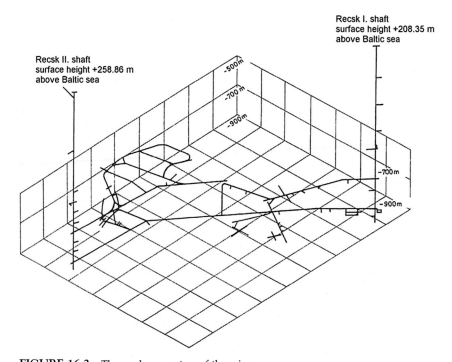

FIGURE 16.2   The roadway system of the mine.

of −900 m there was no problem with the water, because the upper roadway drained the water, its pressure at −900 m depth was 25 bar only.

By the time the roadways had been completed, the price of the copper in the world market fell radically despite forecasts. While the IBRD (International Bank for Reconstruction and Development) prognosticated 6090 USD/ton for 1995, the actual price on the international market was 2000 USD/ton only. Since the price of the copper remained permanently low, the development of the Recsk mine has not continued, since 1981. The Hungarian Council of Ministers ordered the steady interruption of any activity in the mine. The pumping of the water has also been suspended. Thus the roadway and the shafts of the mine are currently flooded. The rise of the water level with time is shown in Fig. 16.3. The shafts are plugged, but a monitoring pipe having a diameter of 250 mm makes possible the measurements of the level, concentration, and temperature of the water.

16.1.3 Geothermal Conditions

High underground temperatures were observed in the Recsk mine during the roadway drifting. Intensive ventilation was necessary in the whole period of implementation. Many temperature data have been obtained in exploratory boreholes. Most the data were measured by mercury thermometers a few days after finishing the drilling operations. Apparently these values are lower at least by 10–20% according to measurements, than the undisturbed rock temperatures. The corrected temperature data of the rock obtained an average value at the upper roadway level 960 m under the surface is 51.8°C. The geothermal gradient based on these data is 0.0435°C/m. The average temperature obtained at the lower roadway level 1.160 m depth from the surface is 59.5°C. The geothermal

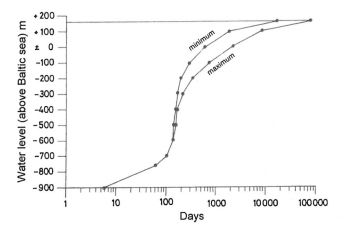

FIGURE 16.3   The rising water level over time.

gradient calculated by these temperatures is 0.0427°C/m. The comparison of these gradients seems to be in rather good agreement. The rock mass around and above the roadway is mainly andesite and limestone. The overall heat conductivity of the cover layers is obtained as 2.53 W/m °C. Thus the terrestrial heat flow calculated by these data is 0.108 W/m^2. The heating of the area is slightly greater than the Hungarian average of the terrestrial heat flow (0.095 W/m^2). The supply of the water flooding the roadway and the shafts is deep water-bearing rock mass around the mine. The temperature of the water is essentially the same as the rock temperature. Temperatures measured on the occasion of water inrushes are in agreement with rock temperatures (Toth, 2007).

The water-filled mine has a large geothermal potential. The volume of the flooded mine is more than 200,000 m^3. At the free surface of the water in the shaft the temperature of the water is 29°C. This temperature increases along the depth. The walls of the roadway are in thermal equilibrium with the water. In the shafts some free convection can occur, deforming the linear geothermal temperature distribution along the depth.

A submersible pump can be lowered to the bottom of the shaft to produce warm water. Assuming 1.2 m^3/min flow rate and 30°C temperature, the obtainable thermal power is 2.512 kW. After utilization the produced warm water can be discharged without any back-pressure into the other shaft.

The utilization may be primarily district heating. It seems necessary to built in suitable heat pumps to increase the temperature of the produced water. Another possibility, to use the large-diameter shaft, is to build in a hairpin-type bore-hole heat exchanger without any water production. Both methods can be economic. The area close to the mine is a wooded recreation area. There are some health resorts with medicinal springs and hotels with medical treatment facilities. The produced water is suitable to supply spas and swimming pools. The clean geothermal heating maintains the clean healthy air. The produced geothermal energy is sustainable for a long time. The heat transfer surface of the roadways and the shafts is more than 150,000 m^2. Assuming a temperature difference of 4°C between the rock and the water, the heat transfer coefficient is 4.8 W/m^2 °C, the thermal power supply of the system is obtained 2.880 kW. Thus the planned thermal power can be enlarged. The geothermal energy is really renewable on this area.

16.2 GEOTHERMAL DEICING OF A MINE TUNNEL

16.2.1 Geological Background

In the southwestern part of Hungary in the Bátaapáti region, a small-to-medium-level underground radioactive waste deposit system was built. The storage space is connected by two 1800-m-long mine tunnels.

The storage area was constructed through these tunnels, and afterwards they were used for the operational traffic and ventilation. If the outside temperature is less than $-5°C$, then to improve the workplace climate and prevent the icing of the entrance section to the mine tunnel, it is necessary to warm the intake air to the tunnels. The original design included the use of traditional oil burners to warm the air. During the construction period 80 m^3/s of air flow was used. This was the maximum air demand. During normal operations 25 m^3/s air flow is utilized. Regardless of the season these air flows had to be heated to between $+2°C$ and $+5°C$. In the case of an extreme weather condition (for example $-18°C$), the air temperature must be increased to between $+20°C$ and $+22°C$. In this latter case, 2316 kW of heating capacity is required. This requires 247 L/h of fuel for the oil burners. The next phase of the development of the waste deposit system would have required the installation of two more oil burners. Instead of this expensive solution, and to decrease the huge CO_2 emission from the oil burners, I recommended a geothermal solution for the floor and the roof heating for deicing the floor of the tunnel and avoiding the formation of icicles on the roof.

16.2.2 Using Geothermal Energy

Tempering the air of the mine tunnels is necessary on the one hand to ensure that the facility's climate is comfortable for the workers, and on the other hand for deicing the floor of the entrance section. Geothermal energy has always had a dominant role in the climate of underground facilities and mine tunnels. The wall temperature of the mine tunnel is about $+17C°C$ to $+18°C$, which is much higher than the inflowing air temperature in winter. The surface of the tunnel walls is a very large heat transfer area. The air intake tunnel walls with the floor and the roof constitute 36,000 m^2 of surface area over the length of 1800 m. The heat flow is constant over this major surface and, thus, heats the air flow in the mine tunnel. This can be determined by a simple calculation.

16.2.2.1 Heating Requirement for the Icing Intake Tunnel

It is enough to heat the first 300 m section of the intake-air tunnels just where the icing appears. The heating system uses hot water and glycol solution being circulated in pipe loops below the floor and on the roof of the mine tunnel. Similar installations like sidewalk, roadway, and bridge deicing systems have been demonstrated in several countries, including Argentina, Japan, and the United States. In our case the heating requirement can be met by two sources. The first is air—water heat pumps utilizing the heat content of the warmed-up air as it flows out of the mine tunnel. The other source is the heat content of the drained water from the deep part of the mine.

Chapman and Katunich (1956) derives and explains equations for the heating requirement of a snow-melting system. Chapman and Katunich (1956) derive the general equation for the required heat output (q) in W/m^2. We can use it for the tunnel floor.

$$q_o = q_s + q_m + A(q_e + q_h) \qquad (16.1)$$

where q_s is the sensible heat transferred to the ice (W/m^2); q_m is the heat of fusion (W/m^2); A is the ratio of snow-free area to total area (dimensionless); q_e is the heat of evaporation (W/m^2); q_h is the heat transfer by convection (W/m^2).

The sensible heat q_s to bring the ice to 0°C (32°F) is:

$$q_s = s \cdot h \cdot \rho_i \qquad (16.2)$$

where s is the rate of taken in snow layer on the floor (0.0025 m/day); ρ is the density of ice (917 kg/m^3); c_p is the specific heat of snow (J/kg °C); h is the enthalpy of fusion for water (J/kg); c_1 is the conversion factor (86,400 s).

It can be assumed that the ice and snow carried by vehicles immediately start to melt, resulting in it not cooling below 0°C. Thus, in the q_s member this can be ignored.

Suppose that the thickness of snow cover entered into the tunnel is 2.5 mm by day. The heat of fusion q_m to melt the snow is:

$$q_m = \frac{s}{c_1} \cdot c_p \cdot \rho = 0.8835 \ \frac{W}{m^2} \qquad (16.3)$$

The heat of evaporation of the molten snow q_e:

$$q_e = \frac{s}{c_1} \cdot c_m \cdot \rho_w = 7.238 \ \frac{W}{m^2} \qquad (16.4)$$

where ρ_w is the density of water (1000 kg/m^3); c_m is the heat of evaporation (2512 kJ/kg).

The heat transfer q_h is between the water film on the floor and the intake air. We can calculate the heat transfer coefficient by turbulent flow. In this case the heat transfer coefficient depends on the Reynolds number.

$$q_h = h \cdot (T_w - T_1) = 39.24 \ \frac{W}{m^2} \qquad (16.5)$$

where T_w is the water film temperature (°C), usually taken as 1°C (33°F); T_1 is the intake air temperature at the entrance cross-section in winter (−5°C).

Summing the fluxes, give:

$$q_o = 47.35 \ \frac{W}{m^2} \qquad (16.6)$$

The heated area in the length of 300 m is 1800 m^2, thus the needed thermal power on the floor of the tunnel is:

$$Q_f = A \cdot q_o = 85.23 \text{ kW} \qquad (16.7)$$

where Q_f is the thermal power for the floor heating (kW).

In winter, where some water appears from the roof of the mine tunnel, the icing occurs as icicles. It appears only in the first 300 m of the tunnel. If we can warm the roof surface over the freezing point (0°C) icicles should not develop, and the water drops could disperge in the intake air flow. If we heat the 6-m arch of the tunnel roof, to a length of 300 m we get 1800 m^2 area. Thus the needed thermal power on the roof is:

$$Q_r = h \cdot A \cdot \Delta T = 70.63 \text{ kW} \qquad (16.8)$$

where $\Delta T = 6°C$ is the temperature difference between the freezing point and the intake air $(1 - (-5))°C$; Q_r is the thermal power for the roof heating (kW).

The floor and the roof heating demand together is

$$Q_T = Q_f + Q_r = 85.23 + 70.63 = 155.86 \text{ kW} \qquad (16.9)$$

where Q_T is the total thermal power demand (kW).

16.2.3 Geothermal Sources for Heating the Mine Tunnel

The maximum inflowing air is 80 m^3/s which occurred during construction. In this case the cross-sectional average velocity is:

$$c = \frac{Q}{A} = \frac{80 \frac{m^3}{s}}{30 \frac{m}{s}} = 2.67 \frac{m^2}{s} \qquad (16.10)$$

where c is the cross-sectional average velocity (m/s); Q is the maximum inflowing air (m^3/s); A is the cross-section of the tunnel (m^2).

The hydraulic radius of the tunnel:

$$R_H = \frac{A}{K} = \frac{30.3}{20.1} = 1.51 \text{ m} \qquad (16.11)$$

where K is the hydraulically active perimeter.

The Reynolds number is

$$Re = \frac{c \cdot 4R_H}{v} = \frac{2.67 \frac{m}{s} 4 \cdot 1.51 \text{ m}}{10^{-5} \frac{m^2}{s}} = 1612680 \qquad (16.12)$$

The Prandtl number of the air is obtained as

$$Pr = \frac{\rho v c_p}{k} = 0.541 \qquad (16.13)$$

where ρ is the air density (1.292 kg/m^3); ν is the kinematic viscosity coefficient (0.024 m^2/s); c_p is the specific heat (1005 J/kg °C); k is the thermal conductivity (0.024 W/m °C).

The Nusselt number can be calculated from the Reynolds number and the Prandtl number.

$$Nu = 0.015 \cdot Re^{0.83} \cdot Pr^{0.42} = 1645.5 \qquad (16.14)$$

The heat transfer coefficient on the wall is:

$$h = \frac{Nu \cdot k}{4R_H} = 6.54 \frac{W}{m^2 \, °C} \qquad (16.15)$$

The temperature distribution along the length can be calculated by the equation:

$$T = T_{wall} - (T_{wall} - T_1)e^{-\frac{4R_H \pi Lh}{\dot{m}c_p}} \qquad (16.16)$$

where L is the length of the air in the tunnel (m); T is warmed-up air temperature after the length L (°C); T_{wall} is the tunnel wall temperature (°C); T_1 is the outside air temperature (°C); $\dot{m}$ is the mass flow rate of air (kg/s).

If the length of the tunnel is 1800 m and the outside temperature is $-5°C$, the air temperature at the end of the intake tunnel can be calculated.

$$T_L = 18 - (18 - (-5)) \cdot e^{-\frac{4 \cdot 1.51 \cdot \pi \cdot 1800 \cdot 6.54}{103 \cdot 1005}} = 15.3°C \qquad (16.17)$$

This calculated temperature was checked by the measured temperature at the end of the intake tunnel, when the outside temperature was $-5°C$. The result was very close. The measured temperature was 16°C.

In an average winter day the air temperature after passing through the intake tunnel is warmed from $-5°C$ to $+15.3°C$. This means: $\Delta T = 20.3°C$.

During the construction period 80 m^3/s (103 kg/s) of air flow is used. This is the maximum air demand.

In this case the maximum thermal power $\dot{Q}_n$ is:

$$\dot{Q}_{max\ power} = \dot{m}c_p(T_L - T_1) = 2101 \text{ kW} \qquad (16.18)$$

It can be seen that the thermal power from the tunnel wall is almost as much as the thermal power of a traditional oil burner at 2310 kW at the maximum air demand.

In order to heat the mine tunnel, it is not possible to use all of the maximum heat power. It can be exploited by the enthalpy difference between the inlet and outlet air of the heat pump: $\Delta T = 15.3 - 5 = 11.3°C$.

In this case the useful thermal power from the air is:

$$\dot{Q}_{airmax} = \dot{m}c_p(T_L - T_2) = 1138.7 \text{ kW} \qquad (16.19)$$

where T_2 is the outlet air temperature (5°C).

During normal operations 25 m^3/s (32.3 kg/s of) air flow is utilized. In this case the thermal power $\dot{Q}_n$ is:

$$\dot{Q}_n = \dot{m}c_p(T_L - T_1) = 652.8 \text{ kW} \qquad (16.20)$$

The useful power in the normal operation is:

$$\dot{Q}_{airmin} = \dot{m}c_p(T_L - T_2) = 357 \text{ kW} \qquad (16.21)$$

Another enthalpy source is the collected mine water. Every day in the mine tunnel about 500 m^3 of water is produced, which means $\dot{m} = 5.79 \frac{kg}{s}$ mass flow rate. After a long-term test the temperature of this mine water is about 15°C. This temperature does not depend on the season, as it is the same in winter or summer. At normal operations this water is collected in a sump under the surface in the mine. From time to time this water is pumped to a creek on the surface. Since the flow rate of this inflowing water is steady, and its temperature is constant, we can use its thermal power as a natural geothermal source. The heat power from the mine water is then:

$$Q_w = \dot{m}c_w(T_w - T_2) = 242.4 \text{ kW} \qquad (16.22)$$

where $\dot{m}$ is the mass flow rate of the collected mine water (57.9 kg/s); c_w is the specific heat of water (4.187 kJ/kg °C); T_w is the mine water temperature (15.3°C); T_2 is the outlet water temperature (5°C).

It can be seen that the heating demand is 156 kW. The heat supply from the air is 357 kW and from the mine water is 242 kW. Thus either heat source is enough to satisfy the heating demand of the mine tunnel deicing.

To collect the mine water, a 100-mm-diameter pipe is used. Every day about 500 m^3 (0.00578 m^3/s) is produced in the mine tunnel. The cross-sectional average velocity in the pipe is:

$$v = \frac{4Q}{D^2\pi} = \frac{4 \cdot 0.00579 \frac{m^3}{s}}{0.1^2 \cdot \pi \text{ m}^2} = 0.737 \frac{m}{s} \qquad (16.23)$$

$$Re = \frac{v \cdot D}{v} = \frac{0.737 \frac{m}{s} \cdot 0.1 \text{ m}}{10^{-6} \frac{m^3}{s}} = 73700 \qquad (16.24)$$

The Reynolds number is high and in this case the flow is turbulent.

A heating loop system is designed on the floor and on the roof for the first 300-m section in the mine tunnel. Present practice is to use plastic pipe, with the typical being polyethylene (PE) according to

Lund (1999, 2000). The relative roughness (the ratio between pipe diameter and absolute roughness) of the PE pipe is $D/k = 10,000$. These Reynolds number and relative roughness values determine the hydraulically smooth behavior of the flow. In this case the friction factor (Karman, 1930) is

$$\lambda = \frac{1}{\left(2\lg \frac{Re\sqrt{\lambda}}{2.51}\right)^2} \tag{16.25}$$

From this implicit equation it is obtained by iteration that $\lambda = 0.01919$.

The total pressure loss is the sum of the pressure loss from the tube and the pressure losses from the resistance of the 1500 elbows by Varga (1970). The tube spacing is 0.4 m in the 750 loops and along the 300 m length. The procedure is similar to running radiant heat in a building's floor slab.

$$\Delta p' = \lambda \frac{L}{D} \rho \frac{v^2}{2} + \sum \xi_k \rho \frac{v^2}{2} \tag{16.26}$$

where λ is the friction factor (0.01919 dimensionless); L is total length of the tube (15,000 m); ρ is the water density (1000 kg/m^3); v is the cross-sectional average velocity (0.732 m/s); D is the pipe diameter (0.1 m); $\xi_k = 0.3$ is the elbow loss coefficient.

$$\Delta p' = 903833 \ \frac{N}{m^2} \tag{16.27}$$

which is about 9 bar.

The temperature distribution along the length is

$$T = T_0 + (T_1 - T_0)e^{-\frac{4UL}{\rho c v D}} \tag{16.28}$$

where T_0 is the wall temperature of the mine tunnel ($^\circ$C); T_1 is the intake water temperature in the tube ($^\circ$C); U is the overall heat transfer coefficient (W/m^2 $^\circ$C); c_w is the specific heat of water (kJ/kg $^\circ$C).

We can calculate the overall heat transfer coefficient.

$$\frac{1}{U} = \frac{1}{h} + \frac{R_{in}}{k_{tube}} \ln \frac{R_{out}}{R_{in}} + \frac{R_{in}}{k_{cement}} \ln \Phi \tag{16.29}$$

where h is the heat transfer coefficient between the flowing water and the tube wall (W/m^2 $^\circ$C); R_{in} is the internal diameter of the tube (m); R_{out} is the external diameter of the tube (m); k is the thermal conductivity of the PE tube (W/m $^\circ$C); ϕ is the shape coefficient.

The shape coefficient can be used for consideration of the asymmetric heat flow pattern around the heating pipe (Bobok, 1993).

$$Nu = 0.015 \cdot Re^{0.83} \cdot Pr^{0.42} = 372.4 \tag{16.30}$$

$$h = \frac{Nu \cdot k}{D} \tag{16.31}$$

$$h = 2234 \frac{W}{m^2 \, °C}, \quad \frac{1}{h} = 0.00045 \frac{m^2 \, °C}{W} \tag{16.32}$$

$$\Phi = \left[2\left(\frac{x}{R_{out}}\right)^2 - 1 - 2\frac{x}{R_{out}} \sqrt{\left(\frac{x}{R_{out}}\right)^2 - 1} \right]^{\frac{1}{2}} \tag{16.33}$$

where x is the distance between the floor surface and the centerline of the tube (m). It is obtained for the shape

$$\Phi = 0.2679$$

Finally the overall heat transfer coefficient is

$$U = \frac{1}{0.0045 + 0.00455 + 0.07901} = 11.90 \frac{W}{m^2 \, °C} \tag{16.34}$$

If the temperature of the wall tunnel is $T_{wall} = 2°C$ and the outlet temperature of the heat pump is $T_{hp} = 24°C$, then the outlet water temperature from the tube is

$$T_2 = T_{wall} + (T_{hp} - T_{wall})e^{-\frac{4 \cdot U \cdot L}{\rho \cdot c \cdot v \cdot D}} \tag{16.35}$$

$$T_2 = 2 + (24 - 2)e^{-\frac{4 \cdot 11.90 \cdot 15000}{1000 \cdot 4187 \cdot 0.737 \cdot 0.1}} = 4.175°C \tag{16.36}$$

The thermal power from the heating loop is

$$\dot{Q} = \dot{m} \cdot c(T_1 - T_2), \tag{16.37}$$

$$\dot{Q} = 5.79 \frac{kg}{s} \cdot 4.187 \frac{KJ}{kg \, °C} \cdot (24 - 4.18) = 488 \text{ kW} \tag{16.38}$$

16.3 CONCLUSIONS

The tunnel floor of the entrance section of an underground waste deposit system in Hungary is exposed to frost and icing in winter. This is rather dangerous for the heavy vehicle traffic. To avoid this danger, an in situ floor deicing heating loop system has been designed. This floor heating system is much more effective than the originally designed intake air heating by traditional oil burners. There are two heat sources of the geothermal energy. One is the heat content of the surrounding rock, which warms up the ventilated air. The other is the heat content of the collected mine water. The temperature of the mine water and the circulated air are the same at about 15°C. The floor heating system from the warmed intake air is a two-stage geothermal direct use. The first stage is the geothermal heating of the intake air by the rock through the huge heat transfer surface of the tunnel walls. The second stage is an air—water heat pump exploiting the enthalpy of the circulated air and transferring it to the heating loop system. Using the thermal power of the mine water makes necessary to apply a water—water heat pump. Both thermal sources are enough to satisfy the deicing heat demand. In this case the geothermal potential of the mine tunnel is proven greater than the deicing heat demand. The benefits of the geothermal solution in spite of the oil burners are lower heating power, elimination of the use of fuel oil, and decreasing the CO_2 emissions radically.

References

Chapman, W.P., Katunich, S., 1956. Heat requirements of snow melting systems, ASHRAE Transactions 62, 359—372.

Földessy, J., 1975. Report on the Geological Mapping Carried Out in the Northern Foreground of the Mátra Mountains (In Hungarian). OÉA Adattár.

Karman, T., 1930. Mechanische Ahnlichkeit und turbulenz. Gott. Mat. Phys.

Kubovics, I., Pantó, G., 1950. Volcanological Investigations in the Mátra and Börzsöny Mountains (in Hungarian). Akadémiai Kiadó, Budapest.

Lund, J.W., 1999. Reconstruction of a pavement geothermal deicing system. Geo-Heat Cent. Q. Bull. 20 (1), 14—17. Klammath Falls, OR.

Lund, J.W., 2000. Pavement Snow Melting. Geo-Heat Center, Oregon Institute of Technology.

Toth, A., 2007. A prospect geothermal potential of an abandoned copper mine. In: Proceedings, Thirty-two Workshop on Geothermal Reservoir Engineering Stanford University, Stanford, California.

Varga, J., 1970. Hidraulikus És Pneumatikus Gépek. Műszaki Könyvkiadó, Budapest.

Zelenka, T., 1973. On the Darno megatectonic zone. Acta Geol. Acadeiae Sci. Hung. 17, 155—163.

Index

Printed in the United States
By Bookmasters